Control of welding distortion in thin-plate fabrication

Related titles:

Self-piercing riveting
(ISBN 978-1-84569-535-4)

Understanding biocorrosion
(ISBN 978-1-78242-120-7)

Handbook of smart coatings for materials protection
(ISBN 978-0-85709-680-7)

Woodhead Publishing Series in Welding and Other Joining Technologies: Number 83

Control of welding distortion in thin-plate fabrication

Design support exploiting computational simulation

Tom Gray, Duncan Camilleri and Norman McPherson

ELSEVIER

AMSTERDAM • BOSTON • CAMBRIDGE • HEIDELBERG • LONDON
NEW YORK • OXFORD • PARIS • SAN DIEGO
SAN FRANCISCO • SINGAPORE • SYDNEY • TOKYO
Woodhead Publishing is an imprint of Elsevier

WP WOODHEAD PUBLISHING

Woodhead Publishing is an imprint of Elsevier
80 High Street, Sawston, Cambridge, CB22 3HJ, UK
225 Wyman Street, Waltham, MA 02451, USA
Langford Lane, Kidlington, OX5 1GB, UK

British Library Cataloguing-in-Publication Data
A catalogue record for this book is available from the British Library.

Library of Congress Control Number: 2013954936

ISBN 978-0-85709-047-8 (print)
ISBN 978-0-85709-932-7 (online)

For information on all Woodhead Publishing publications visit our website at http://store.elsevier.com/

Typeset by Data Standards Ltd, Frome, Somerset, UK

Printed and bound in the United Kingdom

Contents

Author contact details

(* = main contact)

Tom Gray*
University of Strathcylde
UK
E-mail: tom.gray@virgin.net

Duncan Camilleri
University of Malta
Malta
E-mail: duncan.camilleri@um.edu.mt

Norman McPherson
BAE Systems – Surface Ships
UK
E-mail: norrie.mcpherson@baesystems.com

Woodhead Publishing Series in Welding and Other Joining Technologies

1 **Submerged-arc welding**
Edited by P. T. Houldcroft

2 **Design and analysis of fatigue resistant welded structures**
D. Radaj

3 **Which process? A guide to the selection of welded and related processes**
P. T. Houldcroft

4 **Pulsed arc welding**
J. A. Street

5 **TIG and plasma welding**
W. Lucas

6 **Fundamentals of welding metallurgy**
H. Granjon

7 **Fatigue strength of welded structures**
S. J. Maddox

8 **The fatigue strength of transverse fillet welded joints**
T. R. Gurney

9 **Process pipe and tube welding**
Edited by W. Lucas

10 **A practical guide to TIG (GTA) welding**
P. W. Muncaster

11 **Shallow crack fracture mechanics toughness tests and applications**
Conference Proceedings

12 **Self-shielded arc welding**
T. Boniszewski

13 **Handbook of crack opening data**
T. G. F. Gray

14 **Laser welding**
C. T. Dawes

15 **Welding steels without hydrogen cracking**
N. Bailey and F. R. Coe

16 **Electron beam welding**
H. Schultz

17 **Weldability of ferritic steels**
N. Bailey

18 **Tubular wire welding**
D. Widgery

19 **Stress determination for fatigue analysis of welded components: Recommendations of IIW Commissions XIII and XV**
Edited by E. Niemi

20 **The 'local approach' to cleavage fracture**
C. S. Wiesner

21 **Crack arrest concepts for failure prevention and life extension**
Seminar Proceedings

22 **Welding mechanisation and automation in shipbuilding worldwide**
R. Boekholt

23 **Heat treatment of welded steel structures**
D. Croft

24 **Fatigue design of welded joints and components: Recommendations of IIW Joint Working Group XIII-XV**
Edited by A. Hobbacher

25 **Introduction to the non-destructive testing of welded joints**
R. Halmshaw

26 **Metallurgy of basic weld metal**
T. R. Gurney

27 **Fatigue of thin walled joints under complex loading**
T. R. Gurney

28 **Handbook of structural welding**
J. F. Lancaster

29 **Quality assurance in adhesive technology**
A. W. Espie, J. H. Rogerson and K. Ebtehaj

30 **Underwater wet welding and cutting**
TWI/Paton Electric Welding Institute

31 **Metallurgy of welding Sixth edition**
J. F. Lancaster

32 **Computer technology in welding**
Conference Proceedings

33 **Exploiting advances in arc welding technology**
Conference Proceedings

34 **Non-destructive examination of underwater welded structures**
V. S. Davey

35 **Predictive formulae for weld distortion**
G. Verhaeghe

36 **Thermal welding of polymers**
R. J. Wise

37 **Handbook of mould, tool and die repair welding**
S. Thompson

38 **Non-destructive testing of welds**
B. Raj, C. V. Subramanian and T. Jayakumar

39 **The automotive industry: joining technologies**
TWI

40 **Power generation: welding applications**
TWI

41 **Laser welding**
TWI

42 **Fatigue: welding case studies**
TWI

43 **Fracture: welding case studies**
TWI

44 **The welding workplace**
R. Boekholt

45 **Underwater repair technology**
J. Nixon

46 **Fatigue design procedure for welded hollow section joints: Recommendations of IIW Subcommission XV-E**
Edited by X.- L. Zhoa and J. A. Packer

47 **Aluminium welding**
N. R. Mandal

48 **Welding and cutting**
P. T. Houldcrof and J. A. Packer

49 **Health and safety in welding and allied processes**
J. Blunt

50 **The welding of aluminium and its alloys**
G. Mathers

51 **Arc welding control**
P. Julian

52 **Adhesive bonding**
R. D. Adams

53 **New developments in advanced welding**
Edited by N. Ahmed

54 **Processes and mechanisms of welding residual stress and distortion**
Edited by Z. Feng

55 **MIG welding guide**
Edited by K. Wenem

56 **Cumulative damage of welded joints**
T. R. Gurney

57 **Fatigue analysis of welded components: Recommendations of IIW Commissions XIII and XV**
E. Niemi

58 **Advanced welding processes**
J. Norrish

59 **Fatigue assessment of welded joints by local approaches**
D. Radaj

60 **Computational welding mechanics**
Edited by L. E. Lindgren

61 **Microjoining and nanojoining**
Edited by Y. N. Zhou

62 **Real-time weld process monitoring**
Edited by Y. M. Zhang

63 **Weld cracking in ferrous alloys**
Edited by R. Singh

64 **Hybrid laser-arc welding**
Edited by F. O. Olsen

65 **A quick guide to welding and weld inspection**
Edited by S. E. Hughes

66 **Friction stir welding**
Edited by D. Lohwasser and Z. Chen

67 **Advances in structural adhesive bonding**
Edited by D. Dillard

68 **Failure mechanisms of advanced welding processes**
Edited by X. Sun

69 **Advances in laser materials processing**
Edited by J. Lawrence and J. Pou

70 **Welding and joining of magnesium alloys**
Edited by L. Lui

71 **Fracture and fatigue of welded joints and structures**
Edited by K. MacDonald

72 **Minimization of welding distortion and buckling**
Edited by P. Michaleris

73 **Welding processes handbook Second edition**
K. Weman

74 **Welding and joining of aerospace materials**
Edited by M. C. Chaturvedi

75 **Tailor welded blanks for advanced manufacturing**
Edited by B. Kinsey and X. Wu

76 **Adhesives in marine engineering**
Edited by J. R. Weitzenböck

77 **Fundamentals of evaluation and diagnostics of welded structures**
A. Nedoseka

78 **IIW recommendations for the fatigue assessment of welded structures by notch stress analysis**
W. Fricke

79 **IIW recommendations on methods for improving the fatigue strength of welded joints**
P. J. Haagensen and S. J. Maddox

80 **Advances in brazing**
Edited by D. P. Sekulic

81 **Advances in friction-stir welding and processing**
M.-K. Besharati-Givi and P. Asadi

82 **Self-piercing riveting**
Edited by A. Chrysanthou and X. Sun

83 **Control of welding distortion in thin-plate fabrication: Design support exploiting computational simulation**
T. Gray, D. Camilleri and N. McPherson

Preface

The authors of this book hold the view that dealing with welding distortion is as much a responsibility for engineering designers of structures as it is for those engaged in their fabrication. Indeed, there are more opportunities at the design stage to set the agenda for effective control of distortion and, generally, these opportunities are not used. One reason is that there are not many effective and user-friendly tools to assist designers and the main purpose of the book is to provide some.

The research presented here was launched initially at the University of Strathclyde through a project entitled 'Simulation tool for fusion-welded stiffened-plate fabrications'. This was funded by the UK Engineering and Physical Sciences Research Council, with substantial support from BAE Systems (Naval Ships) comprising materials, equipment and technical experience. The project stimulated continuing research in Glasgow and also at the University of Malta and several of the results from these later projects are included.

Every idea has a beginning and in this case it was the accidental discovery, over 40 years ago, of a text in the Andersonian Library at the University of Strathclyde, with the title *The Calculation of Deformations of Welded Metal Structures*. This book was not impressive. It was in the form of a commissioned translation of a Russian text by a researcher, N. O. Okerblom, at the Leningrad Shipbuilding Institute and its presentation was low-budget, typewritten on sub-standard paper. Nor does the title tell the reader that the deformations under consideration are those caused by *welding*, rather than by service loads. Reading and understanding the text therefore was frankly something of a slog.

However, at that time, the only guidance a designer could find on welding distortion was of the qualitative 'good/bad practice' kind, whereas Okerblom's approach offered a way of actually calculating welding distortions – if you could understand the theory and believed the calculations. For that reason of apparent uniqueness, it became the basis of some dozen pages in a textbook written at that time, with a provocative title.[1] It proved to

[1] Gray, T.G.F., Spence, J. and North, T.H., *Rational welding design*, 1st Edition 1975, 2nd Edition 1982, Butterworths, London, UK.

be a popular text and we had a lot of correspondence from readers, but there was never an indication that the section on distortion had been found to be useful by anyone.

With the benefit of some 40 years' hindsight, some of the deficiencies of Okerblom's book, which meant that it did not revolutionise the design of welded ship structures, can be understood. The thermo-mechanical model of fusion welding was so simple that it was difficult to believe that the resulting description was anywhere near reality – and no experimental evidence was provided to bolster the theory. Even if the model and the resulting analyses were accepted, it would have been no small matter at that time to incorporate such a concept into the design process for a ship or similar structure. Perhaps most tellingly, design practices in the industry tended to ignore welding distortion and depended rather on the shop floor to sort it out as best they could.

Nevertheless, one has to admire Okerblom, not least for his uncanny insight into the mechanical processes of distortion and his determination that welding of ship structures should not be left to 'trial and error'. As the years passed and computational capability increased, some of the arguments against using an analytical approach diminished. Small projects to throw some light on uncertain areas of the Okerblom theory were undertaken in 1978 and 1984 at Strathclyde by post-graduate students Wickramasinghe and Cadden respectively (referenced in Chapters 6 and 7) and they had positive outcomes. However, the deciding factor to make a major effort came much later through a local company, Burntisland Fabrications, which posed the question 'if we already have a finite-element model for a ship, why can't we just press another button to find out what happens when we make the welds?' Persisting in answering the question, they supported a preliminary feasibility study towards a simplified computational approach. This provided the starting point for the series of investigations described in this book.

The project noted at the beginning of this preface started at the end of 2001 and, in addition to the three present authors, involved Alex Galloway, who contributed greatly to the practical welding tests, Pierluigi Mollicone, whose work is shown principally in Chapters 9 and 10, Tugrul Comlekci, who provided a very useful data-smoothing tool described in Chapter 6 and Maxine Scicluna, whose work on transient thermal tensioning is included in Chapter 12. The research also benefited from an input of steel property data determined by Corus Ltd.

This research has been an unusually stimulating and rewarding experience for those of us participating, largely because we were learning, instead of seeing what we expected to see. The large-scale experiments described in Chapter 6 were immensely instructive in terms of guiding the modelling and, incidentally, were a lot of fun. The capability of finite-element modelling to provide a transient picture of the welding process also proved to be informative in explaining some of the anomalies in distortion outcomes. But most unexpected of all, Okerblom's simple theory proved, against all scepticism, to be remarkably effective when exploited within a large-displacement, finite-element analysis framework.

As always, many other colleagues contributed to the work that has led to the writing of this book. Although too many to be named, we want to thank them all for their interest, encouragement and input.

Tom Gray, Duncan Camilleri and Norman McPherson

1
Introduction: development of computational welding mechanics approach to welding distortion

DOI: 10.1533/9780857099327.1

Abstract: The opening chapter contrasts a welding design and fabrication strategy based on trial and error with one that is guided by quantitative welding simulation methods. An outline is given of the main contributing factors to the development of computational welding mechanics as a discipline and the possibilities that such studies offer for interpreting welding phenomena such as distortion during fabrication. The importance of establishing simple models is emphasised and a first mention is made of simplified computational methods. The content of the following chapters is briefly described and the relationships between each of the chapters in the sequence are explained.

Key words: computational welding mechanics, heat flow, computationally efficient, reduced solutions.

1.1 Background: control of welding distortion in fabrication practice

Most designers think that welding is related to black magic . . . A few do try to adjust their designs to help the welding engineers, but when they look into the subject, they find it too confusing and so abandon hope.

This somewhat defeatist opinion was offered at an international conference in the early 1970s by a senior metallurgist of the Nuclear Power Group. Power plant construction was at its peak in the UK nuclear power industry at that time and the different disciplines involved in welding large structures tended to be applied in a linear and compartmented manner. Figure 1.1

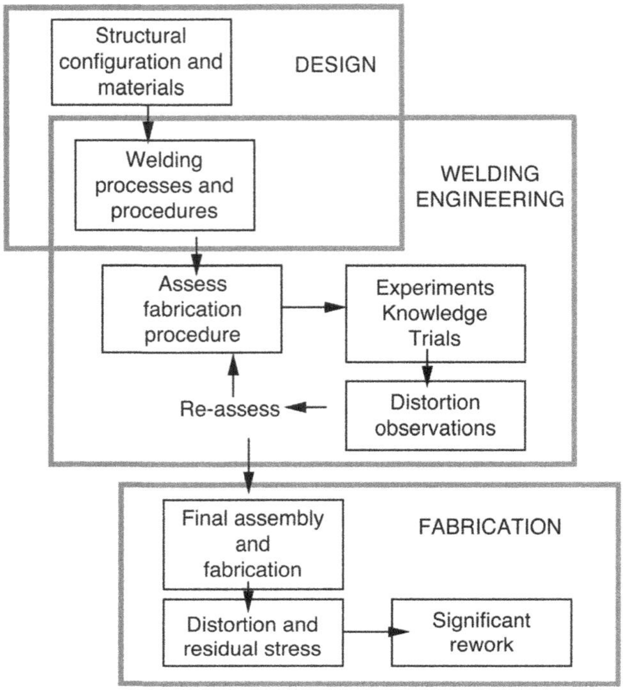

1.1 Former/existing approach to distortion control.

characterises this approach in the context of distortion outcomes. Some engineers operated on the belief that the materials they designed with, including welds, could be treated as homogeneous, isotropic and stress-free.

Metallurgists tended to regard such sweeping assumptions as ignorant and dangerous in the extreme and, on the contrary, sought to elaborate the complexities of metallurgical behaviour around welds. Welding engineers introduced yet another set of new considerations, largely unknown to both structural engineers and metallurgists. There was therefore not much meeting of minds on the subject of how to design welded structures and interaction between structural engineers, metallurgists and welding engineers was largely non-existent.

In this scenario, materials and structural configurations were decided in the first instance by engineers, with specialised input from metallurgists where necessary. There was usually no feedback of information to the primary design process concerning the influence of welding on distortion or residual stress. Such matters *might* be considered at the later 'detail design' stage when the prospective design reached the fabricator, but the approach even then tended to be based primarily on 'rules of thumb' or, in better cases, on experience. Again, the opinions and experiences of fabricators were not consistent, as they depended on the different structural contexts, materials and welding processes with which each fabricator might be familiar. At that stage, welding engineers were presented with the problem of identifying suitable assembly, welding and fabrication procedures to minimise distortion, in the context of a largely fixed design configuration and with no means of quantifying likely distortions. Off-line welding trials of some kind were usually undertaken but, for reasons of cost, these tended to be confined to small test specimens designed to provide information on weld outcomes and metallurgy. Such trials would not reflect the large-scale factors present in distortion mechanisms. Hence, much of the 'experience' in dealing with distortion was often gained in the context of live production with the consequent penalties of scrapped components and significant rework.

The challenge then, as it remains now, was to recognise relevant knowledge from a variety of disciplines and to apply that knowledge in a properly integrated and quantitative manner. The capability to do that has much improved in the last 40 years, based mostly on new tools to model complex descriptions of manufacturing processes, weld metallurgy and structural behaviour. These are based on increased computational power and integrated physical domains covered by different disciplines. For example, the application of finite-element methods for stress analysis of complex geometries is now well established and this has reduced the need for many approximate and sometimes arbitrary 'design rules' previously used in different structural contexts, not least for welds. Typical examples of arbitrary rules might include the concepts of 'joint efficiency' and 'safety factors', which simply assign a given 'strength' to a weld, usually without much quantitative justification. Strength design has instead moved much more towards addressing realistic performance and failure criteria and is now often supported by numerical stress analysis and fracture mechanics modelling.

Analytical treatments of heat flow in welding had certainly been well established by 1970. These treatments are attributed in the literature mainly to the original and independent work of Rosenthal[1] and Rykalin.[2] Available analytical treatments of heat flow underpinned many strategies used by welding engineers and metallurgists at that time to predict fusion zone sizes, cooling rates and other welding features of interest, although the sources of these treatments and their scope may not always have been evident to the users. Increasing computational power has greatly extended the range of welding phenomena that can now be modelled.

The recent development of ambitious simulation treatments of welding can be traced in the transactions of a series of influential bi-annual international meetings with the title *Mathematical modelling of welding phenomena*. These began in 1993[3] and have continued to date. The main topics treated in this series included heat flow, microstructure development, residual stresses and special advanced welding processes. Detailed models of gas flow, weld pool behaviour, solidification and many other phenomena can also be found. The topic of welding distortion appears in the proceedings of the fourth meeting in the series in 1998.[4] A new term, 'computational welding mechanics' (CWM), appears more often in the literature of this period.

Nevertheless, if there is one area of welding technology that still retains more than a whiff of black magic, it might be the subject of distortion. Nothing is surer than that the use of intense heat sources in fusion welding will generate distortion. This might be accepted if the mode of distortion and the magnitudes could somehow be reliably predicted and allowed for. Even if the relative magnitudes of distortion arising from different fabrication routes could be estimated, the problem might be manageable. However, as will be demonstrated in this book, distortion depends typically on a variety of factors arising in the manufacturing phases, some of them apparently so trivial as to be invisible. Success in minimising distortion therefore requires the very close cooperation of designers, welding engineers and metallurgists – much beyond a mere 'adjustment' of designs 'to help the welding engineers'.The difficulty that has persisted in industry is that there has been little commonly accepted quantitative understanding of what is likely to happen when a structure is welded.

Fortunately, several books have been published in recent years to support improved quantitative understanding. At the risk of missing some useful texts, the following publications are of note.

- *Analysis of welded structures* by Masubuchi.[5] This text, published in 1980, pre-dates the development of CWM as we now know it and much of the data are therefore empirical or qualitative. However, it provides a thorough and extensive review of the literature on welding distortion and residual stress, including standards that might typically be achieved in industrial manufacturing and data describing different practical cases. Chapter 7 (Distortion in weldments) and Chapter 13 (Effects of distortion and residual stresses on buckling strength of welded structures) are particularly useful.
- *Welding residual stresses and distortion* by Radaj[6] deals with computational simulation of temperature fields, residual stress development and the distortions that occur during and after welding.
- *Computational welding mechanics* by Goldak and Akhlagi[7] deals very fully with thermal descriptions, heat input modelling and metallurgical interactions. It is, in general, a practical and useful book for anyone building models of weld processes.
- *Computational welding mechanics* by Lindgren[8] is a practical book aimed at 'managers and project leaders who want to know what resources are needed to apply the tools of computational welding mechanics and what competence and improvement of processes/products their application

can lead to... engineers are given detailed information in order to understand phenomena and design issues as well as guidelines for modelling and simulation'.

- *Minimization of welding distortion and buckling*, edited by Michaleris[9], comprises chapters by different authors and covers basic understanding of welding residual stress and distortion. It includes several techniques for minimising distortion in different practical applications.
- *Processes and mechanisms of welding residual stress and distortion* is a similar multi-author book, edited by Feng.[10] It covers principles and applications, contains useful chapters on measuring temperatures and distortions, and provides thermal simulations for different welding processes and strategies for the control of buckling distortions in plates and shells.

There is therefore much more support now, for those who wish to understand and control welding distortion, than was the case in the 1970s.

Notwithstanding the wealth of information available, a challenging question posed earlier by Murakawa *et al.*[11] is worthy of consideration. They ask 'is computational welding mechanics truly attractive for people in industry now?' They conclude that it is not attractive for many and give two reasons to support their claim that 'computational mechanics is not widely employed in industry'. The first reason is 'long computational time and the relatively high cost of the nonlinear transient FEM analysis' (perhaps this is not as serious a problem now as it was even in 1998, when this work was published). The second reason given is 'a lack of consistent bases to interpret the phenomena and to generalise the knowledge obtained by simulations for broader applications'. In other words, simulations can address specific cases, but can one formulate general design and fabrication strategies from these specifics?

The main answer given in the paper by Murakawa *et al.* to their own question is to adopt the simplifying concept of 'inherent strain'. This is in fact very similar to a concept proposed very much earlier (in 1958) by Okerblom, which has not been widely referenced.[12] This earlier idea was also the starting point for the 'computationally efficient' and 'reduced solution' approaches adopted in the present text. As will be shown later, the present authors' opinion on this point is that *some* problems can in fact find good solutions using simplified strategies like the 'inherent strain' approach, but other problems require more rigorous treatment. A central aim of this book is to identify these different cases.

It is also worth asking a second question, namely: what alternatives are there to welding simulation as a basis for understanding and reducing distortion? A forthright answer to that was given by Okerblom as far back as 1958 and it is as relevant today as it was then – not surprisingly, as his research had been based at the Leningrad Shipbuilding Institute over many years previously and was therefore founded firmly on industrial practice. He states (in translation) 'In industry, it is usual to make only a qualitative estimate [of distortion] based chiefly on trial-and-error methods, guess work and past experience'. He then spells out in detail, and with specific examples, what the drawbacks are to this 'usual' approach – principally, the many departures from specified design measurements and the need for correction of dimensions by further expensive cutting, straightening or bending operations. Okerblom points out that the cost of rectification may be more than that of the original welding of the assembly. These themes are discussed further and brought up to date in Chapter 2.

In summary therefore, the weight of opinion in the literature strongly argues that (using Okerblom's words)

> to minimise distortions and departures from the specified design as a result of welding, a quantitative calculation of the deformations produced during different assembly and welding procedures is necessary

The use of CWM, at some level, provides the only obvious basis for such a quantitative calculation. Figure 1.2 provides a characterisation of this approach. The topic of residual stress is added to Okerblom's categorisation, because it is highly relevant to the problem of buckling and, also, some methods used to rectify distortion have the effect of adding residual stress fields. Furthermore, the term 'design' is conceived more broadly than in Fig. 1.1, to cover any feature of the design or the fabrication process that can be decided in advance.

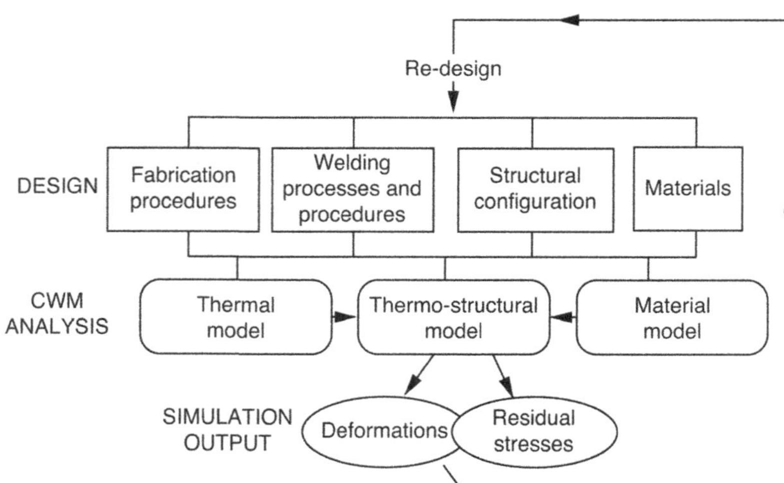

1.2 Integrated design approach based on CWM.

1.2 Aims: integrated design approach utilising computational welding mechanics (CWM)

The main purpose of this book is thus to provide the guidance necessary to encourage all concerned with thin-plate welded fabrication to establish CWM models appropriate to their own situations. If finite-element analysis (FEA) software is already in use for stress and deformation analysis, extension to modelling behaviour of a structure during fabrication should be entirely feasible, especially at what is described in this book as a 'computationally efficient' level. The additional step to do this involves only a separate thermal analysis of the FEA modelled structure to provide a profile of the maximum temperatures caused by the moving welding heat source. A simple algebraic transformation of these temperatures provides an artificial loading profile, which is then applied to a three-dimensional elastic model of the structure, thus providing deformation values in the usual way. One key advantage of this method is that large-displacement FEA can be employed, which is particularly important for realistic modelling of thin-plate structural behaviour.

This kind of treatment should be substantially more effective than the 'trial-and-error and guess work' approach castigated by Okerblom,[12] besides being much less costly. Variations in welding process and structural configuration can be explored economically, to find improvements that reduce distortions. Furthermore, the graphical presentation capabilities of FEA and CAD software can be employed to observe in detail what is happening during and after welding, in the cooling down period. This therefore assists the user to discern the major causes of distortion effects and to find remedies related to particular features such as workpiece support, weld edge preparation, tacking configuration, sequence of assembly and many other factors. It is perhaps worth mentioning at this stage that the major strength of CWM is not so much that it will provide an exact prediction of an outcome (that would be an optimistic claim for many reasons) but more that it can offer an understanding of the causes and mechanisms of the various modes of distortion. As will be demonstrated later in this book, deformations and their development at various stages can be exaggerated in graphical presentation, whereas such movements are barely visible in the case of direct experimental observation.

The material in this book should also be helpful at another more generic level, as many case studies of butt- and fillet-welded structures are included in the chapters, most of them being supported by welding experiments on steel structures at a realistic scale. These results also allow some understanding of the likely outcomes of welding various configurations and, more importantly, the reasons for these outcomes. The approach being recommended therefore is to build specific FEA-based models of structures and fabrication routes as appropriate.

This may not seem to address the second problem raised by Murakawa *et al.*,[11] that the solution of specific problems does little 'to interpret the phenomena or to generalise the knowledge obtained by simulations, for broader applications'. A first response to that would be to point out that in the area of structural analysis, individual FEA treatments of specific artefacts and loadings do little to provide information about the behaviour of structures in general, but they are nevertheless entirely useful and essential to the design process. Moreover, through parametric surveys and parallel simplified analyses, they can also develop generic understanding of behaviour. The same can be said of CWM. At a dimensional level, the quantitative influence of the thermo-mechanical properties of materials involved emerges very clearly, likewise the effects of welding process parameters. In this respect, the computationally efficient or reduced solution approaches (and likewise the inherent strain method) hold a particular attraction in that the connections between the input parameters and the outcomes can be more clearly discerned, albeit the outcomes may be less accurately indicated. The central aim therefore is to provide simple models that can be applied to realistic problems in designing thin-plate, welded structures.

1.3 Structure of the book

The following chapters develop the theme of welding distortion in thin-plate structures in a progressive and inter-dependent manner. Outlines are provided in early chapters and specific detailed aspects relevant to modelling are treated in later chapters.

Chapter 2 describes a typical industrial context, where thin-plate distortion is a central problem and where management of the processes and procedures presents particular challenges. This description is focused on the shipbuilding industry, but not exclusively so. This establishes a practical background against which the application of CWM can be visualised. The early parts of the chapter provide descriptions of the main steps in fabricating ship structure elements and the processes involved. Particular problems such as buckling and the management of known contributory factors to excessive distortion are highlighted. If too much distortion occurs in a given case, despite best efforts, some means to rectify this may be required and the chapter concludes with a brief description of such methods. As many of these techniques depend on local heating and are again mostly based on trial and error, CWM can play a part in understanding how to apply such processes.

Chapter 3 considers several relatively new technologies that can be used to deal with distortion problems. Some of these are currently at a research stage of development, while others have been established in practice to some extent. The artificial neural network (ANN) approach recognises the complexity and sometimes hidden nature of factors involved in welding and the resulting difficulty in relating cause and effect in an industrial context. The ANN process attempts to find empirical relationships between fabrication input parameters and distortion outcomes, using essentially statistical methods to infer functional relationships. The method requires substantial volumes of data from objective observations but is able to reveal influences that might not be recognised in a physical model. This approach contrasts sharply with the CWM approach which depends on knowing *a priori* which physical models are relevant and on having a good material property data bank. However, these two ways of understanding and predicting distortion can often be complementary.

Chapters 4 and 5 provide an initial outline of the physical phenomena involved in the behaviour of structures during welding and cooling. Chapter 4 is based primarily on analytical determinations and Chapter 5 introduces the main features of a computational approach to simulation.

The analytical approach in Chapter 4 is based on well-tried heat flow equations and a simplified and highly approximate thermo-mechanical model. All effects in welding mechanics are driven by the thermal input of the welding process used and the relationships between energy inputs and the resulting transient temperature fields can be clearly drawn using the analytical approach. This provides a basic understanding of the interdependence of heat input, heat source travel speed, material thickness and thermal properties. The thermo-mechanical model again highlights the role of material properties and the basic effects of weld contraction on overall mechanical behaviour and global distortion patterns.

Chapter 5 outlines the basis of more complex multi-physics models, where non-linear material properties are introduced and which are less dependent on simplifying assumptions. This allows more detailed description of transient deformations, which are important in the context of the non-linear, large-displacement distortions occurring in thin-plate structures.

The value of practical experiments is discussed in Chapter 6. It is noted in the first instance that carefully observed experiments are an important route to discerning which features most need to be modelled and which can be given lesser priority. The relationship between the energy input to the welding process and the resulting thermal transients is particularly important to check experimentally, as the process whereby electrical or thermal energy applied is conducted into the workpiece is difficult to model in detail. Practical techniques for out-of-plane measurement, analysis and characterisation are described as this is the form of distortion that is normally problematic in fabricating thin-plate or shell structures. A special procedure to provide smooth representations of multi-point deformations is described and the merits of different parameters for out-of-plane distortion characterisation are discussed.

Chapters 7, 8 and 9 provide the main guidance for computational modelling of welding-induced distortion in thin-plate fabrications. Thermal aspects of modelling are treated in Chapter 7. Simplified or reduced solution approaches to the thermo-mechanical part of the uncoupled treatment are shown in Chapter 8 and methods for full thermo-elastic-plastic treatments are given in Chapter 9.

As will be shown, the thermal (temperature field) stage of CWM analyses and the thermo-mechanical stages can be uncoupled and treated separately. Chapter 7 focuses on the thermal stage and includes both analytical and computational approaches. Various simplifications that can be made (e.g. reduction to two-dimensional treatment) are treated and different methods are tested against experimental determinations of transient temperature fields. Special attention is given in this chapter to the 'thermal efficiency' of different welding processes, that is, the proportion of energy from the process that is converted to heat and is absorbed by the workpiece. Reliable computational models for this ratio have not as yet been proposed for commonly used processes and this step in the overall thermal model has to depend on experimental calibrations of one kind or another. Results from the literature are collated in Table 7.4, although the more satisfactory approach to this problem would be to carry out a live test using thermocouples or other temperature measurement systems. The computed transient temperature field can then be matched to the experimental result, with efficiency as the matching variable. Various strategies for modelling heat input and weld filler deposition are given and their relative advantages explained.

Convection and radiation heat transfer tend to play a minor part in most fusion welding applications. However, the equations normally used to model the heat loss associated with these effects are normally included in computational analyses and appropriate strategies are therefore described in Chapter 7. Treatments to include the effects of thermal property non-linearity are given. A brief study of thermal cutting is also included, showing that the principles used to model heat input in welding can also be applied to model transient heat input and loss of energy through ejection of heated material in the cutting process.

Chapter 8 provides a step-by-step guide to the use of reduced solution strategies, firstly on an analytical basis and then by using FEA elastic models to determine weld-induced deformations. The shortcomings of the simplest computationally efficient approach are indicated and the benefits of hybrid strategies are shown. These strategies also depend on uncoupling of the transverse and longitudinal mechanics to an extent, but the analytically based elasto-plastic approach of the computationally efficient strategy is improved greatly by adopting computational elasto-plastic techniques. The predictions of the various models are compared with experimental results for butt and fillet welds.

Chapter 9 shows implementation strategies for integrated, thermo-mechanical, elasto-plastic structural FEA models and thus completes the portfolio of modelling strategies. These methods make it possible to model non-linear, temperature-dependent dilatation and strength properties. However, the main advantages of such models is that fewer simplifications are made and the methods can be applied to describe complex but important fabrication operations such as assembly by weld tacking and sequence effects in stiffener fillet welding. Case studies and experimental results are again used to draw out the differences between the reduced solutions and the more advanced elasto-plastic computational approaches.

Chapters 10, 11 and 12 focus on application. Chapter 10 deals with butt-welded cases and Chapter 11 with fillet-welded fabrications. Chapter 12 highlights the potential of CWM to assist design and fabrication management by providing simulations that can test alternative designs or fabrication routes.

Butt-weld distortion outcomes can be strongly influenced by fabrication practices and other factors that could easily seem to be of minor significance. These include plate support and the resulting effects of gravity loading on the initial and transient out-of-plane shape, the effects of different procedures such as weld tacking and clamping, and the interactions of multiple residual stress systems in a fabrication with several welds. These factors are explored in Chapter 10 through a variety of case studies, some supported by experimental results.

Various treatments of stiffener fillet welding are shown in Chapter 11. More detail is given on the thermal simulation stage and the relative strengths and weaknesses of computationally efficient and full, transient solution methods. Differences between single- and double-sided fillet-welded configurations are shown, together with studies on multiply-stiffened fabrications, where close spacing of stiffeners can lead to interactions between longitudinal thermal and residual stress systems.

Chapter 12 is concerned with identifying measures to reduce distortion. The first half of the chapter investigates measures that can be applied at the design stage to reduce distortion, including optimising the configuration of stiffened fabrications, the placement of welds relative to other features (including other welds) and the effects of excessive heat input (often determined as a result of design choice) on undesirable outcomes such as buckling. The second half uses simulations to investigate three special fabrication techniques that can reduce distortion. These are transient thermal tensioning, the use of low-transformation-temperature weld material and the use of weld-trailing cryogenic cooling processes. Appropriate computational strategies are used for these studies and they further reinforce the advantages of CWM as a progressive route to understanding the effects of welding processes on out-of-plane distortion and residual stress.

1.4 Conclusion

This book is concerned with quantitative methods to support engineering design and manufacturing strategies for welded, thin-plate structures, where out-of-plane distortion is a troublesome problem. The term 'computational welding mechanics' has been coined to describe such methods, which have many applications in welding technology, beyond providing guidance for control of distortion.

The first section of this opening chapter briefly traced the development of a computational treatment of this problem over the last four decades, whereby it has become possible to improve on the former 'trial-and-error and guess work' approach through simulation of the application of welding processes to manufacturing. Various key texts that support computations of this type were noted. The aims of the book have been outlined, with particular emphasis on the provision of methods that can be simply applied in an industrial context. Another important aspect is the potential of such methods to enable understanding of the driving factors and mechanisms of distortion and the interpretation of phenomena, through exploration of different design and fabrication approaches to a particular case. The remainder of this chapter presented brief summaries of the chapters to follow.

1.5 References

1. Rosenthal, D., 1946, The theory of moving sources of heat and its application to thermal treatments, *Trans. ASME*, 68, pp 849–866.
2. Rykalin, N.N., 1947, *Thermal welding principles*, I.A.N., Leningrad, SSSR.
3. Cerjak, H. and Easterling, K.E., Eds, 1993, *Mathematical modelling of welding phenomena*, Institute of Materials, London, UK.
4. Cerjak, H. and Bhadeshia, H.K.D.H., Eds, 1998, *Mathematical modelling of welding phenomena 4*, Institute of Materials, London, UK.
5. Masubuchi, K., 1980, *Analysis of welded structures*, Pergamon Press, Oxford, UK.
6. Radaj, D., 2003, *Welding residual stresses and distortion: calculation and measurement*, DVS, Düsseldorf, Germany.
7. Goldak, J.A. and Akhlagi, M., 2005, *Computational welding mechanics*, Springer, New York, USA.

8. Lindgren, L.-E., 2007, *Computational welding mechanics: thermomechanical and microstructural simulations*, Woodhead Publishing Ltd, Cambridge, UK.
9. Michaleris, P., Ed., 2011, *Minimization of welding distortion and buckling: modelling and implementation*, Woodhead Publishing Ltd, Cambridge, UK.
10. Feng, Z., 2005, *Processes and mechanisms of welding residual stress and distortion*, Woodhead Publishing Ltd, Cambridge, UK.
11. Murakawa, H., Luo, Y. and Ueda, Y, 1998, Inherent strain as an interface between computational welding mechanics and its industrial application, in *Mathematical modelling of weld phenomena 4*, Cerjak, H. and Bhadeshia, H.K. D.H., Eds, Institute of Materials, London, UK.
12. Okerblom, N.O., 1958, *The calculation of deformations of welded metal structures* (originally in Russian, translated by the Department of Scientific and Industrial Research), HMSO, London, UK.

2
Fabrication of stiffened thin-plate structures and the problem of welding distortion

DOI: 10.1533/9780857099327.14

Abstract: The overall industrial context of thin, stiffened-plate fabrication is outlined. A typical fabrication sequence is described, showing the various stages where distortion issues can arise. Stock quality and flatness have a strong influence on outcomes. Some high-risk factors for distortion are identified, which could be removed by better management of the design and fabrication process. Cause and effect are not always easily related in many cases and application of quantitative modelling tools could often resolve ambiguity and improve distortion outcomes. The chapter closes with a brief description of methods used to rectify any distortion that may have arisen.

Key words: heat-straightening, seam welding, fillet welding, stock flatness, levelling, over-welding, buckling, intermittent welds, sequence welding, tack welding, distortion rectification.

2.1 Introduction

In Chapter 1 it was argued that a major deficiency of typical practice in the design and fabrication of stiffened-plate structures is that the design process is usually not well informed by manufacturing outcomes. This chapter describes various aspects of industrial practice in the fabrication of thin-plate structures and points to several key fabrication issues in relation to welding distortion. These establish the context within which design support tools of various kinds may be deployed.

Typical fabrication sequences used in building naval ships (Fig. 2.1) are described in section 2.3 and attention is given to the various fabrication options applicable and to the problems that may arise at each stage. In fact, some of these problems have their roots in stages before any welding is

2.1 Type 45 Destroyer (courtesy BAE Naval Ships).

applied – during manufacture, transport, storage and preparation of the stock material. These issues are discussed in section 2.4.

A constant criticism, historically and in all countries, has been that many sources of distortion problems are well-known in principle, but the overall management of the design and manufacturing process fails to address these problems and apply the necessary controls. Several of these high-risk factors, as they occur in a shipbuilding context, are listed in section 2.5 and their relevance described. In some cases, there is no argument that the factors have a consistently negative effect, for example, designs featuring particularly thin plate or excessive welding. In other cases, there may be ambiguity. For example, there are differing opinions concerning the use of structural restraint during welding and indeed the outcomes of such use may be positive or negative depending on the circumstances. The use of quantitative assessment tools is particularly critical in these cases.

Finally, section 2.6 includes a brief description of methods that can be used to rectify distortion once it has occurred. However, as these methods are costly, a better aim is to reduce the incidence of distortion in the first instance.

2.2 Welding distortion of stiffened-plate and other fabricated structures

Heat-induced distortion and buckling in welded structures are significant problems in a number of manufacturing areas. Shipbuilding is an obvious case, where distortion due to internal welds is sometimes seen on external surfaces such as hulls. This phenomenon is often described as 'the hungry horse effect', owing to an unfortunate resemblance to the protruding ribs of an under-nourished horse. Although ships are primarily constructed in steel, distortion problems have also been experienced with other metals such as aluminium, titanium and stainless steel. A very obvious example of thin-plate distortion in stainless steel is the Thames Barrier in London, which has external surfaces that have been described as 'quilted' (Fig. 2.2).

Ship plate distortion has been made worse in recent times by progressive moves to the use of thinner, higher-strength steels. This is particularly true of recent Destroyer builds in the USA, UK and Europe and future builds are following this trend with the same basic thin-plate designs. For that reason, a significant proportion of the work and examples described in this chapter are related to the shipbuilding process. However, the discussion and findings are equally applicable to other processes and metals, particularly those with low thermal diffusivities similar to steel (see Chapter 4 (section 4.2) for a definition of thermal diffusivity).

Issues concerning thin-plate distortion have probably never received the correct level of applied research to fully understand the root causes, as it is

2.2 Thames Barrier, London, showing distortion of stainless steel thin plate.

usually feasible for fabricators to fall back on some kind of rectification process. Such rectification, described variously as ‘heat-straightening’ or ‘heat-adjusting’, normally relies on the use of a heating torch, where the level of control can be highly variable. A heating cycle is applied and, as the material cools, it shrinks more than it has expanded, whereupon the out-of-plane distorted height is reduced. This assumes that the process is carried out correctly and that it does not induce further buckling.

Many fabricating organisations assume that a specific level of heat-straightening rework will be incurred in a project, and will factor such cost into the budget. As a result, there has never been a serious drive to resolve the basic problem and, although significant laboratory-based research has been undertaken, adoption of the findings has been low.

Several years ago, the findings of a detailed study on thin-plate distortion rectification were highlighted,[1] based on construction of the Arleigh Burke Destroyer class in the US Navy. The quoted initial rework figures related to heat-straightening were generally in line with data noted elsewhere in Europe. The most significant figure in the statistics was the *consequential* cost of carrying out heat-straightening, which was reported as being ten times the value of the initial heat-straightening cost (see also the statement attributed to Okerblom near the end of section 1.1). An initial figure of £250,000 for heat-straightening rose to around £2.5 million, when factors such as moving equipment, removing insulation, reinstating insulation, reinstating paint damage, effects on schedules, etc. were included. The ten times factor will of course depend on the stage at which heat-straightening is carried out. If it is carried out prior to insulation and final painting, then the ten times factor might drop. However, even if the reduction factor is estimated conservatively to be in the region of five times, the cost remains significant. A recent UK Destroyer build showed that there is scope to reduce the rectification costs of heat-straightening alone by around 75% through adopting certain relatively simple approaches.

An example of distortion problems in another material, namely titanium, has been noted in the manufacture of lightweight howitzer guns. These guns are expected to operate in extremes of temperature. Titanium had a number of attractions as there was an upper weight limit for the guns, due to helicopter transport limitations. The benefits of using titanium, relative to steel, are that it is as strong as steel but 45% lighter and also has a low ductile-to-brittle transition temperature, high strength and high corrosion resistance. It is, however, more expensive and liable to contamination. Distortion was found to be a significant fabrication problem, creating a high risk of rework. As will be shown in Chapter 4 (see Table 4.2), the theoretical potential for distortion in titanium is greater than for steel. This was found to be the case in practice, regardless of the welding process used – whether tungsten inert gas (TIG) welding, plasma arc welding or metal inert gas (MIG) welding. In this

2.3 Out-of-plane distortion of a titanium structure.

case, the remedial strategy was not to tackle the distortion issue directly, as the potential for success was perceived to be low; instead, several of the components in the gun assembly were *cast* in titanium, instead of being fabricated. In one component, a 77% reduction in fabricated parts was achieved, giving a reduction in welding cost and some financial saving, although this was not large. Welding distortion is therefore a significant problem for titanium fabrication, to such an extent that the use of an on-line peening process, carried out behind the weld, has been advocated.[2] Figure 2.3 shows the titanium-clad cosmonaut monument in Moscow where several areas of out-of-plane deformation on the plates can be seen.

Similar problems in fabricating large, complex parts for earth-moving equipment have been reported.[3] Computational simulation was used in that case to optimise the fabrication procedures and to design the assembly jigs so that the required dimensional tolerances would be achieved. In many earth-moving equipment applications, fabricated components manufactured as replacement parts are required to fit correctly to previously manufactured machines in the field, without adjustment. In such cases, distortion has to be very carefully controlled to much tighter limits than would be determined by normal engineering performance. Computational simulation again provides a basis for testing different weld sequences to prevent heat build-up and component distortion. Rigorous implementation of an optimised procedure will then go a long way towards maintaining assembly tolerances. Welding is also a key manufacturing process in the automotive body sector and, in addition to the problem of distortion control, residual stress fields and unsuitable weld profiles may exacerbate potential fatigue problems in service.[4]

Table 2.1 Typical heat inputs for single-wire submerged arc seam welds (single-pass, single-sided, no-gap weld, no arc efficiency allowance)

Plate thickness (mm)	Heat input rate (kJ/mm)	Heat input rate/thickness (kJ/mm^2)
10.0	3.70	0.37
8.0	2.80	0.35
7.0	2.60	0.37
6.5	2.35	0.36
6.0	2.20	0.37
5.0	1.85	0.37
4.0	1.74	0.44

2.3 Outline of a typical fabrication process

The following description of a typical shipyard fabrication process shows the various stages of fabrication and highlights when and where distortion may occur. This example is based on a manufacturing facility where plate and bar cutting form the primary preparation processes. This stage is followed by seam welding, where an automated welding process is used to join plates together to form flat panels. Several different processes can be applied in seam welders, namely submerged arc welding (SAW), MIG welding, hybrid laser welding and friction welding (for future development).

Several variables, relevant to distortion, figure in the seam welding process, such as the gap set between the plates to be welded, the weld edge preparation (e.g. square-edge or V/Y shape) and whether the panel is welded from two sides or from one. Most panel lines operate on the basis that the welding process will be single-sided, reducing the time and trouble required to turn the panel over for the second side. Common industrial practice is to use a square-edge preparation if possible, as that reduces the time on cutting machines. However, the square-edge treatment tends to increase the heat input requirement at the welding stage, which in turn creates more distortion. For thin-plate construction, almost all panels will be welded with a single-sided, single-pass welding process. Table 2.1 shows typical heat inputs for a range of thicknesses. It is useful to normalise the heat input figure for the thickness being welded, and that figure is also shown to be relatively consistent for the particular configuration being used in this case.

There may also be sources of heat extraction within the seam welding machine system. For example, if physical clamping is used, the clamps can in theory provide a path for heat conduction from the workpiece and hence cause heat extraction. However, thermal studies (see Chapter 7) indicate that contact between the surfaces of the plate being welded and any clamp or other heat extraction path has to be very tight to ensure heat conduction. Heat may also be extracted if the seam welder is fitted with a water-cooled

2.4 Typical distortion after seam welding showing lift at end of plate (courtesy BAE Naval Ships).

backing bar, but views on this are contentious and there is little hard evidence. These considerations create uncertainty as to the heat energy conducted into the plate, this being the main driving parameter for welding distortion.

Distortion found after seam welding tends to be according to a very elongated saddle shape, as shown in Fig. 2.4. This is similar in character to the exaggerated schematic views of Fig. 4.7(d) and 4.7(e) in Chapter 4 and this pattern is reiterated at various points in later chapters (notably Fig. 10.17(b)).

The next stage of the process after butt welding is to fit and weld stiffening bars to the panel. In practice, these bars are normally offset bulb plates (OBPs), as shown in end profile in Fig. 2.5. Practice in the Far East tends towards the use of different rolled sections such as T-bars or angles. Attachment of the bars can follow different sequences, as follows.

- Longitudinal bars are tacked into position on both sides and then welded. The transverse bars are then tacked into position and welded.
- All bars, longitudinal and transverse, are tacked into position and fully welded, perhaps using robotic welders to manipulate around inside the semi-enclosed spaces.

Or

- The bars are lifted on a grab system, loaded against the panel and tack welded.

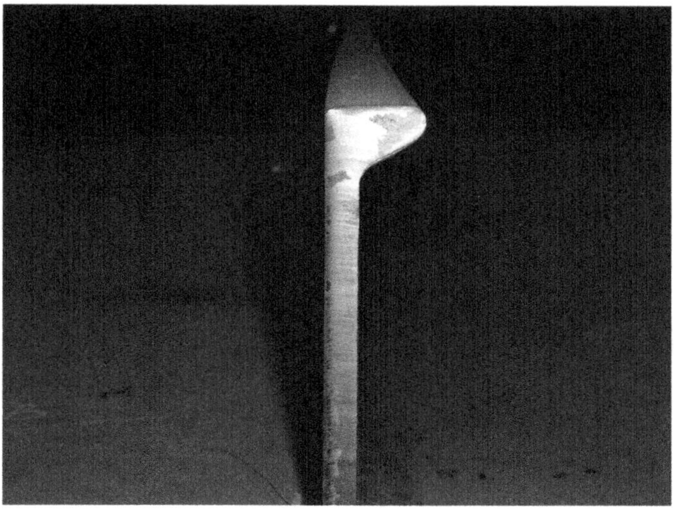

2.5 Offset bulb plate (OBP) (courtesy BAE Naval Ships).

Bars can be welded using several configurations.

- Double continuous-fillet welding: both sides of the bar are fully welded, either consecutively on opposite sides or with two welding heads on opposite sides.
- Intermittent welding: a staggered welding pattern on opposite sides, which could reduce the total welded length by just under 50% (and the heat input similarly).
- Welding on one side of the bar instead of intermittent welding: this is not often used due to the possibility that the bar may be pulled over to one side with the creation of a corrosion-prone crevice.

Several welding processes can be used to carry out fillet welding. These are

- submerged arc welding (SAW)
- MIG welding with solid wire, flux-cored wire or metal-cored filler wire
- hybrid laser welding – from one side or from two sides
- manual metal arc (MMA) welding.

The main function of the stiffener fillet-welding process is to ensure that the bars are attached with adequate weld penetration, but not over-penetrating to an extent that would result in greater distortion. In some instances, full-penetration fillet welds are required for structural reasons and, in these cases, the main aim is to minimise the amount of external weld metal laid down in the fillet, to mitigate against increased distortion tendencies.

It is perfectly feasible to have all the bars positioned on a panel and to carry out the fillet welding to a predetermined welding sequence using a robot. This could then ensure that any areas where a high local volume of welding is required are programmed to cool down before carrying out more welding in that area. This will normally reduce the potential for distortion and help to ensure a consistent outcome. Similarly, welders using a manual or semi-automatic process may be given a predetermined weld sequence to complete the stiffener welding stage. ('Semi-automatic' means that filler wire is fed automatically to a welding gun, but the gun itself is manipulated by the welder, who thereby determines the travel speed.)

2.4 Raw materials and primary process factors

Various aspects related to the supply and stocking of plate and bar materials can figure in the build-up of distortion. The main requirement is that materials should be transported and stocked flat. In effect, this means that there should be frequent supports to ensure that at no time does material bend to the extent that yielding occurs, which would induce residual stress. This is particularly important with very lightweight flat bars or offset bulb bars, which can easily sag between widely spaced supports.

The quality of raw materials is critical to the success or failure of any fabricated structure that demands a high standard of flatness. In shipbuilding, the quality of plate supply with respect to flatness is known to be highly important. The flatter the plate initially, the less liable it is to distort during welding. This statement will be reinforced again and again throughout this book, particularly in Chapters 8 to 12 where the effects of initial out-of-flatness are quantified both by experiment and by simulation. It has also been confirmed directly by experience and observation in a situation where the outcomes of two supposedly similar fabrication contracts, running closely together, were compared. The distortions were significantly different, the crucial difference relating to the fact that different flatness tolerances had been specified for the material supply to the two different contracts. The final distortion on the poorer flatness tolerance material in one contract was significantly worse than on the flatter material in the other, all other factors being the same.

Another benefit of flatter plate is that it is more straightforward to tack the plates together and less local adjustment is therefore required to bring the plates into line. If plate profiles do not match, areas of opposing out-of-flatness have to be forced into alignment, thereby generating a locked-in stress pattern. This problem is normally evident after welding when problematic fit-up areas prior to welding are found to be distorted. A similar philosophy holds for bars. Bars with minimum initial longitudinal bend, or bow, create less distortion. If the bar is initially bent, it has to be pulled against the plate or, in the case of stiff bars, the plate is drawn against the bar. Both of these cases create residual stress systems that interact with the welding thermal stresses and create distortion. Some of the current specifications for permitted initial bend of longitudinal bars are too lax in this respect, in allowing up to 20 mm offset along the length of a 12 m long bar.

In both bar and plate operations, the cutting processes used should be operated at as low a heat input as possible. Current practice is to cut thin plate wherever feasible on a laser machine and, when cutting a bar with slots, this should be carried out on a dry plasma cutting machine. Adopting these actions will generate fewer sites with the potential to trigger distortion. Chapter 7 includes a section on the simulation of thermal cutting operations.

The speed and accuracy of laser cutting machines provides an incentive to include all cut-out operations in the initial plate preparation stage. However, in some cases where the design includes significant cut-outs, there are sound reasons to delay formation of these major openings until as late a stage in the build as possible. For example, in a particular case, window openings in a 4 mm thick structure on a ship's bridge were left uncut until completion of the welding stages, when the structure would be in its most stiff configuration.

The substantially unknown issue concerning stock material supply relates to the possibility that residual stress fields, arising from the material processing stages, may be present in plates and bars. It has been suggested that practices in plate mills are important, particularly with respect to cooling after processing. The use of hot-levelling followed by-cold levelling has been suggested as being beneficial, particularly in the case of thin plate. In general, these are factors which have been poorly researched, not least because process measurement of residual stress is not feasible. The effects on distortion are therefore difficult to quantify, in part because some of the effect is related to the interaction of residual stresses with welding thermal stress and the presence of residual stress also tends to be associated with poor initial flatness.

The process of cold-rolling plates or 'mangling' has been carried out for many years in shipyards. Ostensibly, the purpose is to flatten the plates and make fairing of the structure easier. There appears to be very little science in the understanding of effects, but the process does appear to be beneficial. Based on a limited number of small-scale trials carried out on 4 mm thick DH36 plate processed on shipyard rolls, a 12% reduction in out-of-plane distortion was measured. Of course, many different factors interact in the overall sequence of processes and some care needs to be taken in concluding that any small improvements at this level are due solely to cold-rolling. However, there are enough data to indicate tentatively that there are benefits to be obtained by applying a levelling approach to thin plates. Again, these studies provide no insight into whether the benefit from rolling is due to modification/reduction of residual stress systems or due to improved initial flatness resulting from the rolling process. The experimental and computational studies shown later suggest that the major benefit comes from the improved initial flatness. A similar observation related to earth-moving equipment has been reported on thicker plates (~12 mm), which were strip-cut to witness any effects resulting from roll-levelling the plates. On the levelled plates, the strips stayed relatively flat while on the non-levelled plate there was significant bending of the strips. This suggests that levelling also reduces the effects of residual stress.

2.5 Management issues relevant to thin-plate distortion

The previous section highlighted various important problem areas such as steel mill processing, transportation, stocking, cutting, geometry control and flatness. An in-depth study of distortion[5] clearly identified an over-riding problem that, although the importance of these issues is not a secret, controls to address them are not always put in place. This is essentially a deficiency of management, as it is known that certain factors are likely to create more distortion. The risk areas listed below are dealt with in more detail in this section.

- Designs involving reduced plate thickness.
- High heat input or over-welding.
- Concentration of welding in localised areas.

In addition, there are other factors that are known to contribute to, or indeed mitigate distortion, but there may be a lack of clarity as to the likely effects. These include

- inadequate support of assemblies during welding operations
- restraint of thermal movement
- tack welding of assemblies
- residual stress, due to initial supply or induced by welding
- additional welding applied (possibly unnecessarily) for non-structural purposes
- design choice in frame spacing (a factor in the 'hungry horse' phenomenon).

The last item above is normally determined by the structural design of the ship or other fabricated structure involved. However, it is essential to explore alternative options to the obvious weight reduction approach of simply reducing the plate thickness progressively, to a point where, at around 4 mm, various buckling and large-displacement deformations are very difficult to avoid for reasons shown later.

From a practical viewpoint there are two fabrication issues in the above list that, on the basis of current information, could be managed better. These are additional welding for non-structural purposes and tack welding. The effects of tack welding as an assembly tool are also covered extensively in tests and simulations described in sections 9.3, 10.3 and 12.2.2.

2.5.1 Designs involving reduced plate thickness

Distortion control issues become significantly more severe as plate or section thicknesses are reduced in designs. Figure 2.6 shows data taken from actual shipbuilding cases and demonstrates the marked adverse effects of including 4–5 mm thick plate in projects. Distortion index as used in the figure is a statistical quantity that characterises lack-of-flatness; this is discussed further in Chapter 6.

The main reason for this marked increase in distortion is related to the longitudinal compressive stress induced by contraction of the weld and the increased tendency to unstable large-displacement deformation or buckling behaviour as the thickness is reduced. For example, a steel plate of width a, length $3a$, and thickness t, simply supported on all edges[6] will reach a buckling condition when subject to a longitudinal compressive stress σ_{cr} given by

$$\sigma_{cr} = 3.29 \frac{E}{(1 - \nu^2)} \left(\frac{t}{a}\right)^2 \qquad [2.1]$$

Inserting appropriate values for Young's modulus E and Poisson's ratio ν and setting the plate dimensions as 1 m width and 3 m length, the theoretical Euler buckling strength is given as shown in Fig. 2.7.

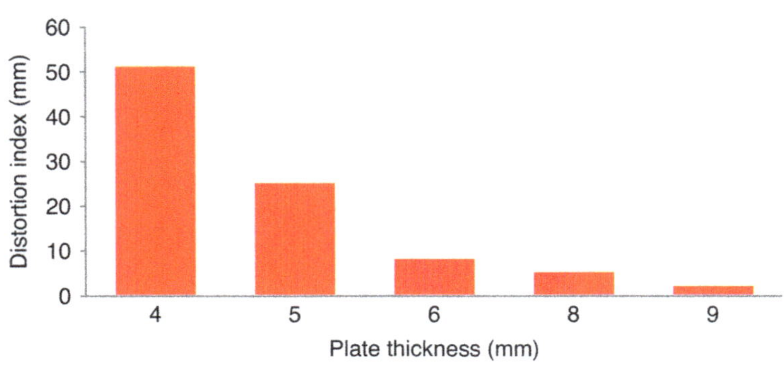

2.6 Plate distortion as a function of thickness, based on actual build data.

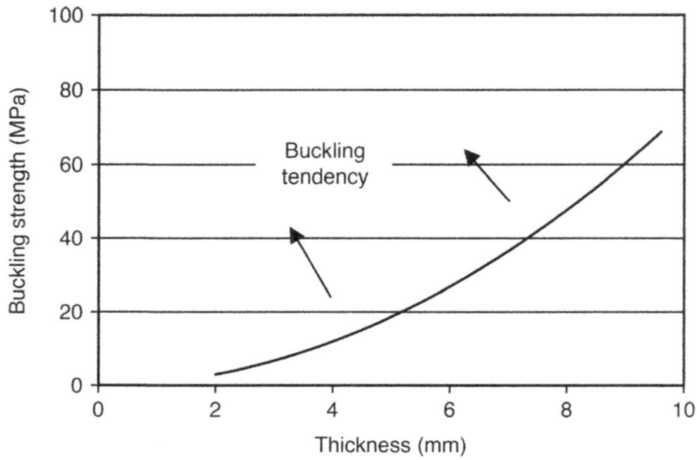

2.7 Typical relationship between plate thickness and buckling strength.

As the contraction force is strongly related to welding heat input, it is possible to derive an approximate relationship between heat input and thickness (see section 4.8) at which a plate of the dimensions given is likely to buckle, as shown in Fig. 2.8. This theory is somewhat pessimistic, as the compressive stress caused by weld contraction is not at a uniform level across the plate width, as assumed in equation 2.1. Nevertheless, comparison with the typical heat input levels of different processes, shown in Table 2.1 and in Fig. 2.9 discussed later,[7] suggests that welding of steel plate less than 6 mm thick must be challenging, as indeed is found to be the case in practice. The buckling strength of stiffened plate is of course much higher, but in severe cases it is nevertheless possible to distort stiffened

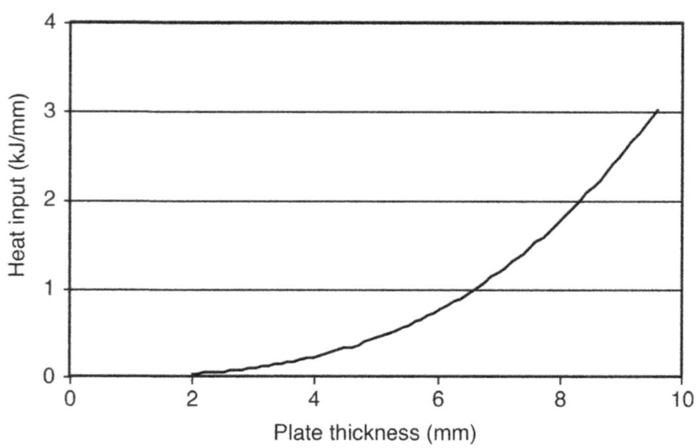

2.8 Relationship between welding heat input and plate thickness.

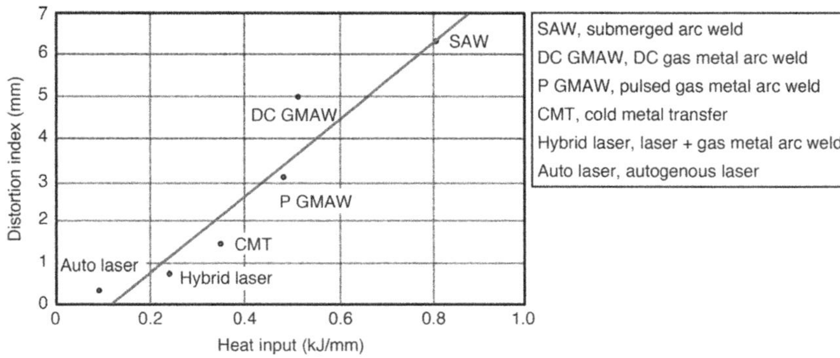

2.9 Effect of welding process and average heat input on distortion (after Colegrove *et al.*[7]).

plates, as will be shown in Chapter 11 on fillet welds and Chapter 12 covering distortion reduction studies.

The question of design choice of thickness is related to yield strength, to an extent, and some of the justification for thickness reduction is often based on the use of higher yield strength material. Elastic buckling strength, as such, is independent of yield strength, but a lower yield strength material will lead to greater permanent deformations and distortions in pre-buckling and post-buckling behaviour. An artificial neural network (ANN) study[8] indicated that lower yield strength (275 MPa) steel was more susceptible to distortion than higher strength (360 MPa) steel. It will be shown in section 4.5 and at other points that, to a first approximation at least, yield strength should not be a strong factor in the development of weld-contraction-induced distortion. However, the differences in material yield strength in such steels are usually related to higher carbon or alloy content and consequential microstructural differences. These will often affect the expansivity of the material and may be an indirect reason for the apparent effect of higher yield strength in reducing distortion. Brust and Kim express a contrary opinion that higher yield strength leads to greater distortion, due to the development of higher residual stresses.[9] The ANN work also confirmed that reducing plate thickness is a significant contributor to distortion. However, there are some options that could mitigate the problems in fabricating thin-plate materials and these will be discussed later.

2.5.2 High heat input or over-welding

The information shown in Fig 2.9 illustrates the effect of heat input on distortion, i.e. the higher the heat input rate the greater the potential distortion.[7] This determination does not entirely allow for heat input or arc efficiency (see section 7.9) as this factor was apparently not varied for the

Table 2.2 Additional weld volume generated by using greater fillet sizes than base case

Fillet leg length (mm)	Over-welding volume (%) compared with volume of 3.5 mm fillet
3.5	0
4.0	31
4.5	65
5.0	104
5.5	147

different processes and the different values should perhaps be considered as relative rather than absolute.

The data shown in Table 2.2 also demonstrate the effect of progressive over-welding of a 4 mm leg-length fillet. During a specific build, it was mistakenly accepted that a 4.5 mm leg length would be the lowest achievable fillet size, when it was actually feasible to produce a smaller 3.5 mm leg length. In this instance, the extent of over-welding resulting from this misapprehension was 28%. In effect, at least 28% more heat was being input to the joints than was required and, as will be shown later, the distortion would increase approximately *pro rata*. Part of the solution to producing a smaller but consistent 3.5 to 4.0 mm leg-length weld was to use a metal-cored welding wire with a MIG welding process. In fully automatic welding, metal-cored filler can be applied at higher welding speeds than a flux-cored filler.

Over-welding can take various other forms. Figure 2.10 shows a case

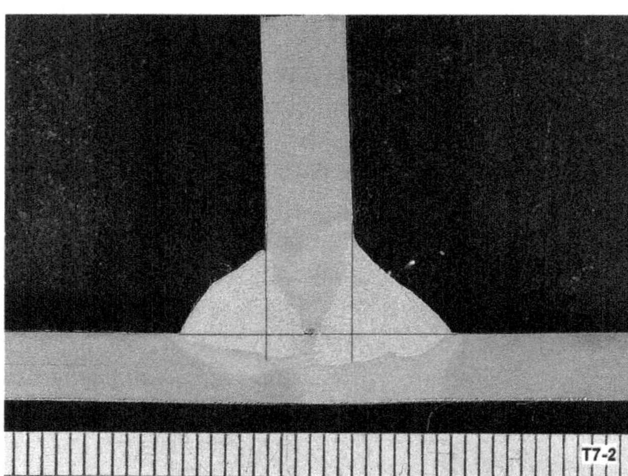

(a)

2.10 Fillet welds showing similar external size but significantly different heat input: (a) significant over-penetration; (b) adequate penetration (courtesy BAE Naval Ships).

(*Continued*)

(b)

2.10 Continued.

where the external size of a fillet weld is seen to be as required in both examples, but excessive penetration seen in Fig. 2.10(a) has resulted in a much greater heat input than the result shown in Fig. 2.10(b). Weld penetration is extremely difficult to assess without extensive trials involving weld sectioning. Statistical modelling processes such as ANNs, discussed more fully in Chapter 3, offer a strategy to optimise process parameters and to improve overall control.

Simple design decisions can have a substantial effect on distortion outcomes, possibly to a far greater extent than designers realise. In the case of fillet-welded stiffeners, 50% of the heat input can be avoided by specifying intermittent welding, as mentioned earlier. Figure 2.11 shows an intermittent weld pattern on aluminium bars welded to plate. Normally, the shear load transfer in a stiffener-web attachment is satisfied by a cross-section much less than the thickness of the web itself and this strength requirement can be met by a reduced total length of weld. However, a significant disadvantage of the intermittent weld strategy is the large number of weld stops. Every weld stop has a potential for crater cracking or pore formation (Fig. 2.12). In addition, if the structure is subject to fatigue loading along the line of the stiffener, each weld end forms a stress concentration, which is a common initiation point for fatigue failure.

In some cases it is permissible to weld only a single side of a bulb bar stiffener in a ship structure, instead of specifying intermittent welding. This reduces the overall heat input to the same level as 50% intermittent welding and has less potential for stop craters and subsequent cracking. However, there is a tendency to pull the bar slightly to one side, off the vertical, which may not be acceptable. An alternative approach to this problem in the case

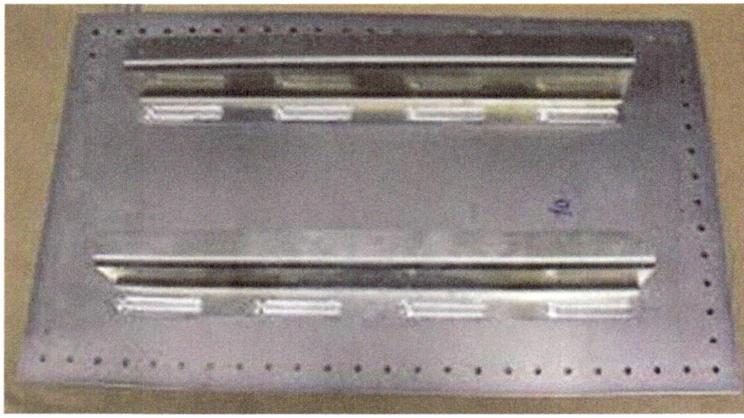

2.11 Intermittent fillet weld pattern.

2.12 Pore defect at fillet weld end – result of insufficient dwell at end of welding.

of bulkheads is to use swedged stiffening instead of welded stiffeners. An example of a swedged bulkhead is shown in Fig. 2.13.

Developments in laser welding have led to the sandwich panel concept – a low heat input low-distortion product that can also be used for bulkheads. Compared to alternatives, this is a higher cost option, but it does offer a very high degree of cosmetic benefit. The use of laser welding in ship structures has gained a foothold in some areas, but probably not to the level anticipated about 10 years ago. Some shipyards that have adopted laser processes have opted for the hybrid laser concept, which is a combination of an arc process and a laser. This approach ensures that good mechanical

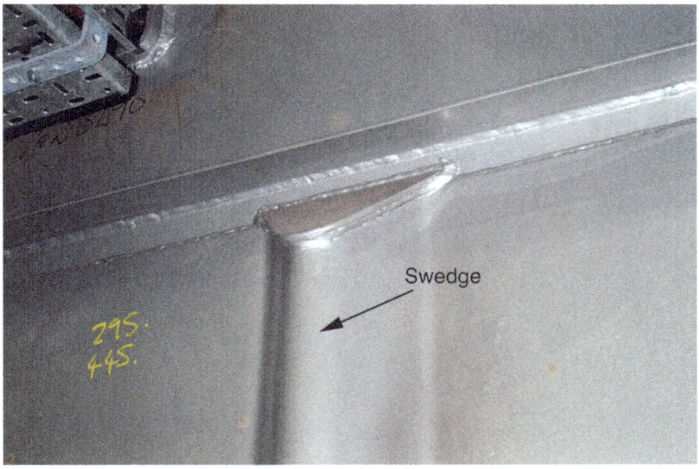

2.13 Example of swedged bulkhead (courtesy BAE Naval Ships).

properties are achieved. In addition, very high levels of panel and sub-assembly flatness have been achieved. However, it is evident that some of these benefits are lost when the sub-assemblies are subsequently combined to construct a three-dimensional structure.

2.5.3 Concentrating welding in localised areas

Concentrating too much welding simultaneously in the same area can create significant distortion by generating a local build-up of heat. If each area is allowed to cool before revisiting for further weld completion, better levels of flatness can be maintained. This is the basic reason for the strategy of weld sequencing. For example, welding from the centre outwards spreads the heat to both ends of the structure. The same concept can be used, even more easily, in fillet welding. In some automatic fillet-welding applications, multiple arc systems are applied, where the arcs are too close together, resulting in high heat build-up. Modelling work (see section 12.2.3) demonstrates the benefits of separating the arcs as much as possible, as this will reduce both distortion and locked-in porosity.

Figure 2.14 shows a simulated fabrication procedure that was put into practice in welding a built-up unit. In the version shown in Fig. 2.14(a), no systematic weld sequence was followed and substantial distortions are seen on some of the internal bulkheads. In the version shown in Fig. 2.14(b), a sequence of weld application was chosen to avoid a concentration of heat, with consequent reduction in distortion.

Some very basic rules for weld sequencing have evolved, although they may not always be put into practice if the need to increase productivity

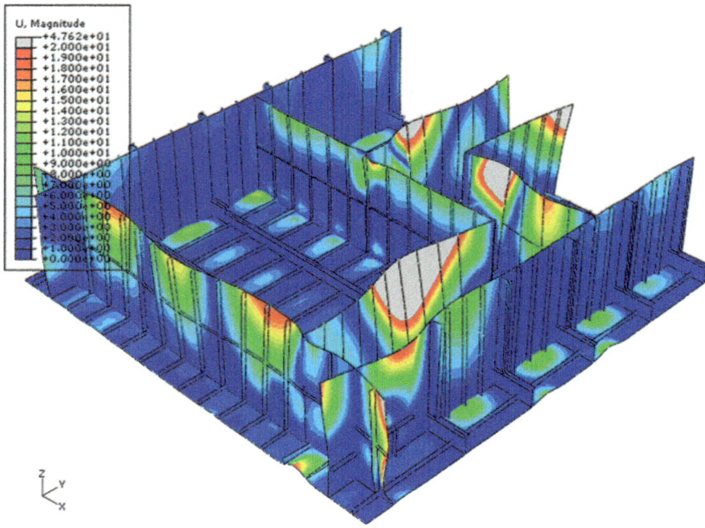

(a)

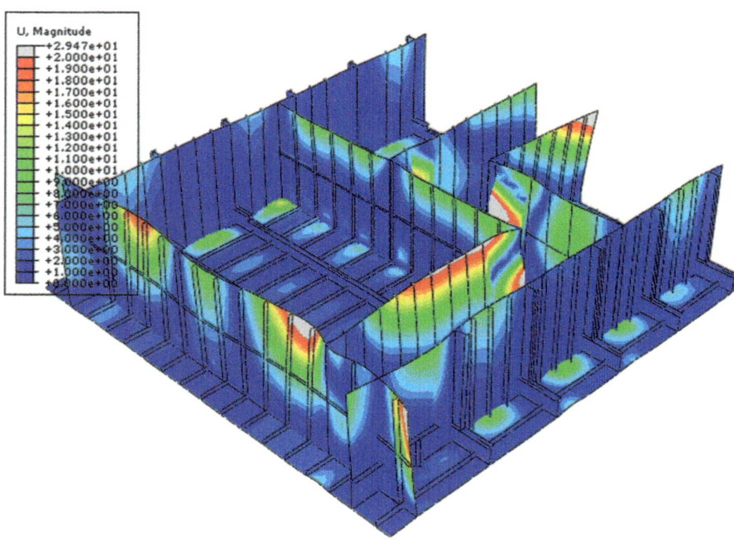

(b)

2.14 Predicted distortion under different weld sequencing strategies: (a) random weld procedure; (b) sequenced weld procedure (courtesy BAE Naval Ships and Dr J. Mawella, MOD).

overtakes the requirement to minimise distortion. This is of course a false economy if additional rectification processes are required. Welding from the centre of a structure outwards is good starting point for a strategy. If two welders are working on the structure simultaneously, then they should work away from each other to drive any heat build-up towards the edges. Thereafter, local experience will suggest further refinements. However, the effects may be difficult to predict and the use of computational modelling can then be useful as an aid to minimising distortion.

2.5.4 Inadequate support of assemblies during welding operations

Reference has already been made to lack of initial flatness as a cause of distortion and, although this may be due in the first instance to poor stock supply or build-up of distortion in consecutive welding operations, it may also be related to lack of uniform support of the structure being fabricated against the out-of-plane force of gravity. The resulting out-of-plane movements are not simply the elastic displacements that would be recovered if uniform support was applied, but may also be the result of non-linear interaction with welding thermal stresses, so that permanent deformations are generated. The gravity and support problem is treated in computational studies as a factor that can be readily included in the boundary conditions of a finite-element model. However, the obvious requirement to support thin flexible plates evenly and uniformly is not as easily achieved as might be imagined, particularly if the plates are not flat in a stress-free state. Unless the supports are sufficiently compliant to follow transient deformations as the welding and cooling process develops, gravity forces will still intervene.

2.5.5 Restraint

There is sometimes a belief that, if the structure being welded is restrained, flatness will be maintained. However, there is little agreement that this is true. Section 10.4 examines a specific case to determine the effects of clamping; it was found in that instance that rigid clamping of a butt-welded plate resulted in a poorer distortion outcome. This question also depends on the configuration of particular build facilities.

For example, in one shipyard (A), the panel line may be based entirely on roller transportation and support, with no solid bed for the structure. In another case (B), the panel is studded down before welding against a steel transportation pallet, and it is therefore restrained to an extent, depending on the stiffness of the pallet. The main difference between the two treatments is that the transverse vertical connections between stiffener bars on line A are not welded out prior to transportation to the next process stage. At this point, the panel is pulled down against a build jig and, as the vertical connections are free in the roller support case, there is a degree of movement as the panel is pulled down. This would not be possible if the vertical connections had been welded prior to applying the restraint. However, on line B, the panel is fully welded out, including the vertical connections, as the panel has already been restrained relatively flat.

There is some evidence from finite-element simulation that welding of the vertical connections is a major source of distortion in panels. In both line A and line B, the different practices are actually carrying out the optimum process but in a different sequence. It is therefore critical to consider the overall process of fabrication and not to blindly follow the practices of one builder or another. In this comparison of practices, it has been found that restraint has been found to be beneficial, although insufficient measurements were made to prove the point conclusively.

Some very basic work carried out on restraint as part of a separate ANN study reached the conclusion that restraint could *not* be classed as being beneficial overall (M.P. Lightfoot, unpublished observations, 2007). For instance, although the 'pattern of restraint' formed one of the variables in the ANN parameters, the 'style of restraint' was important ('pattern' refers to *where* the restraint was applied, whereas 'style' refers to the type of restraint). Furthermore, the level of distortion already present in the area being restrained was an important factor.

2.5.6 Tack welding of assemblies

Tack welding is commonly used as a temporary fixing method in the assembly process. However, any effects arising from tacking are not usually highlighted as contributory factors to the occurrence of distortion. If tacks are recognised for what they are (namely, generators of small amounts of heat input over the length to be joined finally) then, logically, some movement between the items being tacked should be expected. A series of trials on butt-welded plates was conducted and plate shape was measured at various stages using a photogrammetry technique. By comparing the flatness at three stages – prior to tacking, after tacking and after welding – the contribution of tacking to the overall distortion was found to be almost 33%. The various contributions are shown in Table 2.3. As will be shown in later chapters, this form of comparison rather underrates the influence of tacking, as the out-of-flatness at each stage has a non-linear effect on the following stages. However, these results were based on work carried out on 4 mm thick DH36 grade steel and the effects would not have been so marked on thicker material.

Once this effect had been identified, further investigation showed that tack welds typically occupied about 14% of the length to be welded in the final fabrication. This effect is not always readily identified on the shop floor

Table 2.3 Contributions to total distortion of butt welds

	%
Distortion attributed to initial shape	14.4
Distortion attributed to tack welding	32.7
Distortion attributed to butt welding	52.9

due to other contributing factors such as lack of initial plate flatness and unknown inherent residual stress fields, over which no action can currently be taken. Studies on the outcomes of different tacking strategies are described in sections 9.3 and 10.3.

2.5.7 Residual stress

Residual stress is a common factor and theme across the range of distortion problems arising in industry. However, it manifests itself in two aspects of the fabrication process in particular.

The first area of influence is in the delivered material – in both bar and plate form. Mill processing of bar and plate can induce residual stresses into the material, the effects of which will not appear until heat is applied to the product in some form. The stresses can arise from a number of sources, for example differential cooling or stretching during the rolling process. Differential cooling of plates can occur for example on the cooling beds of a plate mill where the plate is not moved far enough backwards across the bed supports. In addition, the plate ends and edges cool faster. Plates may also be stacked unevenly, giving rise to random non-uniform cooling. In the case of a bulb bar used in ship structures, the bulb will cool significantly more slowly than the web and this will set up residual stresses in the material, often manifested as bowing of the bar in the delivered state or progressive bowing during the bar cutting process.

The second aspect of residual stress influence arises from welding processes and these form the main source of distortion and/or buckling in the case of thin plate. The nature and distribution of these residual stresses will be discussed firstly in section 4.7 and at several points thereafter in different chapters.

2.5.8 Additional welding applied for non-structural purposes

The bulkheads in many ship structures are no more than 4 mm thick, but have a high level of welding associated with the 'outfitting' stage of construction, these being attachment points for equipment, piping, etc. In

2.15 Localised non-structural welding of outfit items on 4 mm thick bulkhead (courtesy BAE Naval Ships).

some specifications, the weld lengths for structural welding and outfitting purposes are almost comparable. As in the case of structural welding, it is essential to interrogate the design case for weld sizing in relation to seats, penetrations and other similar attachment welds such as those shown in Fig. 2.15. Any reductions in weld volume from such consideration will be welcome.

In the case of bulkhead penetrations, for example, it is important to ensure that the openings are not oversize, thus creating a need for additional weld volume. A strong case can be made for intermittent welding in such locations unless there is a specific shock loading or other structural condition, for example at certain seats.

2.5.9 Frame spacing design choices

Generally, if stiffeners are placed further apart, less distortion will result. Section 12.2.1 considers the effects of different spacing and configuration options through simulations and experimental comparisons. Thicker stiffeners will also increase distortion, if the fillet-weld sizes are increased to match.

However, this is another area where design decisions should be considered integrally with manufacturing process options to achieve optimum solutions. Some recent work has shown that spacing width was a relatively dominant factor when welding large 5 mm thick panels. The average overall distortion increased progressively as the frame spacing was increased from

500 mm to 600 mm and then to 700 mm. This finding related to panels that were welded in approximately the same sequence using an intermittent welding pattern. However, these tests were not cross-checked against modelling studies and it is not clear that factors other than frame spacing were involved. This is a good example of the need to apply all available tools to a case prior to making decisions that could have an influence on rework levels.

2.6 Rectification of thin-plate distortion

While the main drive should be to eliminate or at least greatly reduce thin-plate distortion in the first place, cost-effective rectification processes also need to be available. Flame heating has been adopted in many areas of industry, particularly where carbon steel is involved. The basic philosophy is to heat the steel up and allow it to cool down, thus creating overall net shrinkage. If this is carried out using a manually operated heating torch, a great deal of latitude rests in the operator's hands and control of time and maximum temperature can be arbitrary. Different heating gases can be used, for example propane or acetylene. The potential benefits of acetylene are greater than propane and acetylene also affords more control. However, past experience in the use of acetylene has to some extent created a negative image from a safety standpoint.

A more controlled version of the arbitrary heating process is to stud weld a long aluminium strongback across any depressed area and then heat the plate up progressively until it moves to be level with the strongback (aluminium is used only for ease of handling). This is a relatively well-controlled procedure, but it generates additional work in fitting the strongbacks, removing the studs and grinding back any scarring created by the studs.

Induction heating has been available for many years and has recently been reintroduced in the form of programmed induction heating coils. The length of plate being heated is set, the upper temperature is set, the time at temperature is set, and a defined pattern is then followed. Experience has shown that this is a highly effective process and reduces the time taken to straighten by 50% compared to the process of heating against a strongback. These processes have many variants, which have been in use for a significant number of years with varying degrees of success.

A point to remember about heat-straightening and in-process tensioning is that these methods rely on generating significant residual stress fields in the components. This may be an undesirable source of dimensional instability in the short term and a factor in fatigue strength reduction in the long term.

2.7 Conclusion

Typical distortion patterns in thin stiffened-plate fabrication have been described, covering different materials and applications. A typical fabrication sequence used in shipbuilding was described and stages highlighted in the sequence where distortion issues typically arise. This chapter considered various aspects of design and fabrication where problematic issues are well-known but steps are often not taken to manage them out. These include reduced-thickness designs, excessive welding that may be avoidable, structural configuration decisions that do not take account of distortion outcomes, residual stresses and fabrication procedures such as the use of supports and restraints and tack welding in assembly procedures. The chapter concluded with a brief description of methods used to rectify distortion once it has occurred.

2.8 References

1. Kirk, M., Conrady, C., DeBicarri, A. and Michaleris, P., 1996, Prediction and prevention of excessive unfairness in Arleigh Burke (DDG-51)-class deckhouse fabrication, in *Welding and weld automation*, DeNale, R., Ed., The Minerals, Metals and Materials Society, Warrendale, PA, USA, pp 67–84.
2. Liu, X., Fang, H., Ji, S. and Dong, Z., 2003, Control of titanium alloy thin plate welding distortion by trailing peening, *J. Mater. Sci. Technol.*, 19, 1, pp 184–186.
3. Chen, X.L., Yang, Z. and Brust, F.W, 2005, Modelling distortion and residual stress during welding: practical applications, in *Processes and mechanisms of welding residual stress and distortion*, Feng, Z., Ed., Woodhead Publishing Ltd, Cambridge, UK, pp 225–263.
4. Barsoum, Z., 2009, Local weld geometry and the influence on the fatigue strength, *Proc. Swedish Conference on Lightweight Optimised Welded Structures, LOST, Borlänge, Sweden* pp 26–33.
5. McPherson, N.A., 2006, Thin plate distortion management, *Welding Cutting*, 5, 5, pp 277–282.
6. Young, W.C., 1989, *Roark's formulas for stress and strain*, 6th Edition, McGraw-Hill, New York, USA.
7. Colegrove, P., Ikeagu, C., Thistlewaite, A., Williams, S., Nagy, T., Suder, W., Steuwer, A. and Pirling, T., 2009, Welding process impact on residual stress and distortion, *Sci. Tech. Welding Joining*, 14, 8, pp 717–725.
8. Lightfoot, M.P., Bruce, G.J., McPherson, N.A. and Woods, K., 2005, The application of artificial neural networks to weld induced deformation in ship plate, *Welding J.*, 84, pp 23s–30s.
9. Brust, F.W. and Kim, D.S., 2005, Mitigating welding and residual stress and distortion, in *Processes and mechanisms of welding residual stress and distortion*, Feng, Z., Ed., Woodhead Publishing Ltd, Cambridge, UK, pp 264–294.

3

Tools to deal with welding distortion: predictive modelling and research on in-process techniques

DOI: 10.1533/9780857099327.39

Abstract: Contrasting approaches to control and reduction of distortion are described. A method based on the use of artificial neural networks (ANNs) uses statistical techniques to interpret observations of distortions and to discover interactions between different input parameters. The section on computational simulation is based on mathematical modelling of the physical processes in welding. Several processes that can be applied along with the welding operation or in the post-welding phase are described. Thermal and mechanical methods include transient thermal tensioning, roller tensioning, intense cooling in the trailing region behind the weld and the use of special weld filler metal.

Key words: artificial neural networks, transformation temperature, tacking, computational simulation, weld process modelling, inherent strain, transient thermal tensioning, roller tensioning, low-temperature-transformation, cryogenic cooling.

3.1 Introduction

Chapter 2 aimed to convey something of practical experience from an industry that is based on the fabrication of thin-plate stiffened elements and where welding distortion is a major problem. Several messages can be found from this experience, including the following.

1. Welding distortion is typically a complex phenomenon. Several factors appear to be involved in any given case and the relative significance of each factor is not always clear. The relationship between cause and

effect is not straightforward to determine and is usually non-linear in thin-plate construction.

2. The management of design and fabrication processes in industry is frequently deficient. The basis of that claim is that, although many of the factors that promote or reduce distortion may be known within an organisation, insufficient measures are put in place to make sure that all that could be done to minimise distortion is done in practice. The complexity noted in point 1 and the difficulty in understanding the relative significance of different factors may form part of the reason for this. It is also clear that distortion is not simply a problem to be addressed by those within the organisation concerned with fabrication but needs to be considered fully at the design stage in an integrated way. This may be difficult to implement in some organisations where the functions are separated.
3. An obvious conclusion to be drawn from the previous two points is that little progress is likely, in terms of producing fabrications of higher dimensional quality, unless active measures are put in place in the first instance to *plan* the fabrication route, to *record* how a given structure has been fabricated and to *measure* the dimensional outcomes, perhaps to an extent beyond the requirements of the build specification. The scale of typical structures in this field is such that trial fabrication exercises are rarely practicable and it is therefore important to extract as much information as possible from ongoing live fabrication experience.

In the light of these three points, the most important tools in dealing with distortion are systematic *observation* of outcomes and rigorous *control* of design and manufacturing processes. Sections 3.2 and 3.3 in this chapter describe some tools that have the potential to assist in understanding the mechanics of welding distortion in a way that provides better strategies for control of the phenomenon. Section 3.4 describes three special process applications, currently at the research stage, which have the potential to reduce distortion.

3.2 Artificial neural networks (ANNs)

Neural network methodology has been applied to a wide variety of tasks in science and engineering. One such task is to infer functional relationships between observations and phenomena. In the case of welding distortion, the aim might be to establish empirical models that will relate the inputs applied to a given fabrication process to the likely deformation outcomes. The process by which the functional relationships are established is essentially statistical, but the methodology has a sophisticated capability to deduce

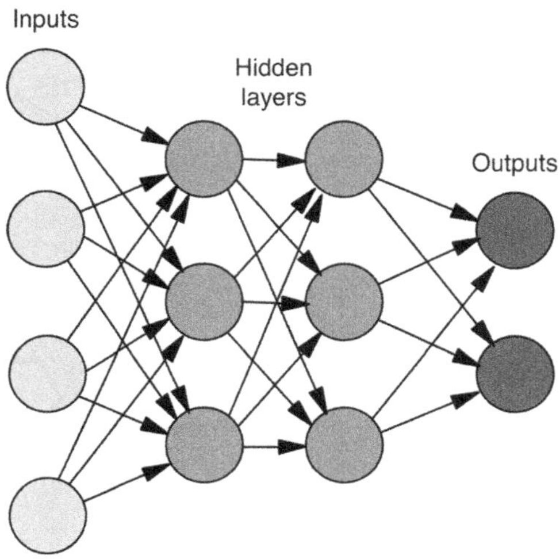

3.1 Schematic representation of a neural network.

interactions between input parameters and to reveal influences that might not be recognised solely through the study of physical models.

A good general introduction to the use of neural networks in materials science[1] notes several applications in welding technology. Lightfoot *et al.* give the only example known to the authors of an application to the welding distortion of thin plate.[2] Figure 3.1 shows a typical schematic explanation of neural network operation. Ideally, input parameters (welding energy input, material thickness, mechanical properties, etc.) should be easily measured and unambiguous. Values that contribute to other parameters (e.g. the welding current value, which contributes to energy input) may also be input independently. It is important to ensure that, for inputs that cover a range of values, there are no significant gaps in the data as there will then be no realistic basis for the fitting process within the range where there are no data.[1] In the case of the example cited above,[2] the distortion output was defined in terms of the predicted out-of-plane deformation at 72 equally spaced points on the surfaces of the two joined rectangular steel plates.

The work of the ANN system is carried out in the so-called 'hidden layers'. Each node in these layers contains a mathematical function, defining a relation between the node input and output. The very simplest of single-layer models, for example, might include a linear regression equation with variable terms or weights[1] which are varied to produce the desired outputs. However, in most ANN software, the functions are non-linear. This feature, together with the multi-layer architecture, allows the system to capture an almost limitless range of non-linear and interactive relationships. The process of finding a relationship that best reflects the behaviour of the real system depends on optimisation algorithms of various types, which progressively

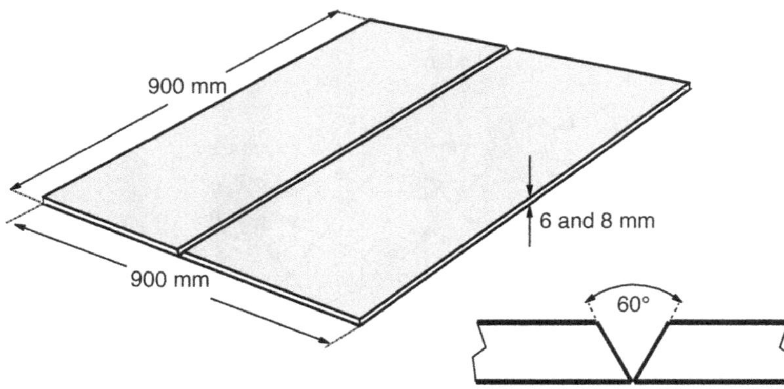

3.2 Test configuration used in ANN study.[2]

Table 3.1 Material input ranges[2]

	Strength and ductility ranges				Chemical composition	
Steel grade	Yield strength (MPa)	Ultimate strength (MPa)	Yield/ultimate strength ratio	Elongation (%)	Individual elements (%)	Carbon equivalent
D-DH36	302–523	422–607	0.716–0.871	24–39	See Lightfoot *et al.*[2]	0.24–0.39

vary the weights iteratively until the desired output is achieved. This process is described as 'training' the system, whereby it is presented with sets of matching inputs and outputs, covering the ranges of input parameters to be analysed. In the work of Lightfoot *et al.*[2] 28 sets of input/output data were used in the training process, with up to 16 000 iterations being required to minimise errors in matching the predicted outputs to those measured. In normal practice of ANN usage, the optimised model and its predictive capability are then 'tested' using unseen sets of data that have not previously been included in the training process. It goes without saying that a very large number of welding trials will be necessary to establish phenomenological trends. Also, predictions on the outcomes of processes and fabrication routes outside the ranges of parameters used to train the system constitute extrapolation beyond valid parameters. For example, if the system is trained on data generated by welding steel, this will give no information on welding aluminium and the likely effects of using intermittent welding cannot be predicted from trials involving continuous welding.

It is of interest to follow through the information presented by Lightfoot *et al.*[2] because almost identical cases, in terms of geometry, material, procedure and process, have been treated experimentally and computationally by the present authors and are shown in Chapters 9 and 10. One of the two

Table 3.2 Ranges of fabrication procedures and processes[2]

Primary processes		Initial shape	Welding process parameters				
Plate treatment	Cutting process	Surface topology at 72 points	Plate thickness (mm)	Welding voltage (V)	Welding current (A)	Travel speed (mm/s)	Heat input (kJ/mm)
Normalised	Laser	Post-tacking					
Normalised and rolled	Underwater plasma arc	Post-welding	6 and 8	21–27	225–275	3.5–6.0	1.013–1.616
As-rolled							

configurations welded in the cited work[2] is shown in Fig. 3.2. The model inputs to the ANN system are given in Tables 3.1 and 3.2 in abbreviated form.

Regarding the chemical composition of the plates (Table 3.1), the percentages by weight of each of the eight elements analysed were matched to the input data for the corresponding plates. 'Carbon equivalent' is itself an empirical relationship, normally used in welding technology to assess hardenability and other features of weld processing.

The individual components comprising the heat input energy (namely voltage, current and travel speed) were also applied as separate inputs. The point here is that, although a physical distortion model based on heat input may be available (see section 4.5) the individual welding process parameters could each have an independent effect on distortion, for example through an influence on the weld shape or some other factor. It should also be noted that there was a difference between the welding procedures for the 6 mm and 8 mm plates – two weld passes were applied in the 8 mm case and a single pass in the 6 mm case. This alters the heat input, as there will be an effective overlap between the stress fields generated by the individual passes. A way of dealing with this problem is given in section 9.4 and also mentioned in section 10.6. Lightfoot *et al.*[2] investigated this point in the first instance by carrying out separate model analyses for the two thicknesses of plate and then by 'lumping' the two weld passes in the 8 mm case into a single pass.

An important point to note is that the tabled input values constituted the only deliberate variations in the fabrication and welding procedures over the range of training samples. Plates were tack-welded together on the underside to a constant procedure and then butt-welded from the upper side. The resulting distortion pattern after initial tacking was in fact very similar to that shown in Fig. 3.3, which was taken from an experimental study by the present authors, to be discussed later in Chapter 10. This deformation is characteristic of tack welding on the underside of the joint. The transverse angular contractions of the tack welds form an inverted V-shape and the longitudinal contractions along the line of the welds bend or bow the plates

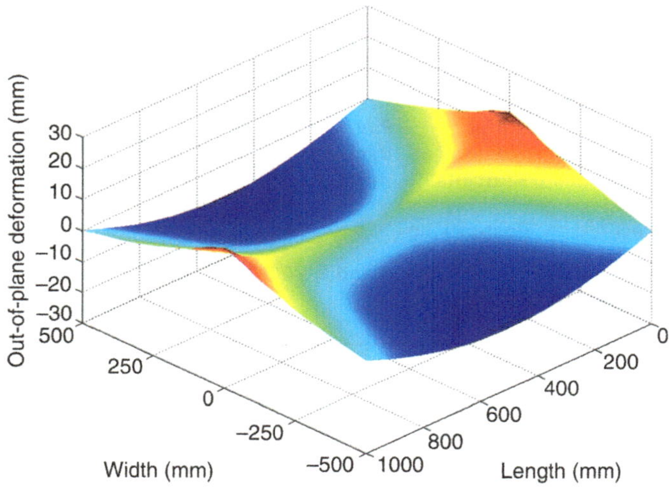

3.3 Typical deformation pattern due to weld tacking on underside of butt joint.

longitudinally into a 'sagging' configuration. The plate edge preparation, tacking pattern, welding conditions and plate thickness in the case shown in Fig. 3.3 were significantly different from those of the previous study[2] but the basic contraction mechanisms described above generated the same overall outcome. During the following butt-welding stage, further longitudinal contraction of the weld had the effect, shown by Lightfoot *et al.*,[2] of increasing the longitudinal curvature, with a relatively small change in angular distortion.

It is therefore important to recognise that the initial post-tacking shape in this ANN study had a critical influence on the subsequent growth of distortion during the butt-welding stage. As no variations in tacking procedure were included in the 'training' process, the model cannot be expected to describe outcomes where other variations may be introduced, such as in the number and placing of tacks.

In the 'test' stage of the model development, predictive results for two 6 mm thick cases were compared with actual experimental results. In both cases, the distorted shape was predicted correctly although, as discussed in the previous paragraphs, this may not perhaps be surprising, given the similar starting points. In one case, the average out-of-plane distortion was predicted as higher than the actual by 13%; in the other, the model predicted an 18% lower distortion than actual (in absolute terms these values corresponded to 'about 2 mm').

Aside from use in the straight prediction of outcomes, a trained ANN model has another capability that may indeed be more useful; namely, as a means of identifying the influence of each input parameter on the specified output. Each input parameter is varied individually, using the final optimised model and the variation in the resulting output is noted. It will come as no surprise that, in the considered study,[2] the strongest influence by far was heat input.

Less expected was a marked influence related to carbon content (ranging from 0.09% to 0.15%), such that distortion was less for the plates showing higher carbon content. Next in significance was yield/ultimate strength ratio (higher ratio giving less distortion), although yield strength, as such, did not seem to have as strong an influence. The authors argue that an increase of the yield/ultimate strength ratio is an outcome of microstructural differences, which may be influencing distortion through a different mechanism. It will be shown later that distortion is directly related to the thermal contraction behaviour of the cooling material and that microstructural changes that lead to lowering of the phase transformation temperature during cooling can have an effect on that property. In this way, ANN analysis has the capability to identify influences that may not have been suspected or for which there is currently no evidence from physical models.

In summary therefore, ANN analysis has to be seen as a research tool at this stage of development. Nevertheless, the underlying strategy is sound, as it is based on careful observation of input and output parameters, whether or not these are known to be involved in physical models of the phenomena. As noted, it requires extensive experimental work in order to provide training data, although it may be possible in some circumstances to use live production information in this context.

3.3 Computational simulation

In contrast to the empirical ANN strategy, the use of computational simulation to understand and predict distortion seems to be at the opposite extreme. Computational simulation relies on physical models of the real situation to describe the complex thermo-mechanical behaviour induced by welding. This has become feasible in recent years through the rapid development of finite-element analysis and the availability of ever more powerful computing resources. However, experimental inputs remain invaluable. As noted by the famous physicist Kirchhoff, 'the task of mechanics consists in describing observable movements in nature – in the completest and in the simplest way'.[3] As will be shown in later chapters, experimental observations of fabrications during manufacture are a primary source of guidance to the simplest form of physical modelling that can describe the problem adequately.

The supreme advantage of the computational route is that the welding is virtual. Full-scale 'experiments' can be undertaken in safety and any parameter can be varied at will. Parametric studies can be undertaken that would be prohibitively expensive in reality. Movements that are not in practice readily 'observable' in a real test can be measured accurately, magnified and speeded up or slowed down for improved visual inspection. These features promote insight and better understanding of the reasons for distortion behaviour in a given instance. To quote Boltzmann, another famous physicist, 'there is nothing so practical as a good theory'.

Most of the following chapters are concerned with describing computational simulation of welding distortion and showing how tools for design support may be formulated. At this stage, though, it is useful to stand back from the detail and consider more broadly what physical effects are being addressed. Most of the welding processes used in manufacturing and structural fabrication fall into the category of fusion welding. The main feature of such welds is that the adjacent edges of the components to be joined are melted, or fused, by an intense heat source that is passed along the joint. The resulting molten bridge then solidifies to form a continuous joint between the components. A wide variety of heat sources can be employed and additional filler material may be added to the molten bridge in some processes. This basic approach provides very convenient and effective manufacturing procedures, but the intense heat source can itself introduce penalties that make it problematic to apply fusion welding to components or structures where dimensional accuracy and freedom from initial stress are premium requirements.

Lindgren[4] notes that computational models may be formulated across a wide range of physical processes in welding fabrication. In particular, he draws a distinction between computational weld process modelling (WPM), which focuses on the physics of heat generation and weld pool formation in the immediate vicinity of the heat source, and computational welding mechanics (CWM), which deals with the resulting effects of heat generation on the mechanical behaviour of a fabrication at a larger scale, which could include a complete thin-plate fabrication. In the classical CWM approach followed in this book, the heat energy input and distribution are prescribed simply on the basis of electrical or other energy specification of the welding process and experimental determinations of the fusion zone shape. Aside from the heat input implications, the local weld process also affects many features of the fusion zone area, notably the microstructure and hence thermal dilatation and strength properties, which will influence the 'welding mechanics' stage. There may come a time when weld process modelling improves to the point where a weld process model can be integrated with a welding mechanics model to provide correct, simultaneous modelling of heat input and fusion zone constitutive properties, but that is currently not feasible for most industrial problems.

The earliest example of CWM known to the writers was attributable to N. Okerblom, whose work was translated into English in 1958.[5] Okerblom's approach was based on an analytical determination of thermal fields, used to construct thermo-mechanical predictions of weld contraction forces, in-plane deformations and angular contraction. This work showed great insight, but there is little evidence that it has been used extensively as a design tool, nor is his work often cited in relevant literature. A computationally based application of the Okerblom approach to in-plane contraction was published in 1987 in the context of thin-plate welding for naval ships.[6]

Improvements in computational capability prompted new approaches to weld process and welding mechanics modelling such as the heat input models of Friedman[7] and the finite-element-based approach of Goldak *et al.*[8] From the beginning it was clear that the general problem of simulating welding in all physical respects would normally exceed normal computing power, especially in a large structure, and much of the research effort in this field has been aimed at simplification and reduction of the problem. Consideration of element types, meshing schemes and sub-structuring have therefore been given considerable attention.[4] Special finite-element software for welding problems has also been developed, e.g. SYSWELD.[9] Uncoupling of the solution domains forms one of the most powerful simplification routes – for example, the heat transfer problem can be solved without reference to the dilatation behaviour or mechanical aspects. The thermal fields resulting from the heat transfer solution are then applied to solve the mechanical problem. This opportunity arises as the two regions of physical behaviour are weakly coupled in practice. This will be discussed further in Chapter 5.

Large thin-section structures present particular challenges and many researchers have specialised in this area, very often in the context of shipbuilding and the fabrication of bridge or other transport-related components. Dimensional reduction (e.g. using a 2D solution for a 3D problem) is inherent in the Okerblom approach (together with a temperature field simplification) and was applied by Iwaki and Masubuchi[10] and Ueda and Yamakawa[11] as early as 1971. Ueda and co-researchers later formulated an 'inherent strain' method whereby a solution obtained from a 3D, moving source, transient analysis is applied to a much simpler elastic model of a large structure, very often containing several welds, such as shown by Murakawa *et al.*[12] This approach and variations of it, have been adopted by Michaleris and others in the shipbuilding field,[13] particularly in the context of buckling deformations. Mention of the buckling problem highlights the warning that one particular simplification that should *not* be applied in the field of thin structures is that of adopting linear strain–displacement assumptions. Out-of-plane deformations are particularly troublesome in thin-plate construction and non-linear displacement is therefore a critical modelling feature.

It is clear from this brief discussion that computational welding simulation is no longer in the category of a research tool but can now be considered a mature technology, capable of industrial application. This view is confirmed by an increasing number of recently published texts, discussed in Chapter 1, which provide an organised introduction and reference source.

3.4 Current research on reduction of distortion

3.4.1 Thermal and mechanical tensioning techniques

A significant amount of effort has been expended on researching various thermal tensioning techniques. The basic aim of this approach is to alter the thermal stress pattern during welding and hence the residual stress distribution around the weld. Various thermal methods are discussed by Michaleris,[14] with the most promising of these based on what has been described as 'transient thermal tensioning' (TTT). The thermal tension is delivered by side heaters that move with the welding heat source(s), parallel to the weld line. An application to twin-arc, double-sided stiffener welding is shown, where transversely oriented heaters are applied, trailing the second arc by 50 mm.[14] Simulation of a twin-arc arrangement is given in section 12.2.3 in Chapter 12, where the interaction with buckling tendency is discussed. The interaction of variables in using this technique is in general fairly complex. Several variables apply to the heaters, including the energy level delivered and the resulting temperatures in the heated zone, their configuration (parallel or transverse), their transverse distance from the weld line and their relative position (in front of or behind the weld heat source). A team at Cranfield University has investigated these aspects experimentally.[15] However, the uptake of this process technology has been extremely slow, as it is primarily applicable to stiffener welding, as illustrated by Michaleris with a twin fixed head.[14] It would not, for example, be easily engineered to work with two independent, oscillating robotic heads travelling along a straight weld line.

In the case of mechanical tensioning, a roller tensioning method has been developed to a certain stage and is the subject of current investigation. In this method, a carefully profiled roll is applied along the weld cap after welding (Fig. 3.4).[16, 17] This produces a slight overall reduction of weld height, but creates a flatter panel. A particular issue in the application of this technology is how to compensate for variations in weld cap shape. However, it does offer some potential benefits in that it is a stand-alone process, separate from the actual welding process. Further work is also needed to move to wider application in order to fully realise the potential benefits, for example involving the treatment of stiffener welds.

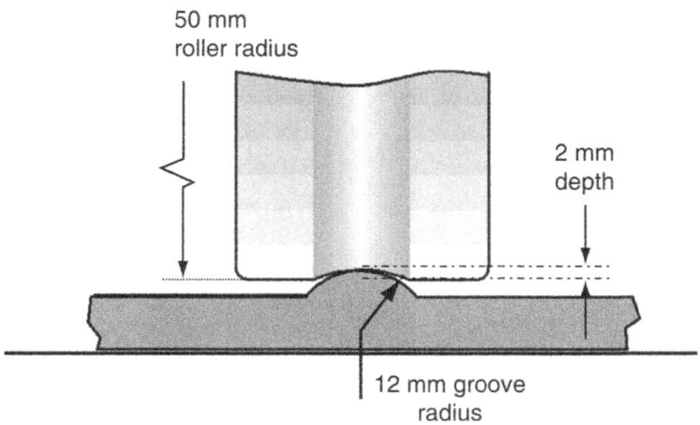

3.4 Configuration of seam weld profile and profiled roller (after Coules *et al.*[16, 17])

3.4.2 Low-temperature-transformation weld metal

The potential application of low-temperature-transformation weld metal is based on the effect of metallurgical transformation on the thermal and residual stress patterns in the central weld zone. The transformation of weld metal to a martensitic phase generates significant volume expansion. The steel cools from austenite, which has a face-centred cubic structure, to a body-centred tetragonal structure via a shear transformation process, which generates a temporary expansion. This develops compressive stresses in the weld region if the temperature is low enough for the transforming material to have significant strength. Initial work, using 12% nickel welding wire, demonstrated a potential for reducing distortion. Nickel has a significant effect in reducing the martensite formation start temperature (M_S) as shown by

$$M_S = 561 - 474\%\text{C} - 35\%\text{Mn} - 17\%\text{Ni} - 17\%\text{Cr} - 21\%\text{Mo} \quad [3.1]$$

This empirical equation has variants, but serves the purpose of illustrating the potential effects of each of the main elements. However, the toughness of the weld metal in the first trials of this technique was not acceptable. Since then, a number of studies have taken place and the problem related to toughness seems now to be approaching a resolution.

A study by Bérĕs, based on a combination of practical welding and modelling,[18] has contributed to the understanding of the overall effects. The conclusions of this study included the following.

- The out-of-plane deformation decreases as the transformation start temperature decreases.
- With conventional welding wire, the central longitudinal residual stress in the fusion zone is tensile and of the order of 300 MPa.

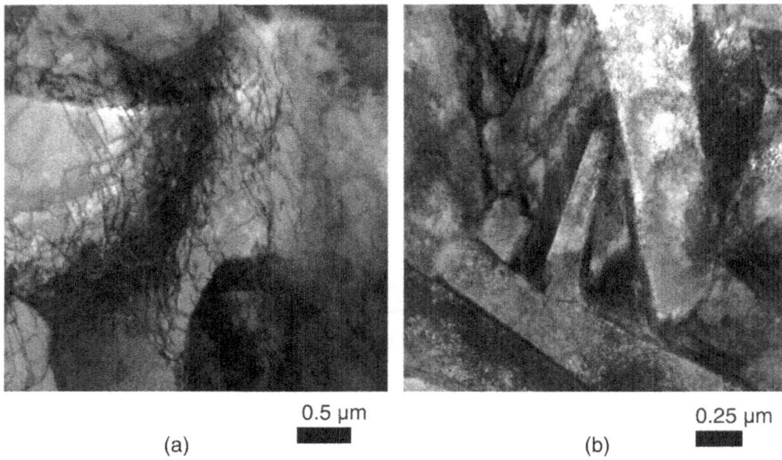

3.5 Structures of weld metal from trial welding wires. (a) Weld metal from conventional welding wire showing regular grain structure with high dislocation density. (b) Weld metal from low-transformation-temperature wire showing more pronounced lath structure.

- With a low-transformation-temperature welding wire, the longitudinal tensile stress is compressive and of the order of −100 MPa.
- The weld microstructure generated by conventional filler wire is characterised by the normal combination of pro-eutectoid ferrite and acicular ferrite, as shown in Fig. 3.5(a).
- The weld microstructure in the low-transformation-temperature wire case is typically all acicular ferrite and a heavily lathed structure (as shown in Fig. 3.5(b)).
- The modelling work predicted that lower transformation start temperatures will lead to lower distortion.

A simulation study of the effects of welding wire composition on distortion is given in section 12.3.2 of this book.

3.4.3 Weld cooling methods

Guan describes a distortion reduction method based on intensive cooling of the weld zone.[19] This technology has the description 'low stress no distortion (LSND) welding' (the *no* distortion claim is perhaps optimistic). Various cooling media, mostly of cryogenic form, have been used by different researchers. The concept of using a gas in solid form, which cools the workpiece due to sublimation, is probably the most adaptable as the cooling is focused on a specific area and in a reasonably well-defined quantity. The process is effective under very well-controlled conditions, but on an industrial scale it is probably difficult to use because the height of the cooling media supply system outlet above the weld has to be controlled very closely. The major process design problem is to achieve consistent sealing of the cooling head against the workpiece, given variations in cap height and out-of-flatness. As in the case of thermal tensioning, uptake of this technology by the major part of manufacturing industry has been low. However, a case in China where the process has been taken forward successfully on an application involving 0.8 mm stainless steel has been reported.[20]

A simulation study of the application of a weld-trailing cryogenic cooling process is given in section 12.3.3.

3.5 Conclusion

Several differing approaches to the observation, prediction, control and interpretation of deformations due to fusion welding have been described. The method based on ANNs (section 3.2) requires extensive data relating generated deformations and associated input parameters and uses statistical methods to reveal empirical connections between the data. The deformations are usually found from experiment but virtual data from computational weld modelling (CWM) could also be used. Most of this book is concerned with approaches based on CWM and the description in section 3.3 concentrated mainly on a broad outline of the origins and major developments in the CWM strategy. Section 3.4 described four different processes designed to reduce deformation due to welding thermal effects. Transient thermal tensioning is based on the simultaneous input of heat energy with the welding process. The roller tensioning method has the same aim, but applies mechanical forces to the weld rather than thermal stress. A method utilising special filler wire was described. This technique relies on a low-temperature metallurgical transformation and the associated volume expansion to counteract the normal contraction forces in the weld zone. Finally, a method based on the application of intense cooling in the trailing region of the weld heat input was introduced. Most of these methods are further discussed in Chapter 12 in the context of computational simulation.

3.6 References

1. Bhadeshia, H.K.D.H, 1999, Neural networks in materials science, *ISIJ International*, 39, 10, pp 968–979.
2. Lightfoot, M.P., Bruce, J.G., McPherson, N.A. and Woods, K., 2005, The application of Artificial Neural Networks to weld-induced deformation in ship plate, *Welding J.*, 84, 2, pp 23s–30s.
3. Timoshenko, S.P., 1983, *History of strength of materials*, Dover Publications, NY, USA.

4. Lindgren, L.-E., 2007, *Computational welding mechanics: thermomechanical and microstructural simulations*, Woodhead Publishing Ltd, Cambridge, UK.
5. Okerblom, N.O., 1958, *The calculation of deformations of welded metal structures* (originally in Russian and translated by the Department of Scientific and Industrial Research), HMSO, London, UK.
6. Cadden, S., 1987, *Prediction of the in-plane contractions in thin butt-welded plates*, PhD thesis, University of Strathclyde, Glasgow, UK.
7. Friedman, E., 1975, Thermomechanical analysis of the welding process using the finite element method, *ASME J. Pressure Vessel Tech.*, 97, 3, pp 206–213.
8. Goldak, J., Chakravarti, A. and Bibby, M., 1984, A new finite element model for welding heat sources, *Metallurgy Trans. B*, 15B, pp. 299–305.
9. ESI Group, Engineering simulation solution for heat treatment, welding and welding assembly, esi-group.com/products/welding_assembly/welding (accessed 20 August 2013).
10. Iwaki, T. and Masubuchi, K., 1971, Thermo-elastic-plastic analysis of orthotropic plate by the finite-element method, *J. Soc. Nav. Arch. Japan* (in Japanese), 130, pp 193–202.
11. Ueda, Y. and Yamakawa, T., 1971, Analysis of thermal elastic-plastic stress and strain during welding by finite element method, *Trans. Japan Welding Res. Inst.*, 2, 2, pp 90–100.
12. Murakawa, H., Luo, Y. and Ueda, Y., 1998, Inherent strain as an interface between computational welding mechanics and its industrial application, *Mathematical modelling of welding phenomena 4*, Institute of Materials, London, UK.
13. Michaleris, P. and Sun, X., 1997, Finite element analysis of thermal tensioning techniques mitigating weld buckling distortion, and residual stress, *Welding J.*, 76, 11, pp 451–457.
14. Michaleris, P., 2011, Understanding welding residual stress and distortion, in *Minimization of welding distortion and buckling*, Michaleris, P., Ed., Woodhead Publishing Ltd, Cambridge, UK, pp 3–21.
15. Nagy, T., Williams, S., Colegrove, P., Ikeagu, C. and Fafiolu, I., 2008, Distortion mitigation in welded ship panels, *Proc. International Workshop on Thermal Forming and Welding Distortion, Bremen, Germany.*
16. Coules, H., Colegrove, P., Cozzolino, L.D. and Wen, S.W., 2013, High pressure rolling of low carbon weld seams: Part 1 effects on mechanical properties and microstructure, *Sci. Technol. Welding and Joining*, 18, 1, pp 76–83.
17. Coules, H., Colegrove, P., Cozzolino, L.D., Wen, S.W. and Kelleher, J.F., 2013, High pressure rolling of low carbon weld seams: Part 2 roller geometry and residual stress, *Sci. Technol. Welding and Joining*, 18, 1, pp 84–90.
18. Bérěs, M., 2010, *Welding of thin sheet steels in marine applications*, PhD thesis, Imperial College of Science, Technology and Medicine, London, UK.
19. Guan, Q., 2005, Control of buckling distortions in plates and shells, in *Processes and mechanisms of welding residual stress and distortion*, Feng, Z., Ed., Woodhead Publishing Ltd, Cambridge, UK, pp 295–343.
20. Li, J. and Shi, Q-Y., 2011, Minimising buckling distortion in welding by weld cooling, in *Minimisation of welding distortion and buckling*, Michaleris, P., Ed., Woodhead Publishing Ltd, Cambridge, UK, pp 214–240.

4

Understanding welding distortion: thermal fields and thermo-mechanical effects

DOI: 10.1533/9780857099327.53

Abstract: A first introduction to welding mechanics is given, based at this stage on uncoupled analytical solutions for thermal and thermo-mechanical behaviour. Rosenthal's well-known quasi-static solution for the temperature field caused by a fast-moving, concentrated heat source is used to demonstrate the influence of different material properties and welding process conditions on the temperature fields. An approximate thermo-mechanical solution due to Okerblom is used to demonstrate the roles of the thermal dilatation of the material being welded and the heat input on welding deformations. This solution is also used to illustrate the development of residual stress fields and the potential for buckling.

Key words: thermal fields, temperature gradients, isotherms, thermal dilatation, thin-plate deformation, uncoupled solutions, Okerblom theory, residual stress, buckling.

4.1 Introduction

This chapter uses simple analytical treatments to provide an overview of solution strategies for thermo-mechanics problems relating to the application of fusion welding processes. The thermal and mechanical parts of the solution are uncoupled. The thermal solution is based on the quasi-static solution of Rosenthal[1, 2] and this is used to show the influence of material thermal properties and welding process conditions on the resulting temperature fields.

The thermo-mechanical overview is based on an analytical application of the approximate treatment due to Okerblom.[3] This treatment will be expanded in Chapter 8, using finite-element computational solution methods. Typical deformations of welded thin-plate structures are described and related to approximate analytical treatments whereby the longitudinal and transverse deformation mechanics are partially uncoupled. An approximate formula for the longitudinal contraction force is derived and related to the thermal dilatation and heat input. The resulting influence of material properties is thereby shown. The chapter concludes with introductory treatments of residual stress fields and buckling, based on the simplified analyses.

4.2 Thermal fields: dependence on welding parameters and material properties

The transient temperature field, generated by the moving heat source, is the main driver for most of the adverse dimensional effects of fusion welding, together with the thermal dilatation of the materials being welded. It is therefore important to understand how the thermal fields depend on the welding parameters associated with specific processes and settings. Detailed modelling strategies for thermal fields will be given in Chapter 7, but a simpler analytical approach is given at this stage to show the main effects of fast-moving heat sources.

Heat dissipation in fusion welding is mainly by conduction from the heat source through the parent components being joined. Dissipation by radiation and convection at hot surfaces also takes place and will be important in fabrications of restricted size, but the main effects can be analysed using two-dimensional (2D) or three-dimensional (3D) conduction theory.

The volumetric thermal capacity of the material, $c\rho$, is an important material property, where c is the specific heat of the material, measured in J/kg K and ρ is the density in kg/m^3. The other important property is the thermal conductivity K, measured in J/ms K. These parameters can be combined to give an important conduction parameter λ, termed diffusivity, where $\lambda = K/c\rho$. The dimensions of diffusivity therefore correspond to m^2/s and, in a sense, diffusivity gives an index of how rapidly heat energy will spread in the material. It is assumed in simplified analysis that these material properties are fixed whereas, in reality, they vary substantially with temperature. Fortunately, the regions experiencing the highest temperatures during welding are normally small relative to the overall size of a typical welded component. Nevertheless, the melted zone has a very high capacity to retain heat due to the high latent heat of fusion. It will then return this heat to the solid regions adjacent to the trailing thermal field.

The welding process and the thermal properties in the regions where material has melted will therefore affect the sizes and shapes of the fusion zones, together with the transmission of thermal energy to the component as a whole. However, the main global thermo-mechanical effects that drive distortion depend more on the thermal fields in the solid regions, where the thermal fields occupy a much smaller range of temperatures, much below the melting point. Therefore the assumption may be applied, with not too much loss, that the significant material thermal properties for welding distortion mechanics are constant, or at worst vary linearly over the small temperature range that is relevant.

So-called 'quasi-stationary' thermal solutions are very useful in the context of simplified analyses. These are based on the observation that, although the temperature at a given point in a structure being welded will rise and fall as the heat source passes, the isothermal contours in most practical configurations stabilise quickly after the start of welding, so that a fixed pattern of temperature contours is established relative to the heat source position. Another way to understand this is to consider the position of the heat source as fixed, along with the observer, while the component, perhaps a large, thin plate, is moved at a steady speed past the heat source. A short time after welding starts, the observer will see a stationary pattern.

An x, y, z coordinate system can be adopted for points in the plate, where x is parallel to the direction of welding, y is transverse and z is measured in a direction away from the surface of the plate, the origin being at the point where the heat source is first applied. A quasi-stationary coordinate system can be defined alternatively, the position of a point relative to the heat source being x', y, z, where the position relative to the true coordinate system is given by $x' = x - vt$, v being the travel speed of the welding head and t the elapsed time. The radial distance from the heat source is then measured by $r = (x'^2 + y^2)^{1/2}$.

Rosenthal provided solutions of the quasi-stationary type which give the temperature T above a base temperature T_o, at a specific set of coordinates in a 2D plate, heated by a fast-moving, intense heat source (2D in the xy plane). The heat source itself is modelled as a vanishingly thin line, which extends uniformly through the thickness of the plate with an energy delivery rate of q/d where q is the power input of the source (in J/s or W) and d is the thickness of the plate (i.e. the length of the through-thickness line). These solutions have the form

$$T - T_o = \frac{1}{2\pi}\frac{q}{d}\frac{1}{K}e^{-vx'/2\lambda}K_o\left(\frac{vr}{2\lambda}\right) \qquad [4.1]$$

in which K_o is a modified Bessel function of the second kind and zero order (lookup tables are normally used). It is also worth noting that this solution implies an infinite temperature at the origin!

Equation 4.1 can be applied to illustrate the basic thermal pattern generated by a fast-moving heat source and the effects of different welding

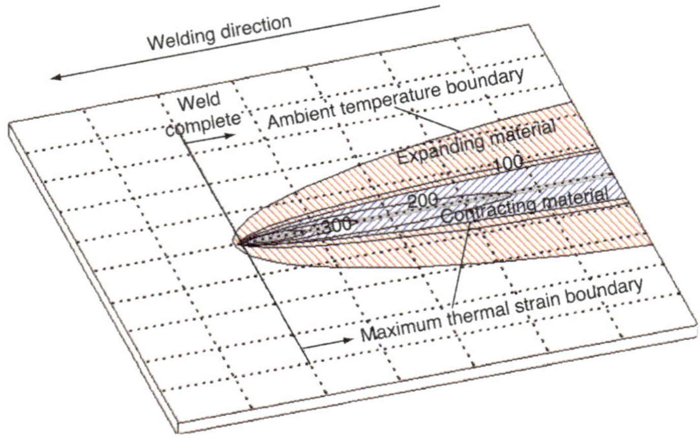

4.1 Thermal gradients developed during welding.

parameters and material thermal properties. Figure 4.1 shows that the temperature rises rapidly ahead of the heat source and falls slowly behind it, and the temperature gradient in the transverse direction is much steeper than in the longitudinal direction. The calculations underlying Fig. 4.1 were based on a welding process electrical input of 5.77 kW when welding a 6 mm thick plate and assuming an efficiency of energy conversion to heat conducted into the plate of 82.5%, giving a notional actual heat input of 4.76 kW. The travel speed was 4.8 mm/s, giving a heat input rate of 992 J/mm.

The forward transmission of heat energy for typical welding conditions is overtaken to a large extent by the forward travel of the heat source. However, in the trailing part of the weld, the material is already at a high temperature and this limits heat transmission in this direction. Hence, the larger part of the heat transmission takes place transverse to the weld direction, where the temperature differential is greatest. The regions experiencing heating and cooling can be identified by drawing a curve that connects the points of maximum width for each isotherm, as shown. Thus, at any transverse cross-section, mechanical strain will be developed in the direction of welding, due to the conflict between inner regions on the weld line that are contracting and the outer regions, which are still expanding.

Equation 4.1 shows that, for a given travel speed, the temperature at a given point depends linearly on power input per unit thickness and the effect of plate thickness on temperature is therefore linear. The thermal patterns shown in Fig. 4.2 confirm that doubling the thickness drops the temperature at a given point considerably, such that, for example, the 600°C contour in the 3 mm thick case (Fig. 4.2(a)) is of a similar size to the 300°C contour in the 6 mm case (Fig. 4.2(b)).

(a)

(b)

(c)

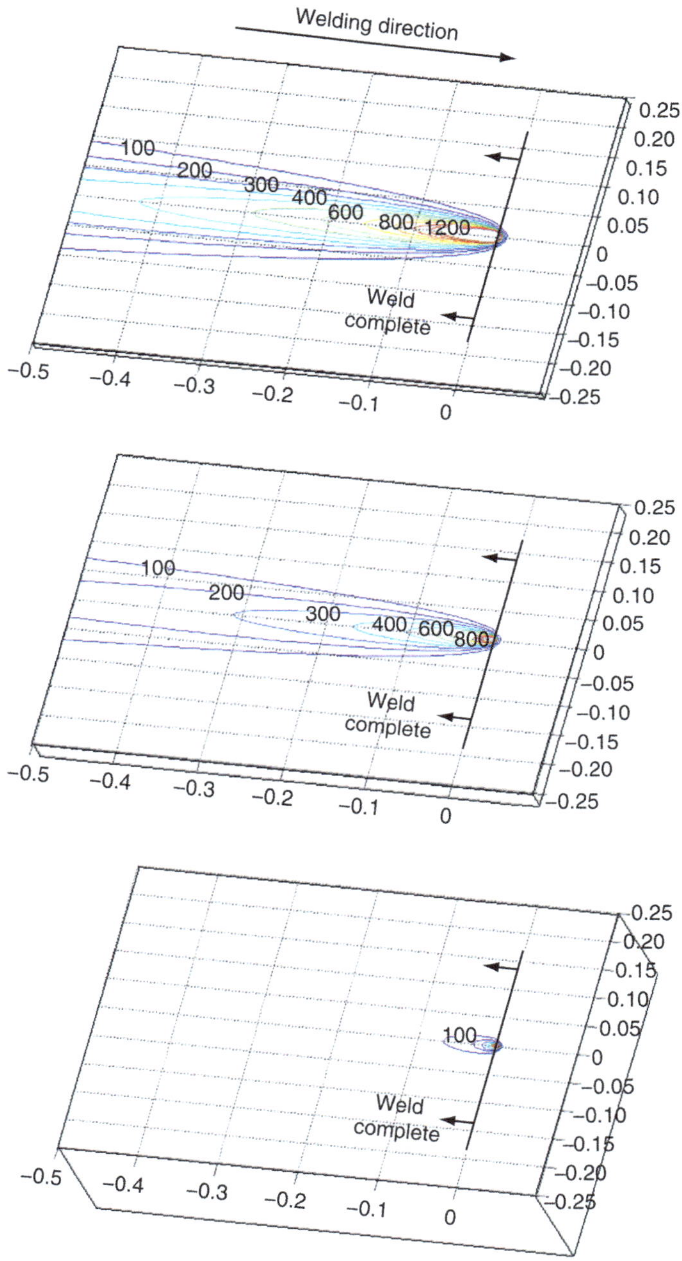

4.2 Isotherms developed during fusion welding for different plate thicknesses but identical welding parameters: (a) 3 mm plate; (b) 6 mm plate; (c) infinite thickness (dimensions (m)).

If the component is too thick to support the assumption that the heat input is evenly distributed along a through-thickness line, 3D quasi-stationary solutions can be applied on the assumption of a point heat source on one surface. In this case, the dimensional system is given by $R = (x'^2 + y^2 + z^2)^{1/2}$ and the temperature at a given point by

$$T - T_o = \frac{q}{2\pi}\frac{1}{K}e^{\lambda v x'}\frac{e^{\lambda v R}}{R} \qquad [4.2]$$

Extending the thickness such that it corresponds to a semi-infinite solid, as in Fig. 4.2(c), results in an even greater shrinking of the isotherm sizes.

The effect of travel speed is illustrated in Fig. 4.3, where the basic heat energy input per unit length assumed in Fig. 4.2 is held constant (obtained by dividing the power input by the travel speed) but the travel speed is reduced, in Fig. 4.2(a) by one-third and in Fig. 4.2(b) by one-half. At higher speeds, the isotherms become more elongated along the direction of welding, but it is interesting to note that the transverse widths of the isotherms are very similar for all three cases. Stated another way, the transverse profile of

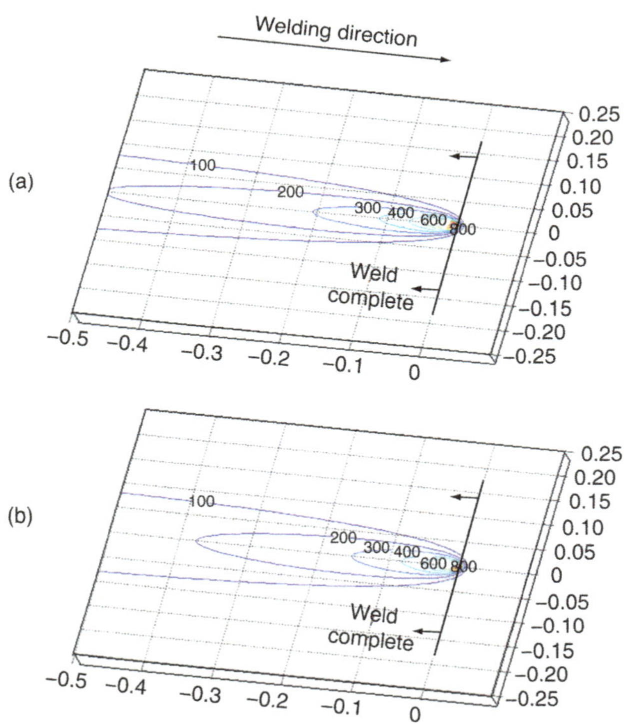

4.3 Isotherms for fusion welding of the 6 mm plate in Fig. 4.2(b), heat input per unit length maintained constant: (a) 2/3 heat input of Fig 4.2(b), 2/3 speed; (b) 1/2 heat input, 1/2 speed (dimensions (m)).

maximum temperatures reached depends primarily on the parameter q/vd obtained by dividing the power input by the travel speed and the thickness. This turns out to be a highly significant point, which will be discussed again at a later stage.

One of the useful features of these two quasi-stationary solutions for 2D and 3D solids is that a given geometry of component, subject to given welding parameters, can be calibrated against each solution to identify whether a 2D treatment would be valid or a 3D treatment is required. Using the cooling rate on the centreline of the weld as a criterion for comparison, Birk-Sorensen[4] applied Rosenthal's equations to find the ratio of 2D/3D cooling rates at a given temperature as

$$d_k^2 = c\rho \frac{v}{q}(T - T_o)d^2 \qquad [4.3]$$

where d_k is considered to be a non-dimensional component thickness. If d_k is equal to unity, then the cooling rates given by both analyses are identical. If d_k is less than unity, a 2D thermal analysis may be applied and the isothermal contours will be larger in area, as shown in Fig. 4.2 and vice versa if d_k is greater than unity. This leaves open the question of the temperature that should be used to index the cooling process, but it would probably suffice to evaluate d_k for the upper and lower values of the temperature range within which the material has enough mechanical strength to drive distortion processes.

As noted earlier, typical material thermal properties vary substantially with temperature and even more so within the range of temperatures normally experienced in fusion welding. In addition to gradually varying properties that may be observed over a limited temperature range, large non-linear variations occur if there are state changes in the materials, such as during metallurgical phase changes in the solid state or during transitions from solid to liquid. Moreover, conductivity will depend on fluid motion, which is often vigorous in the weld pool. It is therefore not easy to measure such properties systematically, and relevant and reliable property data are not readily found in the literature.

Leaving aside these issues, Fig. 4.4 compares the outcomes of welding different materials but maintaining, for the purposes of comparison, the same welding parameters as used in Figs 4.1–4.3. The effects can therefore be compared directly also with Fig. 4.2 (b), which illustrates the welding of CMn structural steel. (Note that it might *not* be possible to apply the same welding conditions to all these different materials in practice and the example is given for illustrative purposes only.) The fixed thermal properties used to generate these graphics are given in Table 4.1.[5–8]

The differing effects of material thermal properties on the thermal fields are dramatic for typical materials. The main effect observed seems to

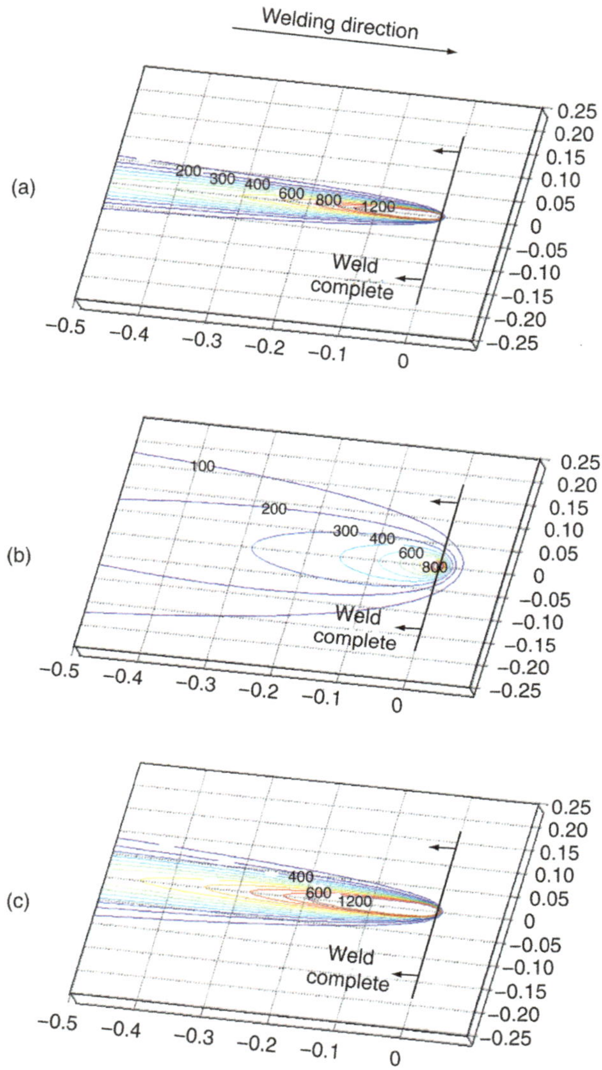

4.4 Isotherm comparison for different metals and identical energy/speed: (a) stainless steel; (b) aluminium alloy; (c) titanium alloy (dimensions (m)).

Table 4.1 Ambient-temperature thermal properties of some weldable materials

Material	K (W/mK)	c (J/kg K)	ρ (kg/m^3)	$c\rho$ (J/m^{3}K)	λ (m^2/s)
CMn steel (DH36)[5]	69.60	782	7800	6.10×10^6	11.40×10^{-6}
Stainless steel prEN3488 (soft)[6]	14.50	490	7900	3.87×10^6	3.75×10^{-6}
Al alloy prEN4202[7]	184.0	920	2700	2.48×10^6	74.20×10^{-6}
Ti alloy DTD 5023[8]	16.95	500	4510	2.26×10^6	7.50×10^{-6}

depend on diffusivity. Lower diffusivity leads to a narrower trailing isothermal field, due to a lower rate of heat spread. The effects interact to a second order with travel speed, as will be shown in Chapter 7.

The examples shown in Figs 4.1–4.4 were constructed using quasi-static, analytical theory. In practice, calculation of thermal fields can be carried out more readily using finite-element numerical programs. Non-linear thermal properties, if known, can often be included in such analyses, together with features of the real welding configurations, to take account of thermal boundary conditions and local variations.

4.3 Thermo-mechanical effects

Having established the nature of the thermal field in a component being welded, the main factor driving the consequent distortion is the thermal dilatation (expansion and contraction property) of the material, as this generates differential strain between the various regions of the component. This property also depends on temperature, especially if metallurgical phase transformations occur within the temperature range of interest. Figure 4.5 shows the results of a dilatometer test where the expansion and contraction of a steel sample were measured during steady heating from 20°C to 1120°C, followed by cooling through the same temperature range. This essentially follows the temperature history of any point in the material that reaches a maximum temperature of 1120°C, although the heating and cooling rates may be different from those occurring in a weld.

The instantaneous coefficient of expansion or contraction is given by the slope of the line at any point and it can be seen that the coefficient is negative during the 700–500°C phase transformation during cooling. Even

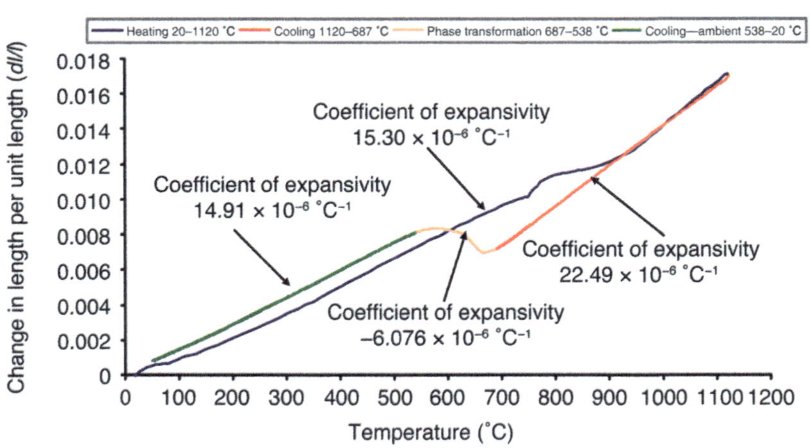

4.5 Instantaneous coefficients of thermal expansion from dilatometer test specimens.

Table 4.2 Coefficient of thermal expansion for various alloys at room temperature

Alloy	Expansivity at room temperature (°C^{-1})	Distortion index $\alpha/c\rho$ (m^3/J)
CMn steel DH36[5]	14.56×10^{-6}	2.39
Stainless steel prEN3488 (soft)[6]	16.80×10^{-6}	4.34
Al alloy pr EN4202[7]	23.50×10^{-6}	9.48
Ti alloy DTD5023[8]	7.60×10^{-6}	3.36

this curve cannot be taken as definitive for the material, as the behaviour can be altered by varying the heating and cooling rates, or by applying stress to the sample. However, it would be possible to approximate a fixed coefficient from these data, either for the whole temperature range or for the part where the component is cooling from 500°C to ambient temperature. Table 4.2 gives typical room-temperature expansion coefficients for common materials (data in the 'distortion index' column will be explained in section 4.5).

Other purely mechanical properties are involved in the mechanics of welding, in particular, elastic modulus and yield strength. If a material were to have a very low modulus and very high yield strength (and perhaps also a very low coefficient of expansion) it is theoretically possible that the differential thermal strains could be accommodated by elastic strain and no permanent distortion would occur. However, this never seems to be the case for typical metallic materials and welding processes. These stiffness and strength properties also depend on temperature. Figure 4.6(a) shows the stress–strain behaviour of samples of steel in tensile tests carried out at various temperatures within the range of typical welding processes and Fig. 4.6(b) shows a simplification or idealisation of the real behaviour.

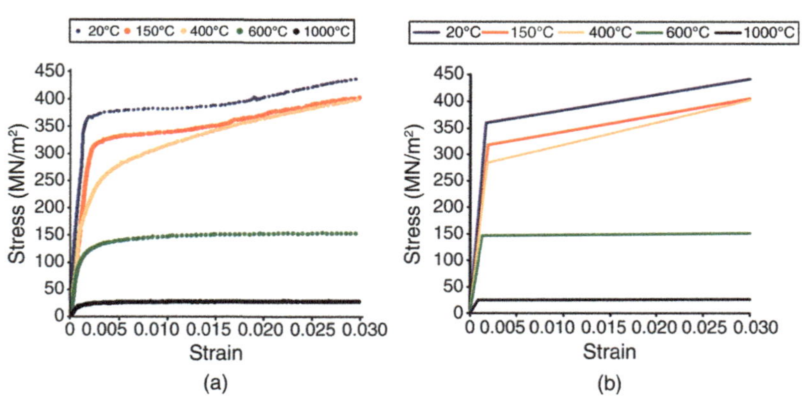

4.6 Stress–strain behaviour of CMn steel at various temperatures: (a) test results on material; (b) bilinear kinematic hardening model.

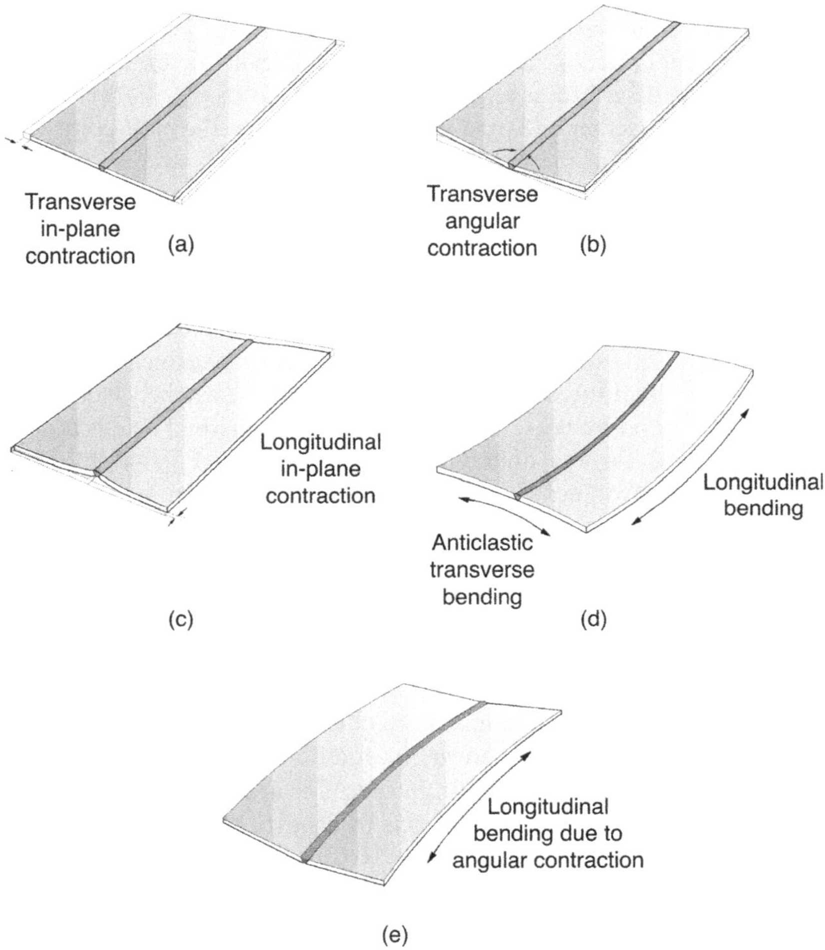

4.7 Typical fabrication deformations in thin-plate welding: (a) transverse in-plane contraction; (b) transverse angular contraction; (c) longitudinal in-plane contraction; (d) longitudinal bending; (e) longitudinal bending and angular contraction combined.

Analysis of the thermo-mechanical movements underlying welding distortion is therefore more difficult than for the purely thermal aspects, due to the complicated interaction of thermal and mechanical strains. It is nevertheless possible to understand some aspects of welding-induced deformation in thin-plate fabrications, at a qualitative level at least, and through such an understanding to develop some simple analytical models. Thin-plate fabrications deform in different directions during welding and for clarity, Fig. 4.7 shows these as separate movements.

Transverse shrinkage of the fusion zone leads to transverse in-plane contraction of the plate, as shown in Fig. 4.7(a). Single-pass fusion zones are typically wider on the surface of the plate adjacent to the heat source and unequal contraction across the thickness therefore leads to rotation of the plates, commonly described as angular contraction, as in Fig. 4.7(b). These two actions occur together and, to a first approximation, the plate contracts as if the whole length of the weld was laid simultaneously and allowed to cool from a temperature somewhat less than the solidification temperature for the material (around 1000°C for steel). The effectiveness of this simple model is more than a little surprising, as one might think that, once some of the completed weld length has cooled enough to possess significant mechanical strength, the resulting restraint would prevent free contraction of the current fusion zone, which should have little mechanical strength. However, it seems that the simple result stems from the extremely elongated nature of the temperature field. In some instances, the thermal field is not so elongated, such as for the aluminium weld shown in Fig. 4.4(b). The question of whether or not the same simple model applies in that case remains unexplored territory, due to the lack of experimental data or published simulations.

It is clear, even from this qualitative description, that transverse contraction will depend directly on the average width of the fusion zone and that the angular contraction will depend on the difference in width between the top and bottom of the fusion zone. Hence, edge preparations and welding procedures that minimise both of these factors will reduce these forms of distortion, as is well known. In addition, increased fusion zone volume correlates approximately with increased heat input and, as will be noted later, this also increases other forms of distortion.

The central region of the weld, including the fusion zone, also contracts in a longitudinal sense, but this is resisted by adjacent cooler regions which are still expanding longitudinally while cooling contraction takes place in the centre, as shown in Fig. 4.7(c). This leads to a central longitudinal contraction deformation and force, of up to yield point magnitude, which develops an overall, non-uniform longitudinal contraction. The picture is slightly different right at the ends of the plate, as force equilibrium precludes longitudinal stress on the end surfaces. The central longitudinal tensile force is reacted in this end zone by a short field of transverse in-plane bending.

If the fusion zone width was uniform through the thickness, i.e. with parallel sides, the longitudinal contraction profile would also be uniform, leading to entirely in-plane deformation in the uncoupled model. However, the typical V-shaped fusion zone means that the line of action of the contraction force normally lies nearer to the upper surface of the plate (i.e. above the neutral axis of the plate) and a sagging curvature would then generated, as shown in Fig. 4.7(d). The resulting longitudinal bending strains and the 2D in-plane response of the material through a Poisson's ratio effect should also lead to transverse curvatures in plate regions near the edges (known as 'anticlastic' curvature) together with transverse stresses in the central flat regions. However, this small plate curvature tends to be swamped in practice by coupling of the individual angular and longitudinal deformation patterns shown in Fig. 4.7(b) and 4.7(c). The V-shaped angular contraction profile establishes a deformed cross-section where the line of action of the contraction force lies *below* the neutral axis for longitudinal bending, resulting in a 'hogging' longitudinal deformation, as shown in the coupled deformation patterns of Fig. 4.7(e). Once more, the differing boundary conditions at the ends of the plate tend to alter the transverse profile at the ends.

Thin structures can also experience non-linear deformations, due to what are described as 'large-displacement' effects. The most commonly known large-displacement phenomenon is transverse buckling of a long, thin strip under axial load. As the strip bends into a bow shape, the moments at points along the length increase at a faster rate than the load increases, resulting in non-linear deformation. This also happens in plates, in areas subject to in-plane compression. Thin-section beams with shallow curved cross-sections also show complex longitudinal/transverse coupling behaviour under longitudinal bending.[9] This will be familiar to users of self-supporting measuring tapes with concave cross-sections, which serve to give the tape some bending stiffness. If the tape is extended and a transverse force applied to the concave side, the resulting longitudinal bending moment has the effect of flattening the transverse curvature. This reduces the bending stiffness non-linearly to a point where the tape buckles. The typical angularly distorted cross-section of a butt-welded plate fits that description well, and later chapters will show examples where such non-linear flattening effects are evident.

The other important factor that may enter into deformation profiles is gravity. The predicted profiles described above correspond to the ideal case where the components are either assumed to be weightless or uniformly supported all over. In practice, an assembly under fabrication will be supported at certain points. Gravity action in the unsupported regions may, as a result, induce out-of-plane deformations, which then trigger further deformations through thermo-mechanical actions. Moreover, the assembly being welded may not be flat in the first place.

4.4 Thermo-mechanical treatment based on longitudinal–transverse uncoupling

As noted in Chapter 1, Okerblom provided the first comprehensive analytical treatment of welding distortion. Firstly, he recognised that the mechanisms controlling transverse, angular and longitudinal contraction are different in principle, as shown in Fig. 4.7, and it could be reasonable to consider them as uncoupled phenomena. So, the transverse and angular contractions could be predicted on the simple assumption that a fusion zone is established simultaneously along the full length of the weld and that, on cooling back to ambient temperature, this molten zone contracts without constraint. The shape and size of the fusion zone are largely determined by the welding process characteristics, the weld edge preparation (if there is one) and the heat input, as discussed earlier. Thus, the fusion zone is often assumed to correspond to the empty cross-section formed by the edge preparation or gap before welding. The shape and size of the fusion zone may also be determined through trials. Alternatively, if some kind of computational thermal prediction of the fusion zone is used, the transverse position where the melting temperature is reached will define the edge of the fusion zone.

The longitudinal thermo-mechanical behaviour is more difficult to simplify, as zones that are expanding longitudinally and have significant mechanical strength are competing with parallel zones that are contracting, as shown in Fig. 4.2. Nevertheless, Okerblom found a simple treatment for the longitudinal mechanics, again assuming that the temperature profile is constant along the length of the weld.

Of course, once the uncoupled transverse and longitudinal deformations (and contraction forces) are established, it is entirely possible to re-couple them mechanically, as implied by Fig. 4.7(e), although there is no indication that Okerblom attempted to predict out-of-plane deformations by doing this.

To predict the longitudinal deformations, Okerblom took as his starting point the quasi-static, 2D, in-plane, thermal analysis of Rykalin[10] with some simplifications appropriate to intense, fast-moving heat sources (this is identical to the Rosenthal solution). He recognised that the major irreversible deformations responsible for distortion are driven by the cooling phase in the region behind the heat source and he greatly simplified the assumed temperature pattern by making use of the typically large differences between longitudinal and transverse temperature gradients shown in Fig. 4.1. What he did was to replace the complex, 2D in-plane, transient temperature pattern by a fixed transverse temperature profile, uniform along the length of the weld. The temperatures in this transverse profile or envelope are assumed to be the maxima reached at each transverse location, regardless of the fact that these maxima occur at different times or locations relative to the longitudinal position of the heat source. The transient aspect of the proper analysis is thus eliminated.

The profile he used, therefore, corresponds to the temperatures on the boundary, shown in Fig. 4.1, between the expanding and contracting material. If the heat dissipation is small and the heat source fast-moving, an inversely proportional relationship between any transverse location y on the cross-section and the maximum temperature $T_m(y)$ reached at that location can be established as

$$T_m(y) = 0.242\frac{q}{vd}\frac{1}{c\rho}\frac{1}{y} \qquad [4.4]$$

where q is the thermal power input to the plate, v is the heat source travel speed, d the workpiece thickness and $c\rho$ the volumetric thermal capacity of the material.

4.5 Plane strain strip: longitudinal deformations and forces

The next step in Okerblom's treatment was to invoke a longitudinal plane strain state, considered either to be along the whole plate or within a transient, transverse slice, leading to suppressed longitudinal thermal strains given by $\alpha T_m(y)$ where α is the coefficient of expansion/contraction of the material. The resulting strain pattern is shown in Fig. 4.8, where the maximum thermal strain at each normal distance from the centreline is non-dimensionalised in terms of the material yield strain ε_y, i.e. $\alpha T_m(y)/\varepsilon_y$ and the normal distance y is non-dimensionalised through equation 4.4 as $y/\left(0.242\frac{q}{vd}\right)\frac{\alpha}{c\rho}\frac{1}{\varepsilon_y}$. The four experimental points shown were taken from actual temperatures in a full-scale test where thermocouples were set at specific distances from the weld centreline.

If the plane strain constraint assumption is accepted, the resulting pattern of longitudinal stress across the section follows directly. In region A of Fig. 4.8, $\alpha T_m(y) > 2\varepsilon_y$ and the heating and cooling cycle within this region generates a final tensile stress of yield point magnitude, which for mechanical or material property reasons cannot be exceeded. In fact region

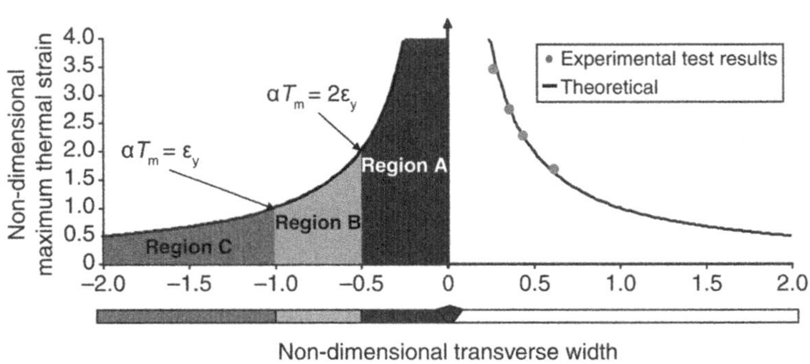

4.8 Non-dimensional maximum thermal strain versus transverse width.

A itself comprises inner and outer zones with different histories. In the inner zone, the material will have melted, whereas the outer zone will have remained below melting point. However, in both inner and outer zones the same limiting yield point stress will be developed during the final cooling phase. In region C, where $\alpha T_m(y) < \varepsilon_y$, the behaviour will be entirely elastic and no contraction stress will be generated. Between these limits, in region B, the tensile contraction stress is given by $E(\alpha T_m(y) - \varepsilon_y)$. Thus, the total stress developed in the shaded areas can be found by integration to provide the overall contraction force as

$$F = 0.335 \frac{q}{v} \frac{\alpha}{c\rho} E \qquad [4.5]$$

This longitudinal contraction force can then be applied to the component being analysed in a simple 'strength of materials' elastic analysis to provide an estimate of longitudinal contraction and, if the cross-sectional shape is known, to estimate longitudinal bending. (This formulation differs from that of Okerblom in that he attributed the overall longitudinal contraction to contractional plastic strains at each point across the width, due to the plane strain suppression of thermal strain. However, the present formulation, in terms of contraction stresses, gives an identical result.)

The accuracy of the simplifying thermal assumption and the plane strain formulation of the mechanical description cannot easily be verified other than by comparison with numerical approaches or experiments. The thermal strain mismatches do not occur simultaneously, as can be seen from Fig. 4.1. However, at each stage, strain mismatches occur between adjacent longitudinal fibres, travelling progressively outwards across the transverse section, until the temperatures eventually converge to a common value. It is therefore possible that the full values of plane strain contraction stress are indeed developed, corresponding to the maximum levels of thermal strain at each point, as shown in Fig. 4.8. For this reason, the analytical treatment arising from this set of simplifications has been designated by the present authors as the 'mismatched thermal strain' (MTS) algorithm.[11]

Equation 4.5 should at least be dimensionally correct and serves to highlight several useful points. Firstly, the most important controlling factors on contraction force are the net heat input rate q/v, as is well known from practice, and the coefficient of contraction, α. For different materials, the important factor is $\alpha/c\rho$ and reference to the right-hand column of Table 4.2 reveals wide differences between commonly used materials. The thermal dilatation may also be heavily influenced by non-linear contractions induced by metallurgical phase changes in the temperature range corresponding to the shaded regions in Fig. 4.8.

Another interesting outcome is that the contraction force is seen to be independent of the yield strength of the material, this being a mathematical consequence of integrating the inverse linear relationship between temperature and transverse coordinate in equation 4.4. Of course, the temperature profile may not be exactly inversely linear in all cases, but in most practical examples examined by the authors in terms of more complex finite, transient and 3D thermal studies, the dependence was indeed fairly near to inversely linear. This suggests that any dependence of contraction force on yield strength is, at most, weak.

The assumption of constant thermal and mechanical properties over the complete range of cooling would seem like a gross over-simplification although it is not difficult to address this problem if a numerical thermal computation is used (assuming that the variations of material thermal properties with temperature are actually known). In the case of the effects of temperature on mechanical properties, principally Young's modulus and yield strength, the significance depends on how much variation there is over the range of temperatures where the contraction force is being developed. For the specific example of 6 mm thick CMn steel welded at around 1.2 kJ/mm gross heat input, as shown in Fig. 4.1 and 4.8, the respective boundary temperatures that define the widths of the inner yielded zone and the outer elastic zone are 259°C and 139°C above ambient temperature. If the model is correct in the implication that distortion is driven by processes occurring in this temperature band, then it is plausible that the assumption of a simple, fixed, room-temperature yield strength will be adequate for the analysis. It is also possible – and much more productive – to implement the analytical MTS algorithm in a finite-element model; this option will be discussed more fully in Chapters 5 and 8.

4.6 Transverse welding deformations

If the weld fusion zone occupies the full thickness of the plate, or any remaining un-penetrated thickness is unconnected mechanically, as in Fig. 4.9, Okerblom's model for transverse in-plane and angular contraction is very simple, as described earlier in section 4.4. The first assumption is that only the fusion zone experiences a net contraction deformation and the surrounding un-fused material simply expands and contracts freely if forces due to heating and cooling are applied, behaving elastically in the transverse direction. (This assumption depends critically on the idea that the longitudinal thermal gradient is small and cooler regions exert no *transverse* constraint.) At any point through the thickness, therefore, the transverse contraction deformation is given simply by

$$\Delta_T(z) = \alpha T_S w_z \qquad [4.6]$$

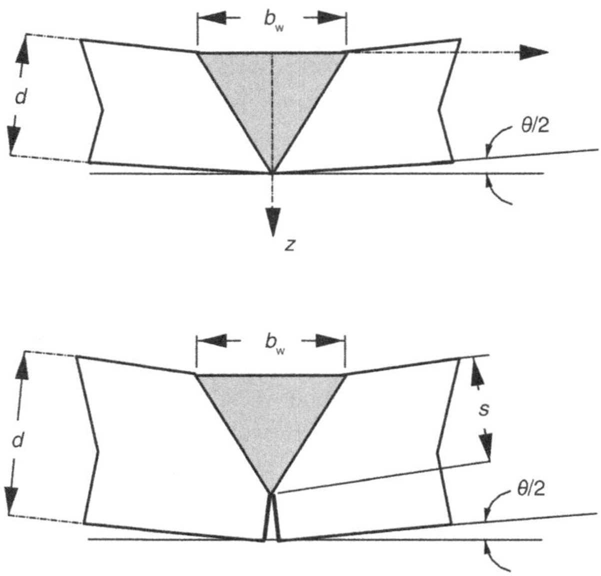

4.9 Transverse contraction model.

where w_z is the width of the fusion zone at coordinate z and T_s is the starting temperature at which the weld metal solidifies sufficiently to establish a fixed shape and this shape then remains as the fusion zone cools to ambient temperature (assumed to be zero for this example). If the fusion zone is assumed to be exactly triangular, as in Fig. 4.9, and there is no other transverse connection between the plates, then it contracts in proportion to the width at each point through the thickness leading to an overall angular contraction given by

$$\theta = \Delta_T/s \qquad [4.7]$$

where s is the depth of the fusion zone and $\Delta_T(z) = \alpha T_s b_w$, where b_w is the fusion zone width on the surface. Note that the mean contraction deformation across the width of the plate, regardless of its width, is given by:

$$\Delta_T/2 = \alpha T_s b_w/2 \qquad [4.8]$$

If the zone is not exactly triangular or some part of the thickness is mechanically connected (i.e. un-penetrated by the weld) then the fusion zone contraction stress, which again cannot exceed the yield strength in tension, is resisted by the assumed elastic response of the un-fused material. The idea is simple, but the resulting formulations for transverse in-plane and angular contraction are complex and depend on the shape of the fusion zone (i.e. triangular, parabolic, etc). Okerblom's formulations were worked through for triangular, parabolic and rectangular fusion zone shapes, leading to nomograms and equations defining the in-plane average and angular contractions for different ratios of penetrated depth to total thickness.[11] This simple analytical treatment was designated by the authors as the 'transverse contraction strain' (TCS) algorithm and this approach can also be used within a finite-element modelling approach, as will be shown in Chapters 5 and 8.

4.7 Residual stress

The residual stress fields generated by welding, although entirely unseen, can often be as troublesome as distortion. Post-welding machining operations to achieve required dimensions will generally not be effective, because mechanical removal of material already subject to residual stress will cause further deformation, sometimes in an unpredictable sense. Failure mechanisms such as fatigue and fast fracture are generally made worse by tensile residual stresses. Buckling, treated in section 4.8, occurs when compressive, residual in-plane loads are applied to thin-section components The residual stresses generated by welding can be sufficient to cause buckling, even in the absence of externally applied loads. Such residual stress fields are described as 'self-equilibrating'. That is to say, the tensile forces generated in a particular direction due to residual tensile stress must be balanced within the assembly by compressive forces in the same orientation.

Like distortion, the highly oriented nature of the temperature fields in welding affects the orientation of the residual stress fields. The longitudinal thermal gradient is small and the effect therefore is that the most-heated strip along the centreline of the weld contracts transverse to the weld line, almost as if it is being cooled simultaneously along its length. There is therefore little restraint and, as a result, little transverse residual stress is generated.

The transverse thermal gradient is large, on the other hand, and therefore the large absolute longitudinal contraction of the central zone is resisted by cooler material on each side which may indeed be expanding. This develops a longitudinal stress field, typically as shown in Fig. 4.10. There is a high central residual tensile stress in a narrow zone, usually reaching a level corresponding to the room-temperature yield strength of the material and this is equilibrated by low-level compressive stress in the flanking regions.

A very approximate estimate of the longitudinal residual stress field may be obtained by applying the plane strain contraction stress field from the MTS algorithm, derived through Fig. 4.8, to a 3D structural model, so that an overall mechanical equilibrium state is achieved in the longitudinal direction, as in Fig. 4.11. The blue curve is the stress pattern developed as a result of the plane strain assumption and the red curve is the stress pattern

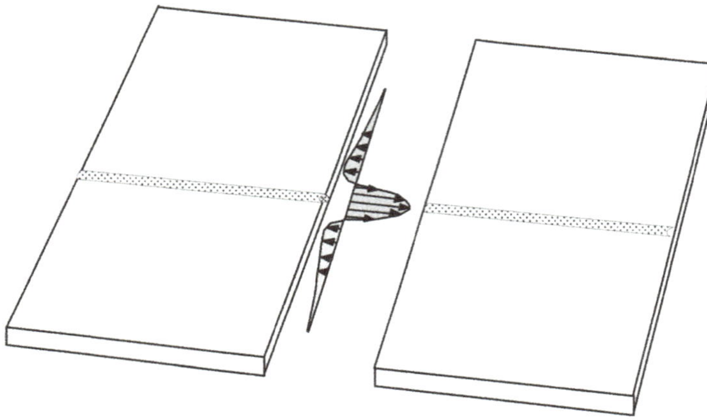

4.10 Typical longitudinal residual stress pattern due to weld thermal cycle.

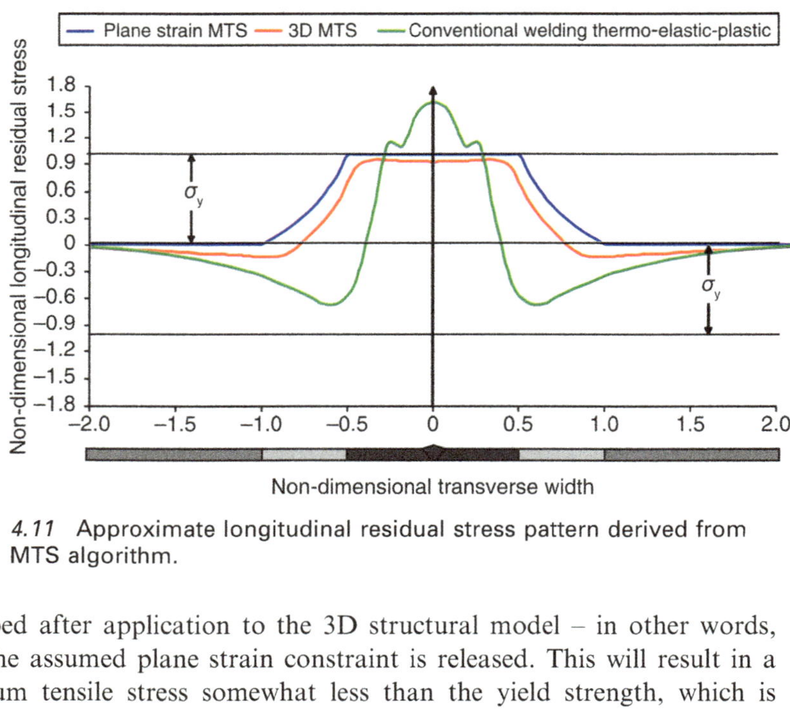

4.11 Approximate longitudinal residual stress pattern derived from MTS algorithm.

developed after application to the 3D structural model – in other words, when the assumed plane strain constraint is released. This will result in a maximum tensile stress somewhat less than the yield strength, which is incorrect, but the method does give some kind of estimate of the width of zone where tensile residual stress exists and of the level of compressive stress in the outer parts of the plate. Figure 4.12 shows the results of this simple treatment for the whole plate, including interactions at each end. However, the accuracy of the residual stress field determined by these simple methods is poor and can only be improved by adopting a more advanced thermo-

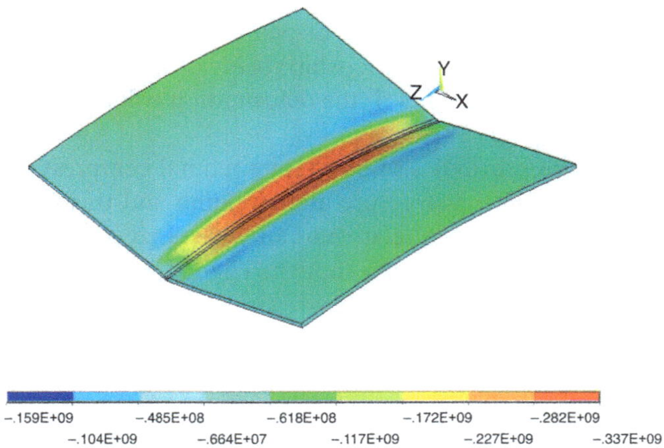

4.12 Approximate longitudinal residual stress pattern from application of MTS algorithm to 3D structural model.

elastic-plastic treatment, together with more accurate modelling of the material properties. This is shown by the third curve in Fig. 4.11 (conventional welding thermo-elastic-plastic) and the methods used to generate this analysis will be treated in Chapter 9.

If the component being welded is thick and, especially if it is being welded by a multi-run procedure, it becomes possible for residual stresses to be developed in any direction. The simple reason is that the contraction in one region is resisted by adjacent material that is much cooler or is indeed expanding. Repair welds in thick components, which may be carried out by excavating a tunnel into the material and then welding outwards, are particularly vulnerable, as shrinkage of the weld bead may then be resisted in all directions, resulting in a dangerous, triaxial tensile stress field within the cooling material.

4.8 Buckling

The buckling problem has already been raised in Chapter 2 as a significant fabrication consideration. In the case of steel panels and the use of high-power welding processes, the possible incidence of buckling should cause an alert for thicknesses less than about 6 mm.

Any long, thin structure, subject to longitudinal compressive force, will tend to buckle. If the structure is perfectly straight and flat, increasing compressive force will lead to uniform compressive stress, until a force level is reached where the force–displacement characteristic becomes unstable and the structure deforms suddenly by transverse bending. This form of buckling is often described as Euler buckling. In practice, structures are normally neither perfectly straight nor flat, and the application of compressive load accentuates the out-of-plane deformations that already exist, the force–deformation relationship being non-linear. If the load approaches the Euler buckling level, the deformation will again be unstable and possibly catastrophic.

As indicated in the previous section, a longitudinal compressive residual stress usually remains in the structure after welding and this will certainly cause out-of-plane deformation relative to an unstressed structure and may also cause Euler buckling if it is high enough. The risk of buckling in simple cases may be judged by comparing the average compressive residual stress due to welding with the Euler buckling stress for given structural shapes and edge-fixing conditions.[12]

For example, consider a square plate of side a and thickness d. the longitudinal force generated may be estimated from equation 4.5 as

$$F = 0.335 \frac{q}{v} \frac{\alpha}{c\rho} E$$

and the resulting longitudinal compressive stress (neglecting the central tensile region) as

$$\frac{F}{ad} = 0.335 \frac{q}{vad} \frac{\alpha}{c\rho} E$$

Camilleri *et al.*[11] showed that if the plate is simply supported on all sides, the Euler buckling stress σ' can be determined using Timoshenko[12] as

$$\sigma' = 3.29 \frac{E}{(1 - \upsilon^2)} \left(\frac{d}{a}\right)^2$$

The point of buckling is reached therefore when

$$\frac{q}{v} \frac{\alpha}{c\rho} = \frac{9.8}{(1 - \upsilon^2)} \left(\frac{d^3}{a}\right)$$

As with the distortion force, the likelihood of buckling increases as the thermal power input increases and notice also the cubic dependence on the plate thickness.

4.9 Conclusion

An introduction to welding mechanics has been given, based on simple analytical formulations. It is noted that uncoupling of solutions and dimensional reductions are powerful tools in terms of reducing the complexity of solutions and retaining insight.

The thermal part of the problem, leading to temperature field results, is based on a 2D quasi-static conduction solution for a moving heat source, with approximations appropriate to high-speed travel of the source. This solution is used to demonstrate

- the influence of material thermal properties (conductivity, volumetric thermal capacity and diffusivity) on temperature fields (Fig. 4.4)
- an approximately linear relationship between welding power input per unit thickness and the temperature at a given point (equation 4.1 and Fig. 4.2)
- an approximately linear relationship between the specific energy input (q/v) and the width of the zone that reaches a given temperature (Fig. 4.3).

The thermo-mechanical part of the problem is based on an approximate elastic-plastic treatment due to Okerblom. The main simplifications in this treatment are that

- the 2D transient temperature field is replaced by an envelope of transverse maximum temperatures, uniform along the length, i.e. one-dimensional (as in Fig. 4.8)
- the longitudinal strain state corresponds to plane strain
- the transverse strains are determined entirely by contraction of the fusion zone from a 'starting temperature'
- the longitudinal and transverse mechanics are treated as uncoupled.

The resulting analysis leads to formulations for the longitudinal force (equation 4.5) and transverse displacements (equations 4.7 and 4.8). The results highlight the key influence of thermal dilatation on the forces and deformations developed. A linear relationship is also shown with the specific energy input (q/v).

The chapter also includes discussion of 3D thermal solutions for thick components (equations 4.2 and 4.3), typical deformations in thin plate (Fig. 4.7), residual stress development (section 4.7) and buckling (section 4.8).

4.10 References

1. Rosenthal, D., 1941, Mathematical theory of heat distribution during welding and cutting, *Welding J.*, 20, 5, pp 220s–234s.
2. Rosenthal, D., 1946, The theory of moving sources of heat and its application to metal treatments, *Trans. ASME*, 68, pp 849–866.
3. Okerblom, N.O., 1958, *The calculation of deformations of welded metal structures*, HMSO, London, UK.
4. Birk-Sorensen, M., 1999, *Simulation of welding distortions in ship section*, PhD thesis, Department of Naval Architecture and Offshore Engineering, Technical University of Denmark, Lyngby, Denmark.
5. Shu Wen, 2003, Data obtained in study on CMn plate steel, private communication.

6. ESDU (Engineering Sciences Data Unit), 2012, *Metallic materials data handbook*, section 10, corrosion resistance steels, prEN3488, sheet, strip.
7. ESDU, 2012, *Metallic materials data handbook*, section 6, aluminium alloys, prEN4202, plate.
8. Section 12, ESDU, 2012, *Metallic materials data handbook*, titanium alloys, DTD 5023, sheet, strip.
9. Rhodes, J., 1991, Some interesting phenomena which can occur in light-gauge steel members, *Proc. international conference on steel and aluminium structures, ICSAS,* 91, pp 128–135.
10. Rykalin, N.N., 1947, *Thermal welding principles*, I.A.N., Leningrad, SSSR.
11. Camilleri, D., Comlekci, T. and Gray, T.G.F., 2005, Computational prediction of out-of-plane welding distortion and experimental investigation, *J. Strain Analysis Eng. Des*, 40, pp 161–176.
12. Timoshenko, S., 1936, *Theory of elastic stability*, Engineering Societies Monograph, McGraw-Hill, New York, USA.

5
Computational simulation of welding distortion: an overview

DOI: 10.1533/9780857099327.77

Abstract: The many different physical phenomena present in welding operations are described. It is shown that the many interactions between these can be simplified to a smaller number, similarly for material properties. An outline of thermal computation approaches is given. Thermo-mechanical computational approaches are described, ranging from 'reduced solution' methods, where substantial simplifications are made, to full elasto-plastic methods that make greater demands on computing resources. The advantages of reduced solutions are exemplified through a butt-welding simulation treated via a 'computationally efficient' method based on the Okerblom approach and two variants of a hybrid method.

Key words: multi-physics, thermal properties, phase change, mechanical properties, transient, steady state, thermo-mechanical, reduced solution, computationally efficient, hybrid methods.

5.1 Introduction

Chapter 4 used analytical, approximate solutions to identify the key driving parameters and material properties that lead to welding distortion. Quantitative prediction of distortion cannot, however, be taken very far using analytical solutions alone and more powerful computational solution methods are therefore required. These are conveniently based on finite-element methods, which are frequently also applied in the structural design process. The treatment in the present chapter stands back from the detail of finite-element implementation for the time being and seeks to outline various ways in which the complex physical features of welding applications can be simplified to a level that permits useful simulation and prediction of the resulting deformations and residual stress fields.

The 'multi-physics' concept is first introduced, describing the complex interaction of several physical phenomena within a typical welding process. It is shown that these interactions can be simplified in various ways to reduce the problem of simulation to a manageable level for a real structure. Modelling of non-linear material properties is discussed, leading to an outline of various approaches for thermal modelling. A range of modelling strategies for the thermo-mechanical part of the simulation problem is given, linking the temperature fields to the deformation and residual stress outcomes. There is then a discussion of the various merits of 'reduced solutions', leading to an illustrative example of butt-welding simple plates. The outcomes of using different approaches are shown and compared with experimental determinations.

5.2 Multi-physics

Many different physical phenomena are involved in the process of welding and each tends to interact with the other, often in a complex way. Figure 5.1 shows this schematically, and extends the scope of a similar interaction diagram[1, 2] through the inclusion of certain metal inert gas (MIG) process features. In the vicinity of the arc and the weld pool, electro-magnetic phenomena and related interactions dominate. Arc plasma and electrical resistance effects generate heat. Magnetic forces influence the fluid behaviour of the molten material and the shielding gas. This influences the pattern of heat conduction into the material surrounding the fusion zone. Material at a distance from the weld pool also responds to the input of heat, but the nature of that response depends on the indirect effects of heat input on the metallurgical state and the microstructure of the material, together with its thermo-mechanical properties. Both heat flow and structural deformation in the fabrication being welded depend on the thermal and mechanical properties, which depend in turn on microstructure, current temperature and rates of temperature change. Plastic deformation also affects microstructure and generates heat.

All this can be appreciated in a qualitative sense and such potentially complex interactions can in theory be simulated through what is called a 'multi-physics' computational approach. Here, the various effects are fully described numerically and the resulting equations are solved simultaneously. However, certain questions are worth asking before embarking on such a course. Is it necessary to simulate all these features and interactions? Is it feasible? Does it add value to the result?

Adopting a multi-physics approach will certainly add to the work of problem formulation. On the question of feasibility, the main difficulty is

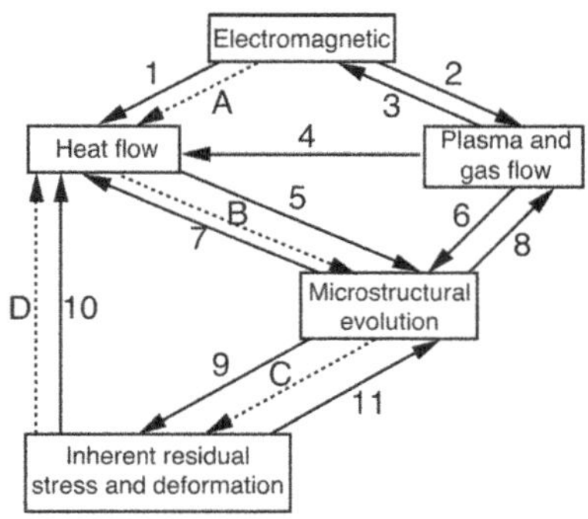

Multi-physics analysis

1. Heat generated from electromagnetic forces
2. Plasma flow, filler wire and shielding gases influenced by electromagnetic forces
3. Efficiency of electromagnetic effects depends on fluid and plasma flow properties
4. Fluid flow interaction influences plasma heat conductivity, heat capacity and thermal boundary conditions
5. Microstructure composition and phase changes depend on temperature transients
6. Microstructure evolution depends on shielding gases and filler material
7. Thermal properties are a function of microstructure
8. Porosity, diffusion and absorption of gases depend on microstructure
9. Thermal expansion, volume changes, elastic and plastic material behaviour depend on microstructure of material
10. Deformation changes thermal boundary conditions. Heat generated due to plastic dissipation, thermal and elastic strain rate
11. Microstructure evolution depends on deformation and residual stresses

Coupled thermo-mechanical analysis

A. Electrical power and efficiency
B. Thermal transients and cooling rates
C. Thermal strains and non-linear material properties
D. Heat generated due to thermal, elastic and plastic strain rate

Uncouple thermo-mechanical analysis

A → B → C

5.1 Multi-physics interaction in welding.

that the dynamic thermal and mechanical properties of the materials involved have to be known and are rarely already documented and readily available. Obtaining the data requires extensive experimental determination, ideally through laboratory simulation of the thermo-mechanical environment experienced by actual materials during welding. This is not a trivial exercise. However, the question of whether or not it is strictly necessary to simulate all these interactions in detail can in the end only be answered by comparing the outcomes of simplified approaches with full multi-physics simulations, or with the results of full-scale tests.

So, what are the alternatives to a full multi-physics approach? Firstly, some of the interactions might be neglected on the grounds that they are considered to be weak. If they cannot be considered to be weak, they might nevertheless be represented in a 'lumped' manner. For example, as shown in Fig. 5.1, the input of electrical and other heat-generating fields is indicated by a single arrow A, which represents a lumped input of heat energy, corresponding to the electrical power delivered to the welding head, allowing for some loss occurring between the measured electrical power input and the thermal energy conducted into the workpiece. This bypasses the real, complicated interactions represented by arrows 2, 3, 4, 6 and 8. The effects of heat on the microstructure may then be represented in a largely temperature-based manner by arrow B. The main outcome is an influence on the temperature-dependent material properties and thermal strains, feeding into the deformation analysis according to arrow C, with a secondary feedback on thermal properties represented by arrows 11 and 7. Some intermediate approaches introduce a rate-dependent material model to reflect the interactions.

If these basic simplifications are accepted, this enables a starting point for a strategy as indicated in box B of Fig. 5.2. The transient nature of the moving heat source is fully recognised and the thermal and structural degrees of freedom in the overall problem are also recognised and solved simultaneously. Detailed input of the temperature-dependent and strain-dependent thermal and mechanical properties of the materials is nevertheless required and this is not a simple task.

Feedback of thermal effects from the deformation fields, as shown by arrow D in Fig. 5.1, could be incorporated. However, many studies have

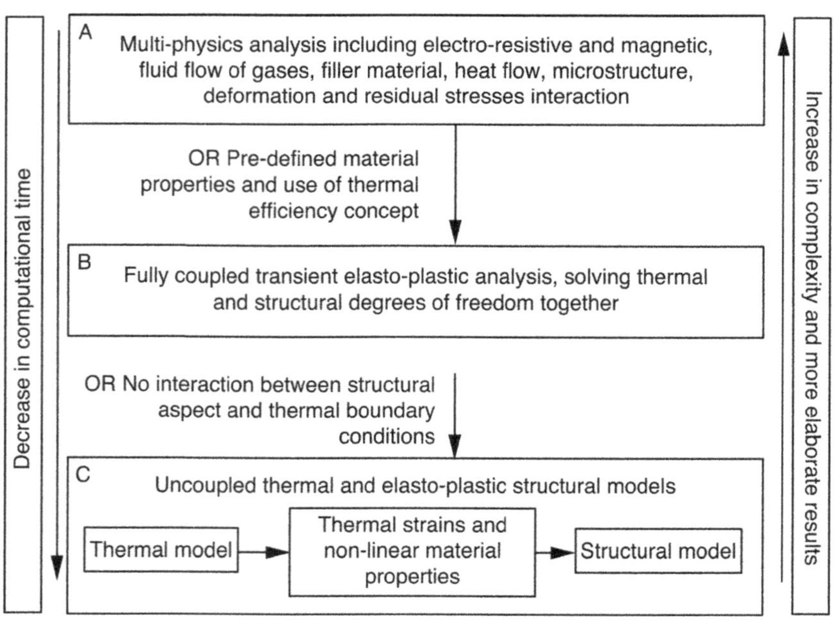

5.2 Computational strategies for predicting thermal transients, residual stresses and distortion in fusion welding.

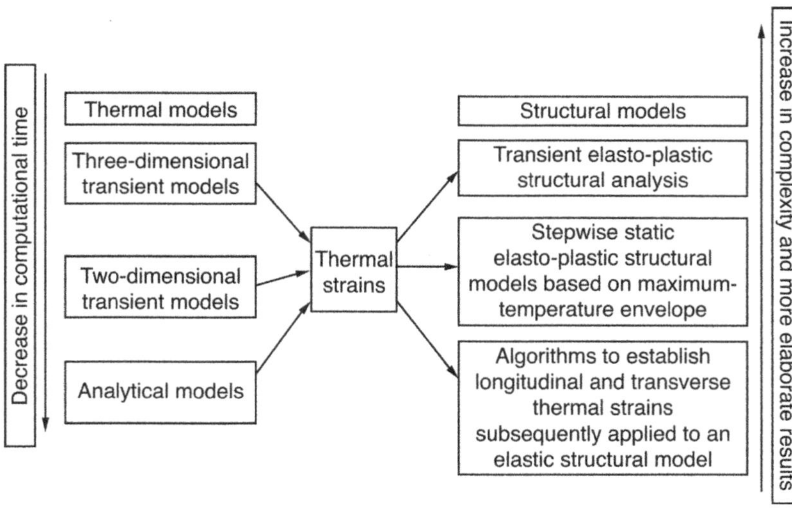

5.3 Different strategies for uncoupled thermo-mechanical analysis.

shown[3–6] that it is realistic to use an even simpler modelling strategy in which the thermal and mechanical phenomena are uncoupled, the interdependence between these two regimes being weak for the process conditions of typical fusion welding. This alternative route is shown in box C in Fig. 5.2.

The heat flow model should, nevertheless, be made to reflect the effects of metallurgy and phase changes on thermal properties (arrow 7) and correspondingly the mechanical model should be set to reflect the effects of temperature and metallurgy on thermal expansion properties and strength (arrow 9). However, uncoupling greatly simplifies the analytical problem and it also becomes more straightforward to validate the predicted outcomes experimentally by measuring temperature fields for the heat flow model and deformation fields for the mechanical model. This then leads to the range of analytical possibilities shown in Fig. 5.3, whereby a thermal analysis of some kind can be carried out and the results, in terms of temperature fields, are then fed into a suitable thermo-mechanical or 'structural' model, with no subsequent feedback of mechanical heat generation.

5.3 Thermal property non-linearity

As stated in Chapter 4, the temperature fields due to a given welding process heat input will depend primarily on the thermal conductivity K and the capacity of the material to absorb heat, given by the product of specific heat and density, $c\rho$. Figure 5.4 shows such data for a typical, weldable carbon

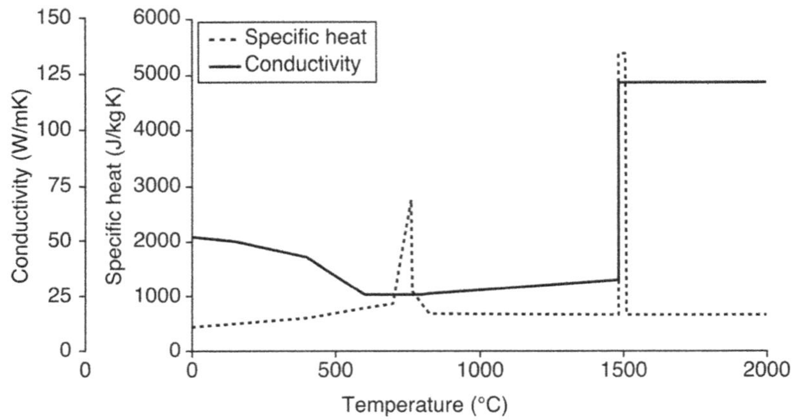

5.4 Typical conductivity and specific heat for CMn steel.

manganese steel.[7] There is a large spike in specific heat capacity, corresponding to the 700°C plus temperature range, where there is a metallurgical phase change. There is an even larger jump in heat absorption capacity at the transition from solid to liquid, directly related to the latent heat associated with melting. A similarly large increase is seen in the conductivity property at the solid/liquid transition, although some of this may be due to the fact that the fluid material can circulate, resulting in greater heat transmission. Here, the conductivity may also be increased to take into consideration the stirring effect of the molten weld pool. The value does not specifically define the conductivity of the molten weld pool as this also depends highly on convection.

It is not at all straightforward to include such extreme variations in computational analyses, due to solution convergence problems.[1] In the case of heat capacity, the heat conduction equations can be cast alternatively in terms of the enthalpy of the material H_e at a given temperature T using the relationship

$$H_e(T) = \int_0^T \rho c\,(\tau)\,d\tau \qquad [5.1]$$

where the specific heat is given as a function of temperature, from Fig. 5.4, for example.

Figure 5.5 shows that the resulting enthalpy–temperature relationship is much smoother and this restores feasibility to any finite-element computational analysis that is used.

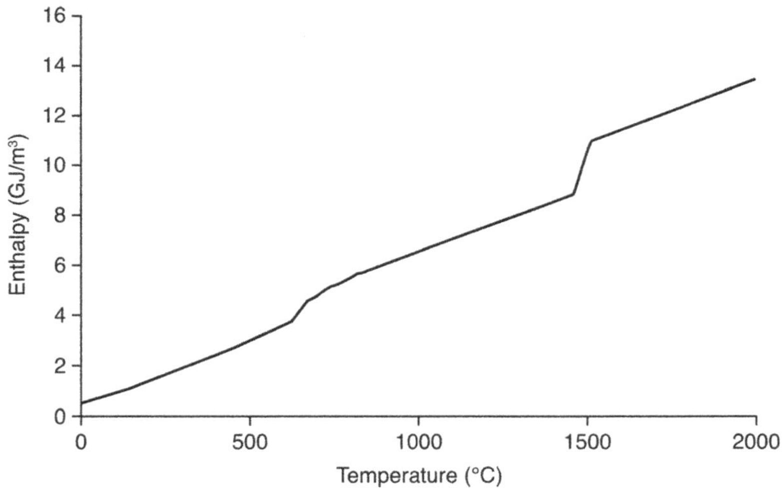

5.5 Enthalpy versus temperature for CMn steels.

5.4 Phase change and non-linear thermal dilatation

In Chapter 4, Fig. 4.5 showed that the contraction behaviour of some materials is highly variable over the range of heating and cooling experienced by a typical weld. In that case, the material is shown to expand briefly at a particular point in the cooling phase, before reverting to the expected contraction behaviour. As the actions leading to distortion are driven primarily by the cooling phase of the welding event, it would be useful to represent this material behaviour accurately. This might be represented in a finite-element simulation as in Fig. 5.6.[8] Various software versions treat this input in different ways and these possibilities will be discussed in greater detail in section 9.2.2, in considering the implementation of finite-element software.

5.5 Mechanical property idealisation

The important temperature-dependent mechanical properties to be incorporated in a computational solution are the elastic and plastic moduli and yield strengths. In the previous chapter, Fig. 4.6 shows a measure of idealisation in that non-linear stress–strain curves measured at a given temperature are represented by straight-line elastic and plastic hardening characteristics. The single values derived from these idealisations can then be represented by temperature-dependent characteristics as in Fig. 5.7.

Such properties are, of course, a function of the mechanical testing conditions and the metallurgy of the material tested. These particular test results were obtained by heating specimens to the specified uniform test

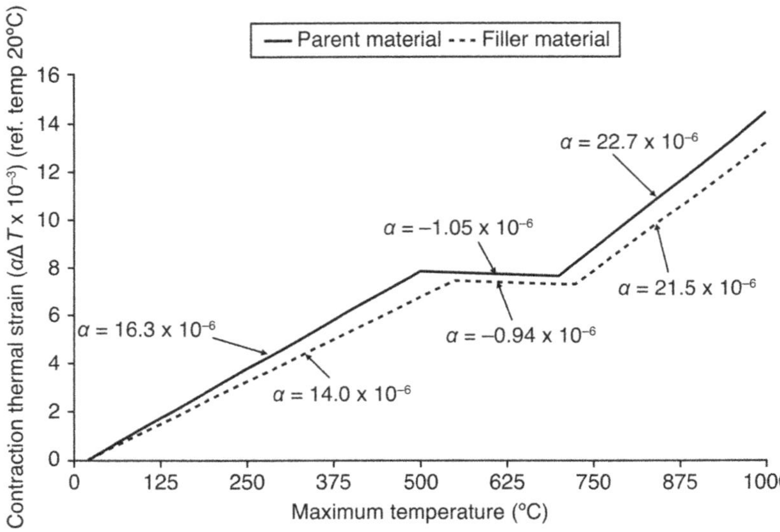

5.6 Thermal contraction strain and dilatation coefficient.

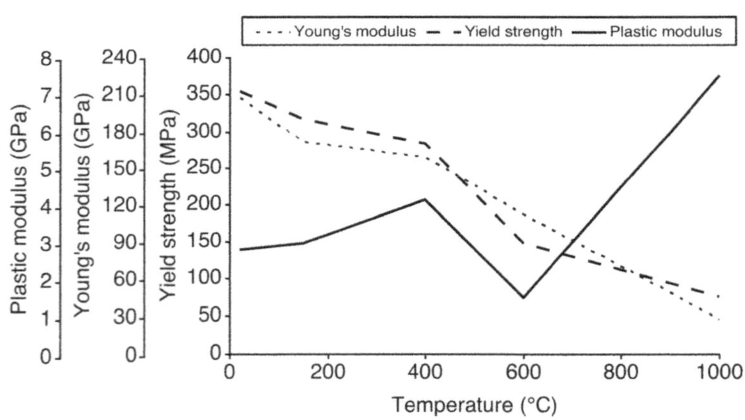

5.7 Non-linear temperature-dependent material properties for CMn steel.

temperature and then straining them at a specified constant rate. This is rather different from the dynamic conditions in a weld zone. In particular, grain growth may occur in the slowly heated tensile test specimens. Phase transformations will also take place differently than in weld zones where temperature changes are rapid and different mechanical conditions will trigger phase changes at different temperatures. Weld simulator testing machines are available that can reproduce many of the transient temperature changes and mechanical conditions in real welds and, if results using such equipment are available, this should improve the quality of data used in the finite-element simulation.

5.6 Thermal computation outline

The ideal approach, ignoring the consequent computational burden, is to apply a three-dimensional (3D), transient thermal finite-element analysis to the problem. This makes it possible to investigate intricate geometry with starting and finishing points for welding process heat input and incorporating finite boundaries. Non-linear, temperature-dependent material thermal properties can be included. Thermal breaks in the component and intimately contacting heat sinks can all in theory be simulated.

Particular aspects of the real process may have to be specially considered in terms of modelling. For instance, how is the heat input from the welding process to be determined and modelled, as this is the major determinant of the temperature fields? (section 7.9 gives more information on the concept of 'weld efficiency'). Temperatures will also be influenced by convection and radiation from the heated surfaces of the workpiece and, although the effects are usually small relative to heat conduction, they need to be accounted for in some way. Furthermore, filler weld material is involved in the majority of industrial welding processes and it may be necessary to account for the change in cross-sectional thermal mass from an empty weld preparation to a section that is filled with weld metal. Modelling strategies for these issues are discussed in more detail in section 7.4.

The following sub-sections present some alternative approaches to modelling heat flow.

5.6.1 Three-dimensional transient thermal finite-element analysis

Figure 5.8 shows time–temperature plots at four transverse locations across the mid-length of a 500 mm long, 6 mm thick plate, MIG butt-welded using a nominal thermal input of 1.19 kJ/mm, corresponding to an arc voltage and current of 23 V and 250 A and at a travel speed of 4.8 mm/s. This form of presentation of thermal fields gives an alternative view to the isothermal contour presentations shown in Chapter 4.

In this example, the arc arrives at the mid-length of the plate at around 52 s and the weld is complete at around 104 s. The maximum temperatures reached at given transverse locations can be readily identified (this is an important parameter in the context of the mismatched thermal strain (MTS) algorithm – see Fig. 4.8) and it also shows clearly how the temperatures across the plate gradually converge after weld completion.

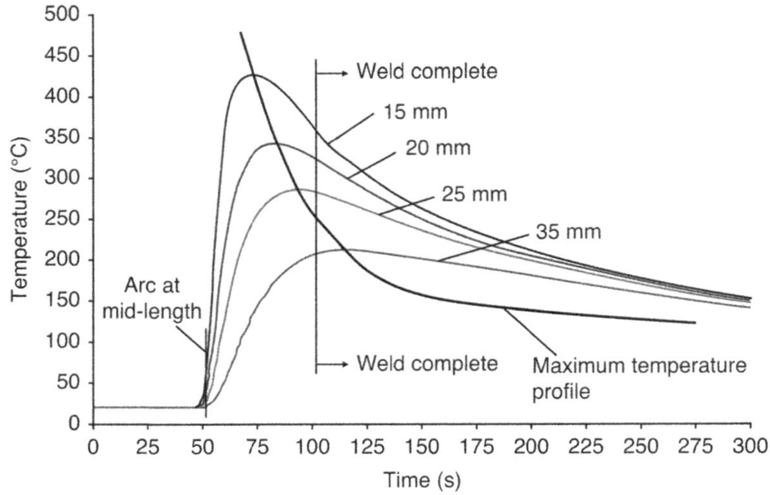

5.8 Three-dimensional analysis showing temperature transients at points across a plate, transverse to the weld line at mid-length.

5.6.2 Three-dimensional steady-state thermal finite-element analysis

Computing economies can be made by formulating the problem in terms of a plate of infinite length, similar to the Rosenthal analytical solution described in section 4.2. In this case, the plate is assumed to have a fixed cross-sectional configuration and 'flows' in a continuous stream past a fixed position where there is a constant weld process heat input. Comparisons of the outcome of such an analysis will not be shown here, but using the same parameters as for the analytical steady-state cases shown in Chapter 4, the temperatures in the computational steady-state case turn out to be slightly higher. This probably reflects the finite transverse dimensions of the computational case.

The disadvantages of the steady-state approach are mainly that heat input starting and finishing cannot be properly simulated, nor can the influence of end boundaries. Also, depending on the software used, it may not be possible to include certain non-linear thermal properties.

5.6.3 Two-dimensional transient thermal finite-element analysis

Given that the heat flow in the case of fast-moving heat sources is predominantly transverse to the weld line, considerable computational economies can be made by basing the thermal analysis on a transverse, cross-sectional strip. Temperatures are assumed to be constant across the

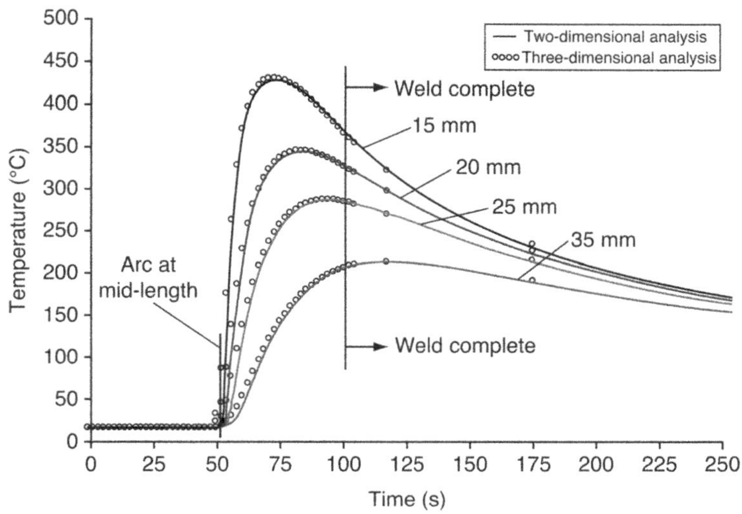

5.9 Comparison of transients calculated using 2D and 3D thermal analyses.

thickness of the strip in the weld direction and no heat flows normal to the strip in either direction, forwards or backwards.

Figure 5.9 shows a comparison of the results from this 2D analysis with the previous 3D transient analysis. The comparison is clearly very good in this instance, at least for points somewhat distant from the fusion zone. The 2D analysis greatly simplifies the thermal computation, the results of which can be used as input to a 3D structural analysis.

5.7 Range of thermo-mechanical approaches available

Having established the thermal transients through one or other of the above routes, the resulting temperature fields then provide the input for the thermo-elastic-plastic structural phase of the analysis, as shown in the right-hand column in Fig. 5.3. Shown at the top of this column is what might be described as the ideal analysis, namely a full, 3D, transient model, complete with non-linear, temperature-dependent, elasto-plastic material properties, together with corresponding descriptions of thermal expansion–contraction and phase-change behaviour. This is a challenging kind of analysis and, other than for relatively small simple configurations, strong computational resources will be required. Figure 5.10 shows results from such an analysis, at various stages of weld completion and cooling. It is worth noting in passing, that results output in the form of a sequence of dynamic pictures (so-called 'movie versions') can be very instructive in terms of understanding the mechanical actions that lead to distortion and residual stress development.

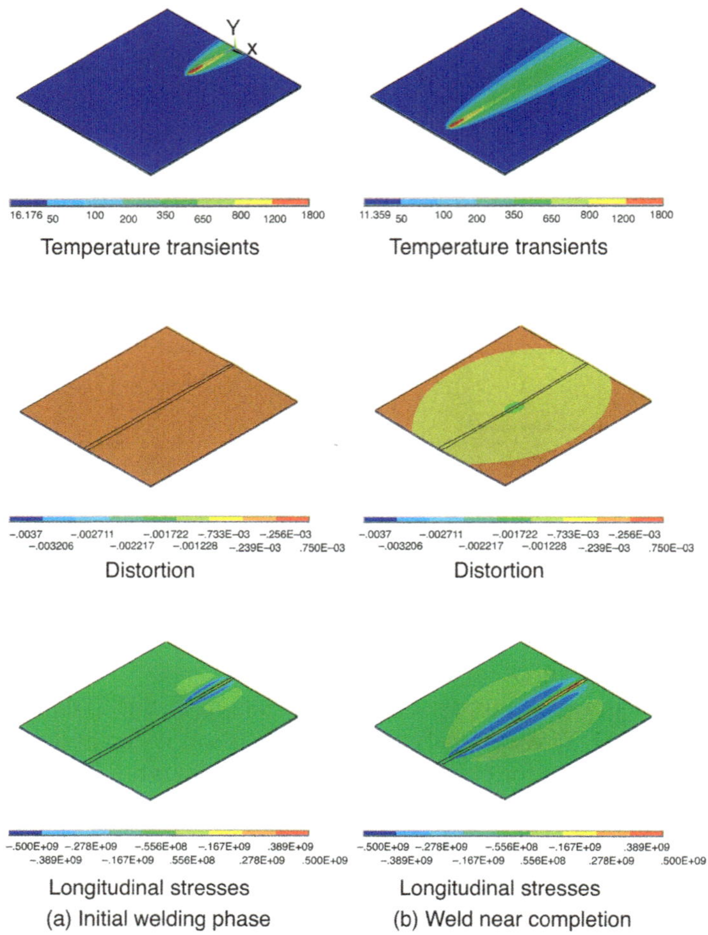

5.10 Transient responses at four stages using full thermo-elastic-plastic analysis. (a) Initial welding phase; (b) weld near completion.
(Continued)

Figure 5.10 shows temperature profiles, distortion and longitudinal stress patterns at four stages ((a) to (d)) during welding of a 4 mm thick plate, 0.45 m long and 0.4 m wide. Welding process parameters were set at 200 A, 20 V and 7.5 mm/s travel speed, with an assumed efficiency of 75% (note that the same contour ranges are used for all the images, so that they are directly comparable). The main mode of distortion during passage of the welding heat source is angular. Longitudinal bowing tends to develop after weld completion and during the long cooling period to uniform ambient temperature.

Comprehensive analyses of this type are remarkably sensitive to what might be thought to be minor modelling factors, such as boundary

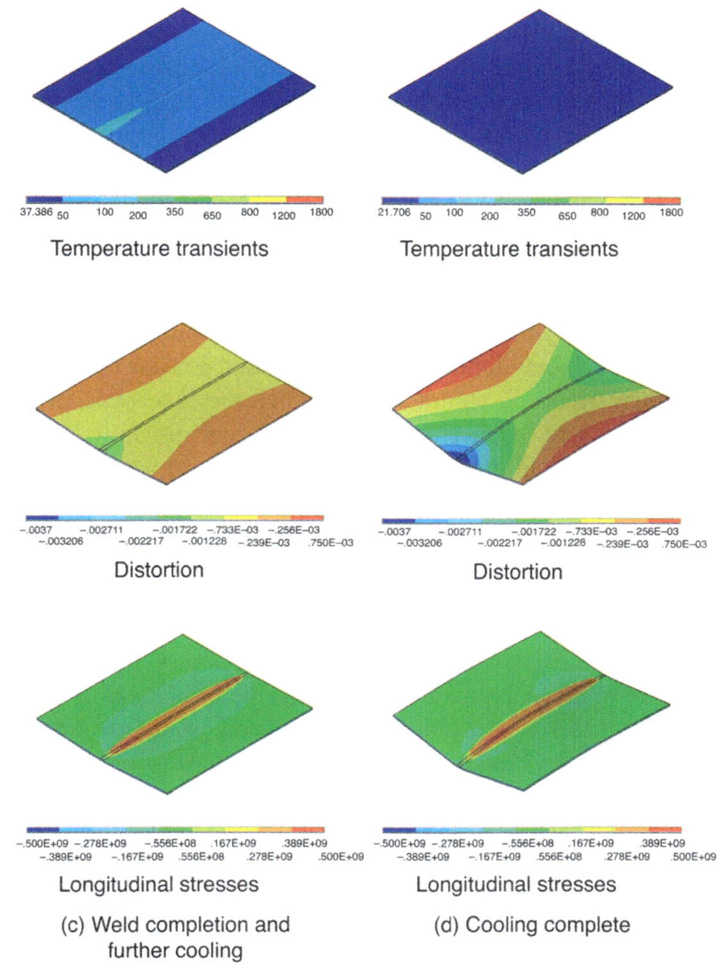

5.10 Continued. (c) Weld completion and further cooling; (d) cooling complete.

conditions used to simulate supports, depth of penetration assumed and simulation of temporary attachments (e.g. tacks) used to hold the plates together for welding. For that reason, comparison of results from different researchers and with different experimental results may be problematic. If the configuration is large and more elaborate than a butt weld, for example a fillet-welded stiffener, the modelling and computational effort multiplies to the extent that a full thermo-elastic-plastic model becomes impractical.

The analytical approach described in sections 4.4 to 4.6, together with the computational model based on the MTS and transverse contraction strain

(TCS) algorithms, represent the opposite extreme from the full, 3D, transient models just described and are therefore located at the bottom of the column of structural model options in Fig. 5.3. These approaches, described as 'computationally efficient', are based on approximate descriptions of the longitudinal and transverse thermal strains. The resulting contractions are applied to a simple 3D, static, elastic model, thus achieving a very large saving in computational effort relative to the full analysis.

However, a price is paid for simplicity and there are many situations where this description is inadequate. The problems lie mainly in the fact that the calculated longitudinal and transverse strains in the Okerblom approach[9] are each one-dimensional and assumed constant along the length of the structure. Hence, inaccuracies arise when these methods are applied to configurations where the mechanical response depends on transient factors that develop as the weld progresses along the length, or where the stiffness in one plane (i.e. transverse or longitudinal) is affected by constraints in the other plane. A particular example of this arises due to supports or clamps, where the developing displacements result in changes in structural boundary conditions. This can only be treated by some form of transient analysis.

Nevertheless, the simple approach models most cases surprisingly well and, apart from the massive computational savings they introduce, they also allow the user to develop greater insight into the mechanics of distortion in more complex cases, for example where several welds are applied in an assembled structure. This benefit has led the authors to develop hybrid methods that combine the best features of the simple and the more complex analysis methods. These are based on stepwise application of separate analyses, whereby transient effects are incorporated to a limited extent. This option is shown in Fig. 5.3 between the two extremes of structural modelling.

Hybrid models overcome many of the deficiencies of the simple transverse description (the TCS model) by applying 3D thermo-elastic-plastic analyses to the angular (and possibly also to the longitudinal) deformations of the weldment, albeit in a more simplified and uncoupled manner than would be the case in a full transient thermo-elastic-plastic analysis. This treatment can also incorporate improved simulation of expansion/contraction and high-temperature strength properties. However, one simplification – which is a key part of the TCS algorithm – is carried over to the elasto-plastic analysis in the hybrid application. Only the elements inside the fusion zone are assumed to be active in terms of cooling from the cut-off temperature. This analysis is carried out in a stepped fashion (i.e. like a coarsely graded transient analysis) and it can therefore reflect any changing 3D constraint as the weld progresses. The longitudinal action can be analysed in different ways, as will be discussed in the next section.

5.8 Reduced solutions and their advantages

Given the range of solution strategies outlined in Figs 5.2 and 5.3 and described in more detail in sections 5.6 and 5.7, a prospective user may be uncertain about the best approach to adopt in a given context. It is evident that solution methods that are more rigorous in terms of description are also more complex and costly to implement – as reflected in the side bars in Figs 5.2 and 5.3. The more simplified strategies will be described here as reduced solution methods, reflecting the fact that certain physical aspects of the full problem are not included on the grounds that they are believed to influence the outcome of the analyses only to a small extent. Such simulation methods have the advantage that they are simple and quick to apply in terms of formulation, solution and interpretation, but they may miss important aspects of the real problem in some instances. It is therefore useful to briefly consider the typical design context where support through simulation strategies may be helpful and thus identify where reduced solutions could be appropriate.

The attractive feature of fusion welding that is most exploited in design is that it provides a powerful and versatile capability to create very large built-up structures by joining stock metal components such as plates, rolled sections and other components. However, the very versatility of the method can lead to trouble if the fabrication implications of the process are not understood at the design stage. The same structural design objective can be achieved in most applications through a potentially infinite variety of configurations and starting from a variety of stock material inputs. As a simple example, a ship's hull plate or deck structure, which is required to support certain superficial or in-plane loads, may be formed by attaching various stiffeners, in various configurations, to thin plate. Assuming that light weight is a design objective, structural and/or cost analysis may be used to optimise the combination of plate thickness and stiffener size and pitch. However, more frequently spaced stiffeners will mean more frequent welds, and thinner plate usually gives rise to greater welding distortions. Moreover, a variety of welding and fabrication procedures might be applied, driven by cost and availability constraints, each with their own thermal advantages and disadvantages. The variables in the design equation are not therefore simply determined by material cost or structural mechanics and some care is required to ensure a realistic prediction of possible distortion outcomes.

It is therefore useful to illustrate the application of different reduced solutions through an example that represents a basic building block for thin-plate structures, namely a built-up, extended plate, formed using a simple butt weld. Later examples in Chapter 11 will treat stiffened plates, where upstanding strips or rolled sections are fillet-welded to plates. In both cases, matching experimental determinations are available to provide some cross-reference on the effectiveness of the computational solutions.[10, 11]

5.8.1 Example: butt-welded plates

Figure 5.11 shows the common configuration for the 6 mm thick steel test plates used in this study. The plates were held together initially by 100 mm square, end-tab plates, tack-welded to the edges of the main plates. Complete scans of out-of-plane deformation were taken before and after welding (the deformation scanning procedures are described in Chapter 6).

The corresponding symmetrical-half finite-element model is shown in Fig. 5.12. Support positions were modelled, with gravity body forces applied, through the use of bar elements having zero stiffness in tension and specified stiffness in compression to simulate contact (or loss of contact). Interaction between the end-tabs and the plate was simulated by including the tack welds in the model. However, the end-tabs were assumed to be at ambient temperature throughout. This offers maximum restraint on the structure but does not take into account the real initial and final thermal strain patterns in the end-tabs due to starting and stopping.

Material expansivity was modelled in different ways, but always based on the dilatometer test results shown in Fig. 5.13.[8] Heating and cooling rates of the same order as in the real weld were used for these determinations. The corresponding instantaneous coefficients of expansion, when assuming

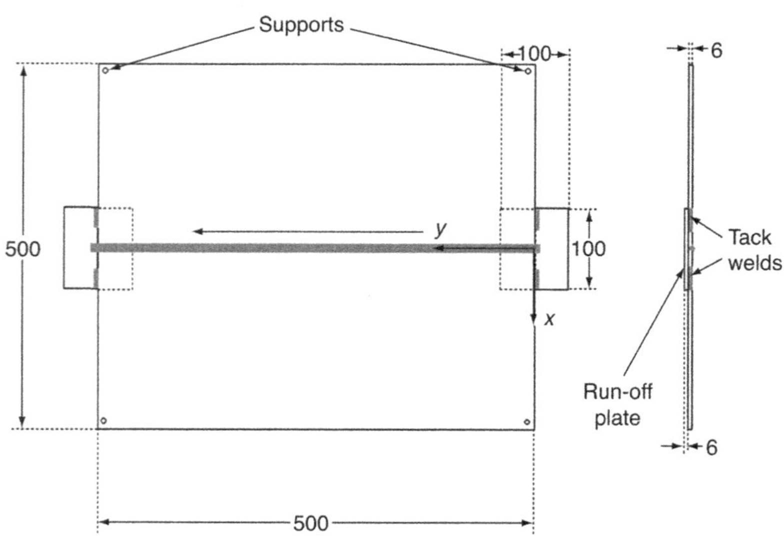

5.11 Schematic illustration of butt-welded specimen (dimensions in mm).

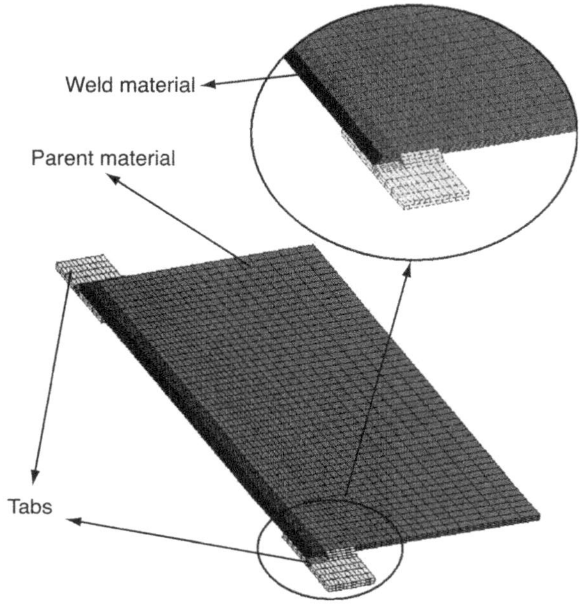

5.12 Finite-element model of butt-welded test specimen.

constant or temperature-dependent material properties, are shown in Fig. 5.14.

Computationally efficient (CE) model

Two-dimensional transient thermal analysis was used for the first part of this simulation, according to the reduced thermal solution strategy outlined in section 5.6.3. A simple structural model, based on TCS and MTS algorithms was then applied, using fixed average expansion and contraction coefficients for parent and weld material as given in Fig. 5.14. This is the finite-element version of the analytical treatment described in sections 4.5 and 4.6, and further explanation of this development will be given in Chapter 8. Note that this method can also be fine-tuned to match the initial out-of-plane deformation pattern of the real assembled and tacked components, if this is known. Such deformations are used as the starting point for the analysis. However, the initial deformation patterns varied substantially over the ten separate specimens welded in the study and therefore a zero out-of-plane initial profile in the models was assumed for simplicity. The CE model for the test structure shown in Fig. 5.11 can be solved in about 8 minutes.

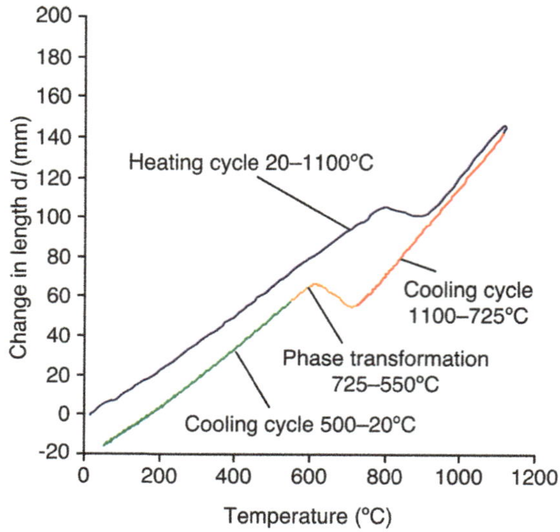

(a)

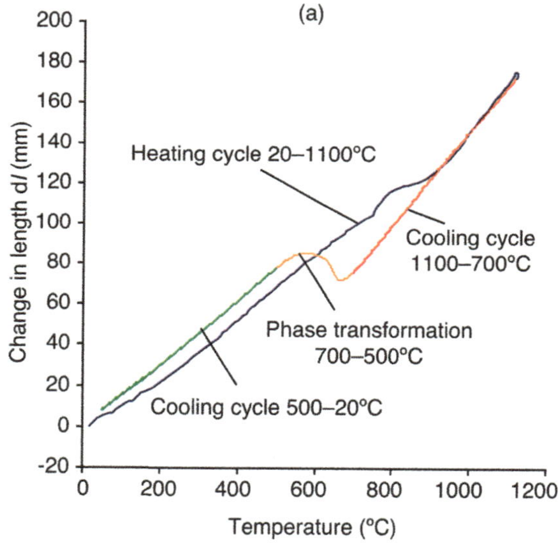

(b)

5.13 Dilatometer test results: (a) weld metal; (b) parent material.

Sequential hybrid (SeqH) model

This model treats the transverse and longitudinal actions as somewhat independent and seeks mainly to improve on the TCS model by replacing it with a non-linear finite-element analysis applied to a transverse strip. This is advanced in a step-wise mode, with improved simulation of the non-linear, temperature-dependent material properties. The transverse strip or plate section is attached at all times to the rest of the plate, including the end-tabs.

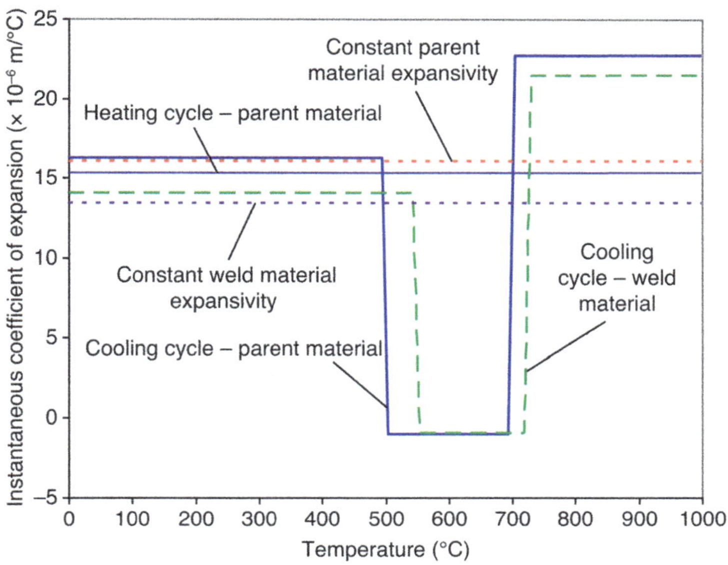

5.14 Instantaneous coefficients of expansion used in various computational strategies.

This therefore makes it possible to simulate more of the angular deformation mechanics in the active section, caused by the restraint of parts of the structure ahead of and behind the weld zone. Weld filling is simulated by 'killing' the fusion zone elements before and during heating and then reactivating them during the subsequent cooling cycle (see section 7.4 for further explanation).

Instantaneous coefficients of expansion and contraction, derived from Fig. 5.13 are as shown in Fig. 5.14 and non-linear, temperature-dependent strength properties are incorporated as given previously in Fig. 4.6. The transverse actions are de-coupled from the longitudinal in this stage, by the simple expedient of setting zero coefficients of expansion and contraction in the longitudinal direction.

Note that, as in the simple CE model, thermal analysis is used to establish maximum temperatures across the transverse strip (a 2D analysis is adequate) and these are then applied and followed by slow cooling to ambient temperature. Some further features of the CE model are retained, such as the assumption that the thermal strain is entirely due to contraction of the fusion zone elements from a reference cut-off temperature. The transverse expansivity of the parent material elements is set to zero to achieve this.

The resulting step-wise loading sequence is shown in Fig. 5.15 for a model divided into three 167 mm wide transverse sections (i.e. relatively coarsely divided) and 'welded' from left to right. The key outcome of this stage of the

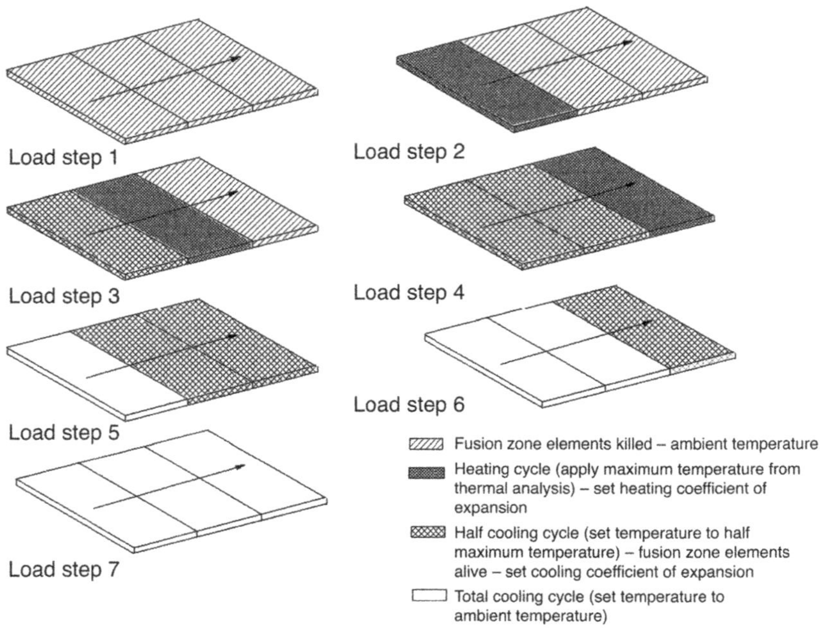

5.15 Step-wise non-linear thermal strain modelling strategy.

analysis is the out-of-plane displacement profile over the whole area of the plate, due to transverse angular deformation. The difference between this profile and the profile that resulted from the TCS stage of the CE model is that the SeqH result reflects more realistic non-linear material properties and analysis and, for SeqH analyses based on more than one or two longitudinal sections, it becomes possible for the transverse profile result to vary along the length.

In the second step of this analysis (the longitudinal stage), fictitious loads derived from the MTS algorithm are applied to the angularly distorted plate in a single step, corresponding to the out-of-plane profile derived from the previous load steps. In this case, elastic material properties and fixed coefficients of expansion are used in the longitudinal direction, but transverse thermal strains in both the weld and parent material are now set to zero. Note that, although in the first transverse stage the plate is subdivided into several sections, the longitudinal contraction algorithm is applied in a single load-step, embracing the total length.

The time taken to solve this model increases substantially relative to the CE model, principally due to the transverse thermo-elastic-plastic analyses.

Simultaneous hybrid (SimH) model

Having moved to a more rigorous thermo-mechanical analysis for transverse actions of the strip in the SeqH model, this prompts the possibility of doing the same for longitudinal actions, simply by resetting the longitudinal coefficients of expansion and contraction to realistic levels (whereas in the SeqH model the longitudinal coefficients were set to zero for the transverse stage). This is a more rigorous approach than using the MTS algorithm and does not require a significantly greater computational time than the SeqH model – for this simple joint configuration at least. Formulation is also more straightforward and robust. In the SeqH model the analysis is split into two stages where the first-stage deformation profile has to be extracted and applied to form the starting point of the second stage, whereas in the SimH model the transverse and longitudinal actions at each section are analysed together at each step.

This model retains the computational economy gained by using the maximum temperature profiles as input, instead of the true transient profiles. The stepping strategy makes it possible to reflect some of the transient mechanical constraints without incurring the full computational burden of a true transient analysis.

Comparison

Figure 5.16 compares the results for the three different thermo-mechanical strategies with the experimental findings. The description of distortion is presented in three ways, corresponding to angular, longitudinal and overall out-of-plane deformation shown in Figs 4.7(b), 4.7(d) and 4.7(e). All of these deformations vary along the length and across the width but are expressed for convenience by a single parameter, namely the root mean square (RMS) averages of the transverse angular, longitudinal bending and overall out-of-plane deformations. Welding distortion simulations were carried out for the three forms of analysis (CE, SeqH and SimH); in the latter two cases, models having one, three, ten and twenty five sections were analysed (the CE model is carried out as a single section).

Experimental issues on measuring welding distortion will be discussed more fully in Chapter 6, but it is worth noting at this stage that it is usually very difficult to control the conditions in live tests with sufficient rigour to ensure that the results provide accurate validation or otherwise of simulations. In particular, the models assume perfectly flat, stress-free components, whereas real test pieces are usually neither flat nor stress-free as a result of the assembly tack-welding procedures needed to hold them together during the weld pass. In the case of the ten test welds studied here, three of them acquired uncharacteristic longitudinal final curvatures relative

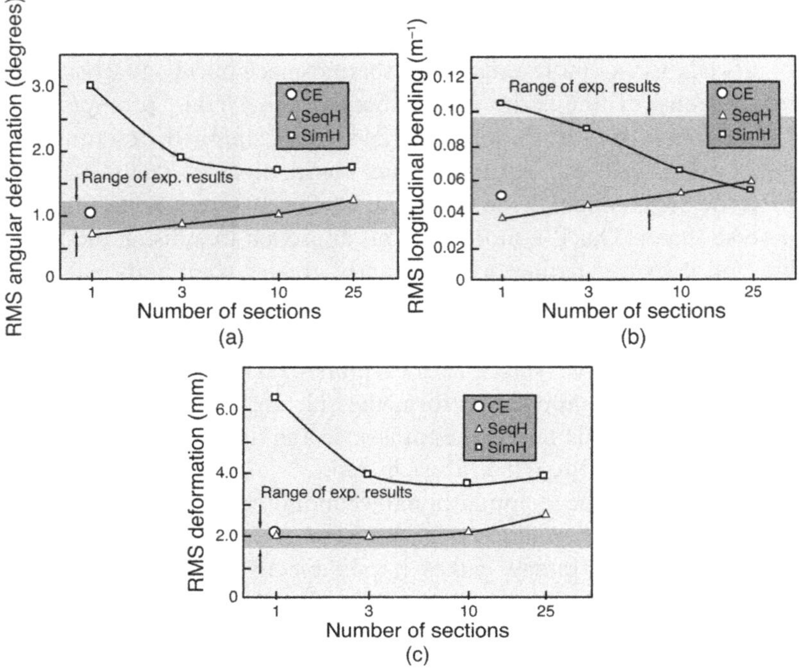

5.16 Comparison of thermo-mechanical analyses with tests: (a) angular deformation; (b) longitudinal bending (hogging); (c) out-of-plane deformation.

to the others – sagging (i.e. positive) curvatures in the case of specimens 5 and 8, and nearly flat in the case of specimen 7 (see Table 6.1). This was due to the fact that initial welding of the end-tabs generated a slightly negative initial angular deformation, which meant that the longitudinal contracting zones were positioned rather higher than the others relative to the neutral plane of the cross-section. This led to sagging curvature results relative to the hogging curvatures of the remaining seven test results. (This pattern of deformation is similar to the experimental results of the artificial neural networks research described in section 3.2 and the reasons here are similar.) The majority of the results, which were hogging, conform to the mechanical description given for Fig. 4.7(e). The three outlier results are not therefore included in Fig. 5.16, despite which there remains substantial variation in the experimental longitudinal curvatures.

The differences between models with respect to computational budget are shown in Table 5.1. The time taken for a full transient, thermo-elastic-plastic model of a similar but not identical model[12] is included for comparison. The principal 'reduction' shared by all of the reduced solutions lies in the simplification of the thermal input to the mechanical analysis.

Table 5.1 Computational time (in hours) for alternative models

Model type	Number of sections				
	1	3	10	25	Transient
CE	0.14	—	—	—	—
SeqH	0.21	1.00	3.11	6.68	—
SimH	1.20	2.19	3.64	9.31	—
Thermo-elastic-plastic[12]	—	—	—	—	37.5

This is based on the transverse maximum-temperature profile rather than the more complex transient temperature pattern.

The results in Table 5.1 confirm that the CE model is remarkably effective for this simple case. The SimH model does not incur a large time penalty and might well be preferred from the point of view of rigour, provided that enough sections are taken. It has been noted[10] that this model also provides a much more realistic prediction of the residual longitudinal stress state, as it does not rely on the MTS algorithm.

This case study demonstrates the usefulness of reduced solutions, especially in terms of generating quick robust results that draw attention to the main outcomes of a welding procedure without necessarily providing accurate predictions of these outcomes. The computational time for the CE model was several hundred times shorter than that needed for the full transient elastic-plastic version and it is also clear that the algorithmic approach is simpler to formulate in model terms.

5.9 Conclusion

Welding involves many complex physical phenomena that are challenging to model rigorously. Simplification through 'lumping' of phenomena and dimensional reduction is therefore necessary. Material properties also tend to be complex and non-linear. Numerical approaches to modelling are therefore particularly appropriate.

A comparison of results from three different thermal computation methods showed only small differences for the example problem of a butt weld in carbon steel. Transient solutions are probably preferable to steady-state solutions, as the computational resource needed for the former is not prohibitive and the transient formulation has the capability to model boundary conditions and local thermal features. Two-dimensional solutions are also accurate for the example problem and provide a worthwhile reduction in computational resource requirement.

Stage-by-stage results from a transient elasto-plastic simulation, starting from a perfectly flat plate (Fig. 5.10), show that a V-shaped angular distortion forms immediately during passage of the heat source. Further angular distortion takes place during the longer cooling period to ambient temperature and this is accompanied by longitudinal bending or bowing in a 'hogging' sense.

Three reduced solution methods have been described that are more feasible for large-scale structural problems than transient elasto-plastic computations. The common factor in these methods is that they are all based on the Okerblom simplification of the temperature field, whereby the true transient field is replaced by the transverse envelope of maximum temperatures, assumed to be constant along the length of the model. The largest computational resource saving is attributed to the computationally efficient method, based on longitudinal/transverse uncoupling and the MTS and TCS algorithms outlined in Chapter 4. Two forms of hybrid model were also shown, in which a measure of transient behaviour can be introduced by dividing the model into short stepped lengths longitudinally. In the sequential hybrid model, the transverse deformation behaviour is solved first using elasto-plastic methods. The MTS algorithm is then used to solve the longitudinal aspect in a single step, applied to the transversely deformed shape from the first stage. In the simultaneous hybrid model, elasto-plastic methods are used to solve both orientations simultaneously, but in an uncoupled manner. This simplification nevertheless achieves a substantial reduction in computing resource.

5.10 References

1. Lindgren, L.E., 2001, Finite element modelling and simulation of welding. Part 1: increased complexity, *J. Thermal Stresses*, 24, pp 141–192.
2. Lindgren, L.E. and Josefson, B.L., 1999, Welding residual stresses and distortions simulated by the use of simplified methods, *Keynote Lecture, Symposium on Recent Advances in Welding Simulation*, Institution of Mechanical Engineers, London, UK.
3. Lindgren, L.E., 2001, Modelling for residual stresses and deformation due to welding –knowing what isn't necessary to know, *Math. Model. Weld. Phenom.*, 6, pp 491–518.
4. Jiang, W., Yahiaoui, K., Hall, F.R. and Laoui, T., 2005, Comparison of sequentially and fully coupled generalised plane strain FE modelling of multi-pass welding, *NAFEMS World Congress, Malta.*
5. Abid, M., Siddique, M. and Mufti, R.A., 2005, Prediction of welding distortion and residual stresses in a pipe-flange joint using the finite element technique, *Model. Simul. Mater. Sci. Eng.*, 13, pp 455–470.
6. Radaj, D., 1989, Finite-element analysis of temperature field, residual stresses and distortion, in *Weld. Res. Abroad*, 35, pp 31–38.
7. Goldsmith, A., Waterman, T.E. and Hirschhorn, H.J., 1961, *Handbook of thermophysical properties of solid materials*, Revised edition, Vol II: Alloys, Macmillan, London, UK.

8. Wen, S., 2003, Corus data obtained in study on CMn plate steel, private communication.
9. Okerblom, N.O., 1958, *The calculation of deformations of welded metal structures*, HMSO, London, UK.
10. Camilleri, D. and Gray, T.G.F., 2005, Computationally efficient welding distortion simulation techniques, *Model. Simul. Mater. Sci. Eng.*, 13, 8, pp 1365–1382.
11. Camilleri, D., Mollicone, P. and Gray, T.G.F., 2006, Alternative simulation techniques for distortion of thin plate due to fillet-welded stiffeners, *Model. Simul. Mater. Sci. Eng*, 14, 8, pp 1307–1327.
12. Mollicone, P., Camilleri, D., Gray, T.G.F. and Comlekci, T., 2006, Simple thermo-elastic-plastic models for welding distortion simulation, *J. Mater. Process. Technol.*, 176, 1–3, pp 77–86.

6
Experimental investigation of models of welding distortion: methods, results and comparisons

DOI: 10.1533/9780857099327.102

Abstract: Experimental strategies and techniques underpinning this book are described and the role of practical studies is highlighted. Thermal investigation of welding heat input and conduction in the workpiece is based on thermocouple arrays and thermography. Welding and measurement rigs to create industrial-scale structures are described. The arrangement integrates welding with thermal and deformation measurements to minimise workpiece handling errors. A special deformation-smoothing procedure is given that ensures appropriate interpolation between point deformation measurements and leads to profiles that can be differentiated to provide accurate surface slopes and curvatures. The use of different statistical measures of out-of-plane deformation is discussed.

Key words: thermocouples, thermography, emissivity calibration, laser displacement transducers, transient displacements, out-of-plane deformation, smoothing, differentiable profiles, flatness and distortion characterising measures.

6.1 Introduction

This book is based on the outcome of a series of research projects focused on the theme of the book title. The aim was to provide strategies for the prediction of welding distortion in order to support the design and fabrication of thin-plate welded structures. Very few experimental studies in this class of fabrication could be found in the literature at the outset of the projects and there were even fewer studies that could be readily related to generic computational treatments. It is perhaps not surprising that this is the case. Welding processes cannot easily be scaled down in size. It is therefore not feasible to build physical-scale models of typical fabrications to a size that can be accommodated readily in a typical laboratory situation. Furthermore, many features of fusion welding processes, such as intense heat sources, strong magnetic fields, hot components and weld spatter, constitute a hostile environment for measurement transducers and experimental systems. Such considerations also apply to making measurements during normal production fabrication, where there are other constraints related to production priorities and free access to fabrications in manufacture.

Nevertheless, it seemed important to undertake experimental investigations in parallel with the planned computational development. This would, at the very least, form a basis for confidence that the outputs of given theoretical models could be relied upon to provide a reasonable description of the real behaviour of typical fabrications assembled using normal production techniques. To that end, supporting experiments were undertaken to include the welding of fabrications that as far as possible reflected materials, component thicknesses and the overall magnitude of structures typically found in shipbuilding fabrication shops where stiffened, thin-plate structures are key components. The welding equipment and process parameters used in the experimental studies also matched those applied in real production situations.

The research programme was advanced in three phases – 'pilot studies' using specimens with relatively small in-plane dimensions, 'full-scale studies' utilising a large-area rig of 10 m × 3 m in-plane dimensions and, finally, an 'intermediate study' rig with improved control and instrumentation and capable of welding 1.2 m square samples.

Steel was used for all test plates and stiffeners, conforming to normal shipbuilding supply. All surface preparation and other preliminary processes applied were identical to normal shipbuilding practices. This introduced troublesome considerations in the research sense as test specimens were not flat at the outset and almost certainly contained unknown residual stress systems from prior metalworking stages. However, the advantage of this strategy was that, if any of these factors were to have significant influence on the outcomes, there was less risk that the studies would overlook important factors and fail to reflect the real situation.

This chapter includes discussion of thermocouple application and thermography techniques used to determine transient thermal fields and estimate the heat input of welding processes. Methods to measure out-of-plane distortion are described, together with a special procedure developed to construct complete and continuous deformation profiles from measured point values. Different characteristic measures of distortion are described and exemplified.

6.2 Importance of experimental observations

The value of experimental observations in validating a theory or strategy is obvious enough. However, well-observed tests can also provide insight into the behaviour of the test cases and understanding of the reasons for a particular result. Such tests can point to the factors influencing the outcome and the features that need to be addressed in the theoretical model. This was particularly true for the present project, where a wide variety of assumptions could have been made concerning the significance or otherwise of different material properties or alternative modelling assumptions.

The accuracy of thermal and thermo-mechanical computer simulations is highly dependent on material properties, which are in general not readily available. A slight change in chemical composition can influence the thermal conductivity, specific heat and latent heat of fusion, together with associated mechanical properties. Experimental work is therefore important in validating material properties adopted. Phase-change phenomena in steels are particularly sensitive to material composition. A small shift in phase-change temperatures due to chemical composition or the use of particular welding filler wires can influence thermal stress systems greatly, as this will change the point at which mechanical expansions and contractions occur, relative to the hot strength characteristics of the cooling material. It should be clear from earlier chapters that thermal dilatation is a major driving parameter in welding distortion and specific experimental determinations of this property may be required, through dilatometry, to ensure good simulation results.

Given the basic strategy outlined in earlier chapters whereby an *uncoupled* theoretical approach is used, leading to separate thermal and mechanical analyses, this uncoupled strategy was reflected in the experimental approach. This aimed to validate the thermal aspects of the process through thermocouple determinations and the mechanical aspects through deformation measurements, during welding and after cooling to ambient temperature. This separation of the observation strategy meant that any divergence between theory and experiment could be narrowed down to the thermal or mechanical stages of the analysis and the associated assumptions. In particular, the thermal investigations were important in relation to the models used for weld heat input, material properties, filler material addition, the effects of radiation or convection and the alternative use of two-dimensional (2D) or three-dimensional (3D) idealisations. Deformation measurements, on the other hand, were important in terms of validating the mechanical analysis outcomes and also assisted in understanding the dynamic development of distortions. Factors such as material property idealisation and the need for 2D or 3D and transient modelling could also be investigated through the deformation outcomes.

6.3 Welding process application in test work

The most important requirement for the welding process to be used in validation test work is, as explained above, that it should be realistic relative to the processes that may be used in the practical situation of interest. That accepted, it is critical that the process is closely controlled and monitored with respect to the energy input and in terms of orientation of the welding head and the heat source. That will generally mean using an 'automatic' process application in which the welding head is traversed mechanically at a constant rate along the weld line. Control and monitoring equipment is required to ensure constant welding parameters with independent monitoring. The important process parameters will normally be welding current, voltage and travel speed, together with wire feed rate, if filler wire is used.

In the present case, a gas metal arc process was used, where a gas mixture of 80% argon and 20% carbon dioxide provides the arc-ionising atmosphere between a continuously fed filler wire electrode and the workpiece. The filler wire also incorporated a flux core that contributed to the shielding atmosphere and the metallurgical quality of the weld. Tests showed that this welding process was relatively 'efficient' in that a high proportion of the measured electrical energy input is translated into heat energy, which is conducted into the plate, as tracked via thermocouple measurements. Some heat is normally lost to local heating of the shielding gas and surface surroundings and this will vary according to the welding process and the configuration of the welded component surrounding the arc plasma heat source. Other minor variations may be expected depending on the angle of the electrode relative to the molten weld pool. Variations from the idealised thermal model may also be encountered in smaller test specimens due to electrical resistance heating of certain regions, depending on the welding current flow path. In some industrial applications, the workpiece is surrounded by jigs or clamping devices that can act as heat sinks, to an extent that is often difficult to determine or model accurately.

6.4 Thermocouple arrays

At present, thermocouples provide the most accurate and convenient method to determine transient temperatures in welded components. The best-known thermocouple in the chromium–nickel–aluminium group is the so-called 'type-K' thermocouple, which has a potential measurement range from −200°C to 1100°C with an accuracy of ±1.5°C. This is more than adequate. Guidance should be sought for the application of thermocouples within electronic data acquisition systems, in particular concerning signal insulation and interference screening. Compromises often have to be made between thermocouple wire sizes and cable lengths. Protection of thermocouples and cables from impacting hot weld spatter is also a challenge.

Thermal gradients in welding are steep and therefore small-diameter thermocouples are preferred to track such gradients. In the present case, 0.315 mm, type-K thermocouples were used. Techniques were developed and improved in the series of projects, leading to a preferred arrangement where mineral insulated thermocouples were used. By this means, the thermocouple junction was electrically insulated from the workpiece and the process equipment. The thermocouples were placed in interference-fit drilled holes in the test plate.

Preliminary theoretical studies of temperature fields were undertaken, as such information is always useful in planning layouts of discrete, point measurement systems such as thermocouples or strain gauges. This is especially true for situations where the rate of change of the measured parameter varies steeply. In this case, as discussed in earlier chapters, it was conjectured that welding distortion is driven mainly by thermo-mechanical actions within a temperature range where the material exhibits reasonable mechanical strength, which for steel might be in the range from ambient temperature to 500°C. The transverse disposition of the thermocouple array layout shown in Fig. 6.1 was designed to give good coverage within this temperature range. Identical positions for thermocouples on the lower and upper surfaces provided information on any through-thickness variation of the temperature field. The successive rows A, B and C provided a measure of duplication in the event of a transducer malfunction and also some indication of temperature field variation as welding progressed along the length of a weld run.

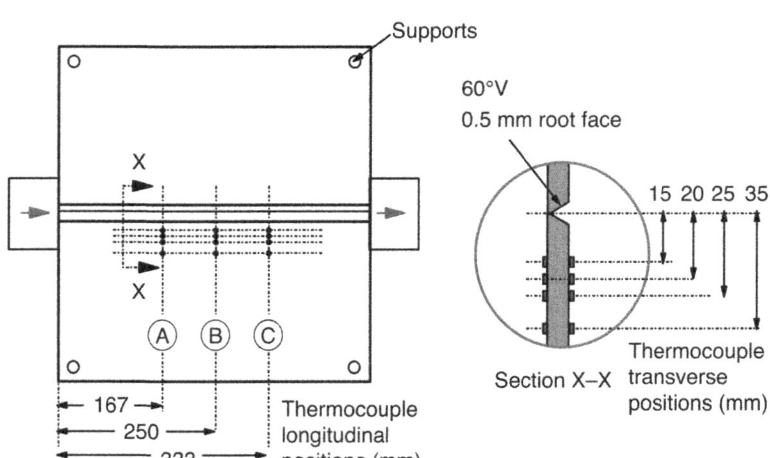

6.1 Thermocouple positions.

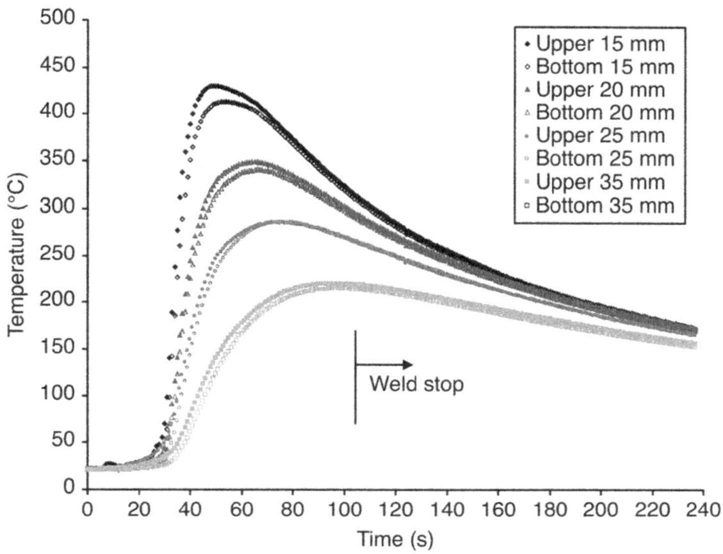

6.2 Row A thermocouple results.

Figure 6.2 shows results for thermocouples in row A of the pilot study series. This shows a very small time lag between lower and upper face temperatures at a given position and suggests higher temperatures on the upper face. However, the irregular shape of the curve in the highest temperature region at the 15 mm location suggests that the upper transducers may have been receiving more heat energy than the lower ones as a result of direct radiation heating from the arc or hot shielding gas. It is also important to note that a small inaccuracy in thermocouple position will lead to a significant temperature error and that is where duplication of position within the array configuration provides greater confidence that any phenomenon observed is not a spurious effect.

Figure 6.3 shows results for the lower thermocouples (hence shielded from direct radiation effects) in successive rows. This demonstrates the effect of thermal boundaries at either end of the specimen, with a slightly greater build-up of heat towards the end of the weld in the central region near the weld.

6.5 Thermography

The thermographic principle is based on capturing and imaging infrared emissions from the subject under investigation. The imaging system produces a whole-field, moving, thermal picture of the subject, as distinct from the point location, transient values provided by thermocouples. The advantage therefore is that the thermal state of all parts of the target

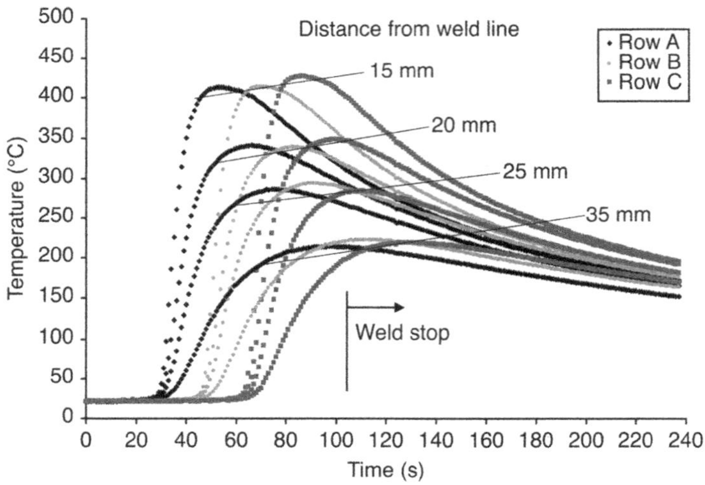

6.3 Lower-surface thermocouple results.

structure is revealed and the influence and effects of boundaries, junctions and all other real features of the case can be examined directly, through a non-contacting and non-destructive system. These advantages assume, of course, that an unobstructed view of the structure being welded can be obtained at all times, which is not always feasible in an industrial context.

In theory, the output data from such a system could be used directly as input to a thermo-mechanical computational analysis, thus bypassing the thermal computation stage of the uncoupled approaches described in earlier chapters. This would be of value in industrial configurations where the thermal fields would be too complex to model computationally. Thermography has its own complexities, however, as the temperatures corresponding to the infrared outputs depend on the emissivities of the imaged surfaces, which in turn depend on temperature and wavelength. Emissivity thus has to be determined through a process of careful calibration. Moreover, as the input temperatures have to be applied at specific locations in the thermo-mechanical model, any spatial distortions of the imaging system will need first to be corrected. Thermography has not been widely used in welding studies, possibly due to the complications noted above. Cheng *et al.*[1] used thermography to cross-reference finite-element thermal simulations and other researchers[2, 3] have used thermography for weld process control.

A FLIR Systems ThermaCAM® SC500 camera was used in the present study. This system detects infrared radiation at wavelengths from 3 to 12 μm, giving a temperature range of 10–650°C with an accuracy of ±2°C. Images are captured by a focal plane array of 320 × 240 pixels.

Emissivity calibration was carried out by relating camera image results to

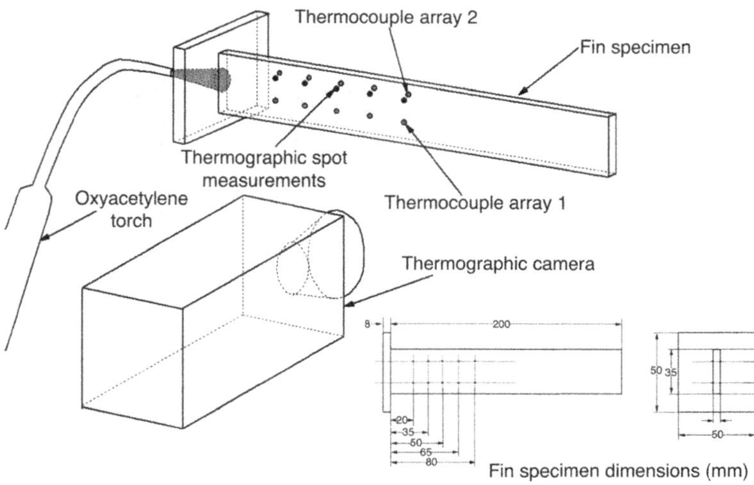

6.4 Experimental set-up for determination of emissivity.

thermocouple-measured temperatures on a heated fin made from the weld test material, having a surface preparation identical to the distortion test plates (see Fig. 6.4). Spatial errors of the image caused by aberration of the wide-angle lens were corrected through a transformation matrix, generated in the first instance by imaging a resistance-heated rectangular grid.[4] Thus the images captured by the camera in the experiments were corrected for spatial distortion and adjusted in temperature values through the emissivity calibrations. These calibrations enabled the production of isothermal contour figures following a similar pattern to those previously computed and shown in Chapter 4 (e.g. Fig. 4.4). Figure 6.5 shows a calibrated thermographic isothermal plot of temperature fields in a butt-welded plate. The specimen here was from the pilot study, at a stage near the end of the weld pass. The effect of the finite boundary at the left-hand edge is shown in the spread of the 100°C contour. The finite boundary at the right-hand end has little effect, as forward transmission of heat is overtaken by heat source forward travel. The emissivity of the weld bead surface at various temperatures was unknown and is clearly less than in other areas of the plate, as indicated by the 'notches' in the trailing ends of the contours.

The thermographic results showed higher temperatures than the thermocouples or the computational determinations, especially nearer the weld line. This is thought to be due to extra radiation from the arc, as the computational results (and the lower-face thermocouple results) are derived from the conduction component of heat energy, whereas the thermographic results will include reflected radiation. The two sets of results also diverge at the starting edge of the plate where the thermographic results correctly

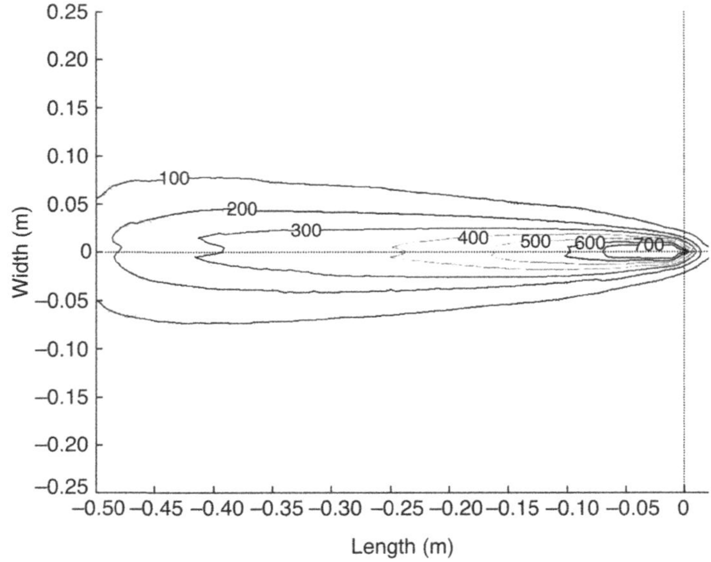

6.5 Thermographic temperature contour plot as weld nears end.

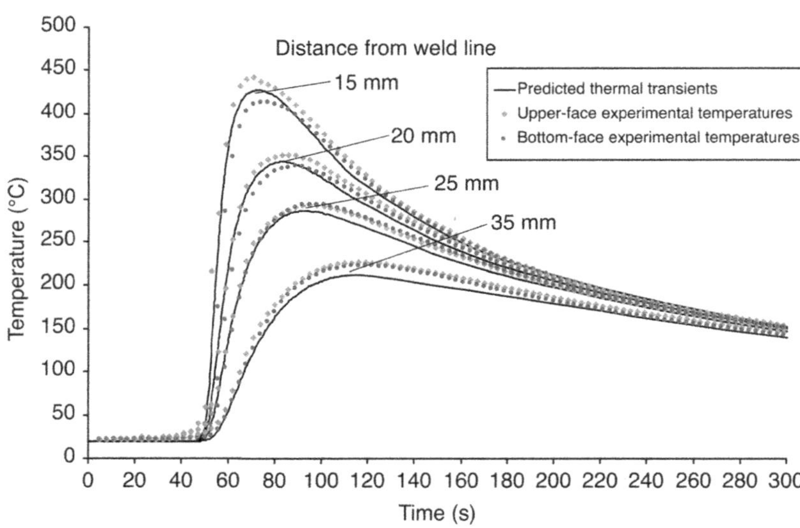

reflect the loss of heat through the run-on plate, which was not modelled in the theoretical analysis.

A comparison of thermocouple and 3D computational results is shown in Fig. 6.6. It should be noted that the computational transients did not show a thermal gradient across the plate thickness at the chosen locations and upper and lower computed temperatures followed the same thermal history. The 3D determination also depended on an assumed weld efficiency, derived by matching the overall computational results to the temperatures recorded by the thermocouple array.

6.6 Deformation measurement

From the point of view of the industrial user, out-of-plane deformation is the major problematic distortion of interest and was therefore the focus of the deformation measurement strategy in these projects. An important consideration is that out-of-plane movements can be very local in thin structures, due to buckling, and, if the final shape is to be accurately determined, the measurement map needs ideally to be based on finely spaced measurements. In-plane weld-induced deformations may nevertheless be important is some applications, in terms of the finished sizes of welded assemblies[5] but this seems to cause less difficulty in practice.

Michaleris and DeBiccari[6] and Tsai *et al.*[7] used linear variable displacement transducers (LVDTs) and a laser displacement sensor to verify computed displacements. Huang *et al.*[8] employed a Lidar sensor system to measure out-of-plane deformations. Various purely experimental studies of angular distortion include those of Suresh and Dattrajan[9] and Parmar and Arya.[10] Various points emerge from these studies – in particular, the advantages of measuring deformations through the use of non-contacting systems and the need for a fine in-plane grid of measurements, together with robust data-smoothing algorithms.

Such considerations led to the use of an experimental arrangement for the 'full-scale' and 'intermediate' size rigs in which the welding head and thermographic camera were mounted in a fixed overhead position and the test piece was traversed smoothly below the welding head on a moving support and reference frame as shown in Fig. 6.7. Vertically adjustable point supports were used to locate the test plate to a fixed plane initially. A non-contacting, two-axis laser scanning system mounted on a fixed bridge above

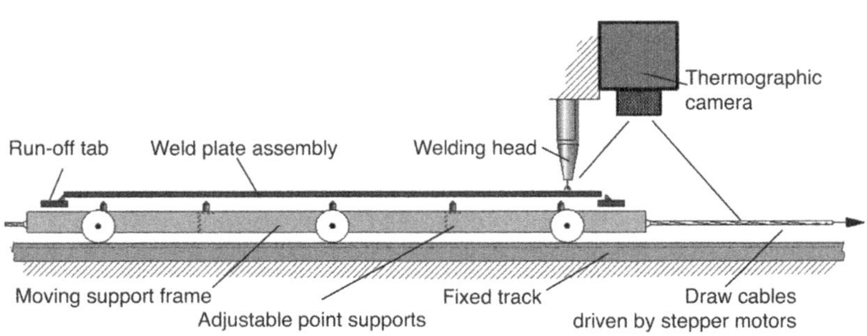

6.7 Schematic illustration of welding rig.

the support frame was then used to determine static out-of-plane deformations before and after welding. This arrangement therefore made it possible to determine accurate differences between the initial and final out-of-plane profiles, after complete cooling of the test plate to the original ambient temperature. The initial shape patterns are also very important as they provide some clue to the existence of any pre-assembly residual stress systems in the plate and trigger various features of the final deformation patterns in the welded assembly. Many of the deformation patterns are unstable, due to buckling effects, and it is therefore important to understand the influence of initial out-of-plane deformations on the outcomes.

It is also very useful to monitor deformations *during* welding and subsequent cooling, as these assist understanding of the way in which the final deformations develop, through interaction of changing shapes and contraction forces. In the present project this was carried out using a local array of out-of-plane measurements in an area placed approximately one-third of the length from the weld starting position (see Fig. 6.8). Four LVDTs mounted on the moving support frame were set to maintain contact with the plate during welding and to provide a dynamic record of the local displacement behaviour. Processing of the outputs provides an indication of the angular deformation, from the relative displacements of LVDTs 1 and 3, and local curvature of the plate edge can similarly be derived from LVDTs 2, 3 and 4.

Figure 6.9 shows a typical processed output from the dynamic deformation array. This demonstrates, for example, that the angular distortion develops immediately as the weld passes the point of measurement, whereas most of the longitudinal hogging curvature develops during the long cooling period, after welding is complete. This is the basic pattern of transient deformation shown earlier in the computational simulation in Fig. 5.10. Note also that a temporary *sagging* curvature develops as the weld passes. The plate in this local area has already taken up a shallow V-shape at

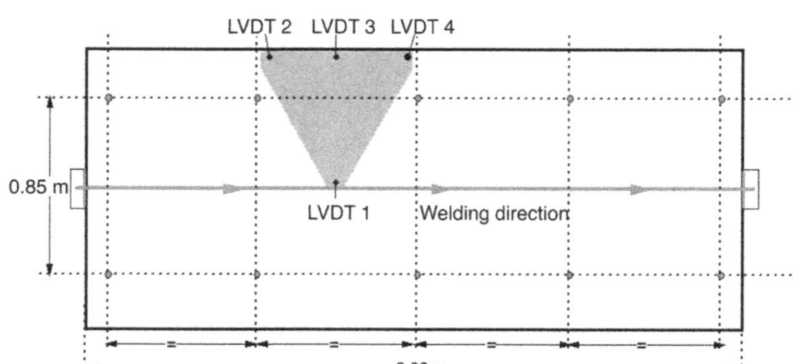

6.8 Layout of supports and LVDTs.

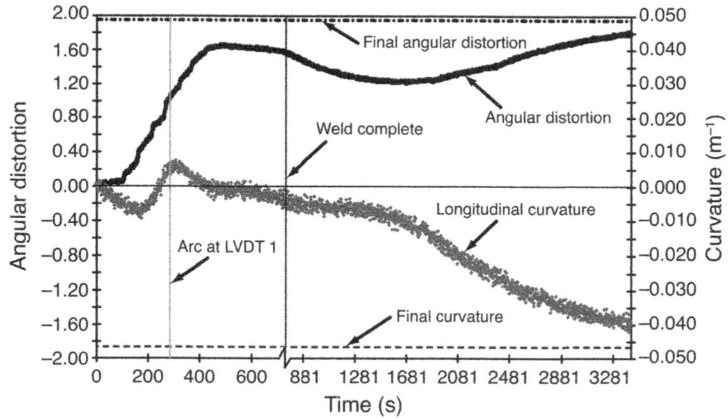

this stage and longitudinal expansion of the hot weld zone has the effect of curving the plate locally into a sagging mode. In an ideal case, it would be useful to generate complete maps of the test plate at every stage of the process, but this would present some practical difficulties.

6.7 Completion and smoothing of measured deformation profiles

In the case of the large-scale butt-welded specimens, the laser scanning system described in section 6.6 was typically set to capture 8668 point deformations on a 394 × 22 point grid. This provided a detailed record, but for the purposes of analysis and comparison, a complete and continuous field of deformation is needed, which will fill in the areas between and across the measured points. It is likely also that such measurements will contain random errors, both in the out-of-plane values themselves and in the x and y measurement positions (a description of accuracy and repeatability studies undertaken is given by Camilleri[11]). There is therefore a requirement to construct, from the measurements, a full-field, continuous, 2D profile that will be faithful to the measured points, while rejecting errors, and which will also provide smooth transitions between the measurement points.

There are many possible approaches to this problem. The most obvious is to use standard interpolation, regression or spline-fitting methods, such as provided by MATLAB[12] or similar software. Such an approach is readily available and will often provide rapid and satisfactory results. The disadvantage is that the interpolation functions used are normally unrelated to the physical situation. Local slopes and curvatures derived from the profiles may therefore be incorrect. The method chosen in this research used a finite-element based technique developed by Feng and Rowlands,[13] which combines smoothing capability with the finite-element concept. The complete surface of the welded plate model is discretised into finite elements and a minimisation procedure is then used to find a continuous profile for the assembly of elements. This approach was configured by Comlekci *et al.*[14] to address the present requirements in terms of constructing profiles that would best match the point measurements. The main advantage of the method is that the finite-element model reflects the physical situation in the assembly of elements. Slopes and curvatures derived by differentiating the deformation profiles are therefore more likely to be realistic.

The out-of-plane distortion is represented in the first instance by an approximate, smooth, scalar function z in the 2D space R representing the plate. This function, although smooth, does not necessarily agree exactly with the measured values of distortion z_i. A quadratic, positive definite functional $\phi(z)$ is introduced and then minimised over the finite-element space. This relationship is given by

$$\phi = \phi(z) = \frac{1}{2}\sum_{i=1}^{T} \omega_i[z(x_i, y_i) - z_i]^2 + \frac{\zeta}{2}\int_R \left[\left(\frac{\partial^2 z}{\partial x^2}\right)^2 + 2\left(\frac{\partial^2 z}{\partial x \partial y}\right)^2 + \left(\frac{\partial^2 z}{\partial y^2}\right)^2\right] \mathrm{d}R \qquad [6.1]$$

where z is the predicted twice-differentiable piecewise function defined in region R, z_i is the measured input value at data point i at location x_i,y_i T is the total number of data points in region R, ω_i is a weighting parameter that can be used with particular data points and ζ is a smoothing parameter. The smoothing parameter ζ influences the *interpolation* characteristics of the function. If it is set at a value near to zero, the predicted smoothing function z will approach the measured values z_i, giving a pure, least-squares fit to the experimental data. This may be undesirable if there are errors or 'roughness' in the data. At very large values of ζ, the procedure will fit a flat plane to the data by least-squares. The optimum value of the smoothing parameter is therefore problem-dependent. Likewise the point-weighting function ω_i is available for use, if there is a desire to fit the function more closely (or less closely) to specific points in certain regions. However, in the present case, with equally spaced data and no anomalies in particular regions of the model, the weighting parameter was not used.

The element used for this application is described as isoparametric, nine-noded, Lagrangian-type, as shown in Fig. 6.10. The scalar and shape functions of the nine-node element are given as

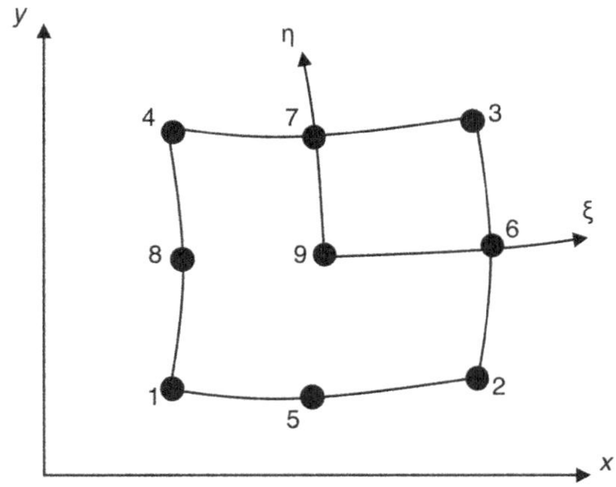

6.10 Nine-node Lagrangian-type element and coordinate systems.

$$N(\xi, \eta) = a_1 + a_2\xi + a_3\eta + a_4\xi^2 + a_5\xi\eta + a_6\eta^2 + a_7\xi\eta^2 + a_8\xi^2\eta + a_9\xi^2\eta^2 \quad [6.2a]$$

$$N_1 = \frac{\xi^2\eta^2 - \xi^2\eta - \xi\eta^2 + \xi\eta}{4} \quad [6.2b]$$

$$N_2 = \frac{\xi^2\eta^2 - \xi^2\eta + \xi\eta^2 - \xi\eta}{4} \quad [6.2c]$$

$$N_3 = \frac{\xi^2\eta^2 - \xi^2\eta + \xi\eta^2 - \xi\eta}{4} \quad [6.2d]$$

$$N_4 = \frac{\xi^2\eta^2 + \xi^2\eta - \xi\eta^2 - \xi\eta}{4} \quad [6.2e]$$

$$N_5 = \frac{-\xi^2\eta^2 + \xi^2\eta + \eta^2 - \eta}{2} \quad [6.2f]$$

$$N_6 = \frac{-\xi^2\eta^2 - \xi^2\eta + \xi^2 + \xi}{2} \quad [6.2g]$$

$$N_7 = \frac{-\xi^2\eta^2 - \xi^2\eta + \eta^2 + \eta}{2} \quad [6.2h]$$

$$N_8 = \frac{-\xi^2\eta^2 + \xi^2\eta + \xi^2 - \xi}{2} \quad [6.2i]$$

$$N_9 = \xi^2\eta^2 - \xi^2 - \eta^2 + 1 \quad [6.2j]$$

Representing the shape functions in the matrix form $[N] = [N_1 N_2 \ldots N_9]$, the global coordinates within the element are defined as

$$x = [N]\{C_x\} \text{ and } y = [N]\{C_y\} \quad [6.3]$$

where the nodal coordinate vectors of the elements are

$$\{C_x\} = [x_1 \;\; x_2 \;\; \ldots \;\; x_9]^{\mathrm{T}} \text{ and } \{C_y\} = [y_1 \;\; y_2 \;\; \ldots \;\; y_9]^{\mathrm{T}} \quad [6.4]$$

The scalar function within the element is defined as $z = [N]\{z_e\}$ where $\{z_e\} = [z_1 \;\; z_2 \;\; \ldots \;\; z_9]^{\mathrm{T}}$ represents the vector for unknown nodal values.

To minimise the quadratic function 6.1 with respect to the unknown nodal values, equation 6.5 is applied

$$\left\{\frac{\partial\phi(z)}{\partial z}\right\} = 0 \quad [6.5]$$

Expanding equation 6.5 and using the global coordinates results in the relationship

$$\sum_{j=1}^{M}\left[\begin{array}{l}\sum_{i=1}^{J}\left[\omega_i\left\{[N(\xi_i,\eta_i)]_j^{\mathrm{T}}[N(\xi_i,\eta_i)]_j\{z_e\}_j - [N(\xi_i,\eta_i)]_j^{\mathrm{T}} z_i\right\}\right] + \\ \zeta \int_{R_j}\left\{\left[\frac{\partial^2 N}{\partial x^2}\right]_j^{\mathrm{T}}\left[\frac{\partial^2 N}{\partial x^2}\right]_j + 2\left[\frac{\partial^2 N}{\partial x\partial y}\right]_j^{\mathrm{T}}\left[\frac{\partial^2 N}{\partial x\partial y}\right]_j + \left[\frac{\partial^2 N}{\partial y^2}\right]_j^{\mathrm{T}}\left[\frac{\partial^2 N}{\partial y^2}\right]_j\right\} \mathrm{d}R\{z_e\}_j\end{array}\right] = 0 \quad [6.6]$$

where M is the number of finite elements in region R, J is the number of input data points applied to element j and ξ_i,η_i are the local isoparametric coordinates of input data point i.

Equation 6.6 can be simplified in the form

$$\sum_{j=1}^{M}[k]_j\{z_e\}_j = \sum_{j=1}^{M}\{p\}_j \quad [6.7]$$

where

$$[k]_j = \begin{array}{l}\sum_{i=1}^{J}\left[\omega_i[N(\xi_i,\eta_i)]_j^{T}[N(\xi_i,\eta_i)]_j\right] + \\ \zeta \int_{R_j}\left\{\left[\frac{\partial^2 N}{\partial x^2}\right]_j^{T}\left[\frac{\partial^2 N}{\partial x^2}\right]_j + 2\left[\frac{\partial^2 N}{\partial x\partial y}\right]_j^{T}\left[\frac{\partial^2 N}{\partial x\partial y}\right]_j + \left[\frac{\partial^2 N}{\partial y^2}\right]_j^{T}\left[\frac{\partial^2 N}{\partial y^2}\right]_j\right\}\mathrm{d}R\end{array}$$

and

$$\{p\}_j = \sum_{i=1}^{J} \omega_i [N(\xi_i, \eta_i)]_j^{\mathrm{T}} z_i$$

Assembly of the contributions of each element gives equation 6.8, where $[K]$ is a pseudo-stiffness matrix and P is the pseudo-load

$$[K]\{\sigma\} = \{P\} \qquad [6.8]$$

The pseudo-stiffness $[K]$ is independent of material properties or the thickness of the plate but involves the shape functions and the total geometry, through the element layout. It is of interest to note in passing that features of the real case, such as holes or cut-outs may be included in the finite-element model. This option is not applicable in simple fitting routines. The pseudo-load P also involves the shape function as well as the measured displacements.

Equation 6.8 is now in a form that can be solved using a standard finite-element solver to provide the unknown nodal out-of-plane displacements.[15] The derivatives of the resulting scalar function z can then be evaluated and averaged at element nodes such that $\tan^{-1}(\partial z/\partial y)$ is the transverse slope of surface at each point and $(\partial^2 z/\partial x^2)$ is the longitudinal curvature. The relationship between the Cartesian coordinate system of the plate and the isoparametric coordinates is obtained through two Jacobian matrices, both making use of a number of chain rule derivatives.

As noted earlier, the choice of smoothing parameter is problem-dependent and the performance of the smoothing strategy when applied to out-of-plane deformation was investigated by Comlekci *et al.*[14] This showed that the main critical factor was to ensure that each element encompassed a reasonable number of measured points. (It is also advisable, of course, to reduce measurement errors as far as possible.) In the case of the large-scale test plates, 16×9 plane elements were used, giving approximately 60 measurement points per element. It was then found that the smoothing parameter could be reduced to zero for typical problems, meaning that the smoothed profile was basically a least-squares fit to the data points, although based fundamentally on the finite-element shape functions, rather than on arbitrary piecewise polynomials.

In practice, the ANSYS system pre-processor was used to generate a finite-element model of the plate (Fig. 6.11). The global x,y coordinates of the measurement points were transformed to local element coordinates using inverse isoparametric mapping. These coordinates were then input to the finite-element based smoothing program, yielding the nodal values of the

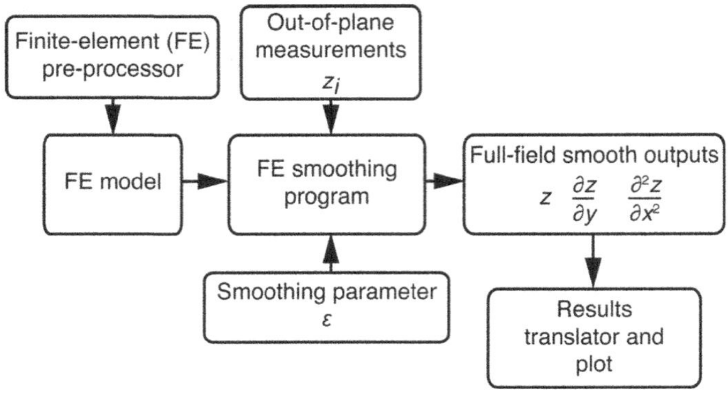

6.11 Flow chart for smoothing program.

smoothed profile and the relevant nodal derivatives. The results could then be displayed using the ANSYS post-processor.

This finite-element based approach demands substantial effort to program and, as noted earlier, it is not strictly necessary that this method be used. It does, however, produce robust results and gives confidence that the smoothed values reflect the actual profile of the plate and the local slopes and curvatures.

6.8 Characterising out-of-plane deformation

Some consideration needs to be given on how to analyse and present out-of-plane deformation data. The most informative way, from a practical point of view, might be to provide a full isometric view as in Fig. 6.12, which shows the initial profile of a full-scale test specimen after assembly and tacking. Such a representation also reveals any gravitational effects in relation to the support positions and dimensions (compare with Fig. 6.8). Transient local deformations during welding, for a similar test specimen, have in fact been shown previously in Fig. 6.9 and the overall net deformations due to the welding operation are shown in Fig. 6.13. This form of presentation helps to identify correspondences between the initial shape and the deformations caused by welding, as shown in the transient results and the net deformation profile. A full isometric view of the final shape (i.e. initial deformations plus those due to welding) should also be helpful from the point of view of judging final fitness-for-purpose. However, if a single characterising number is required rather than a 3D graphic, for example to compare the outcome of one fabrication route with another, there are many possibilities, each with their own advantages and disadvantages.

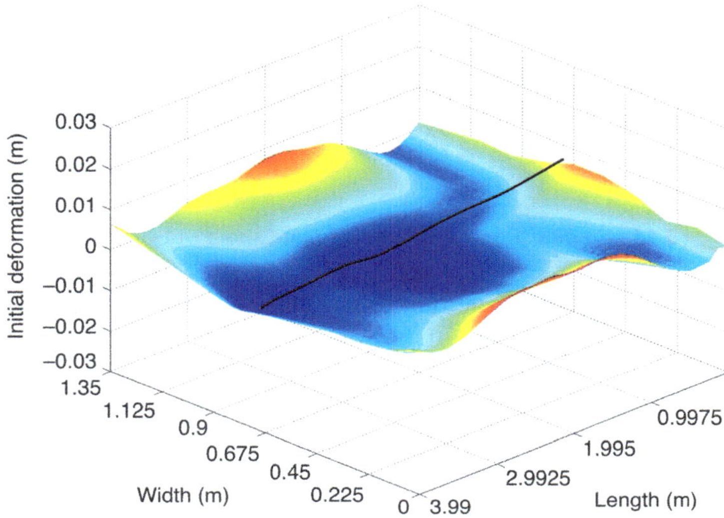

6.12 Initial profile of 4 m × 1.35 m × 5 mm test plate.

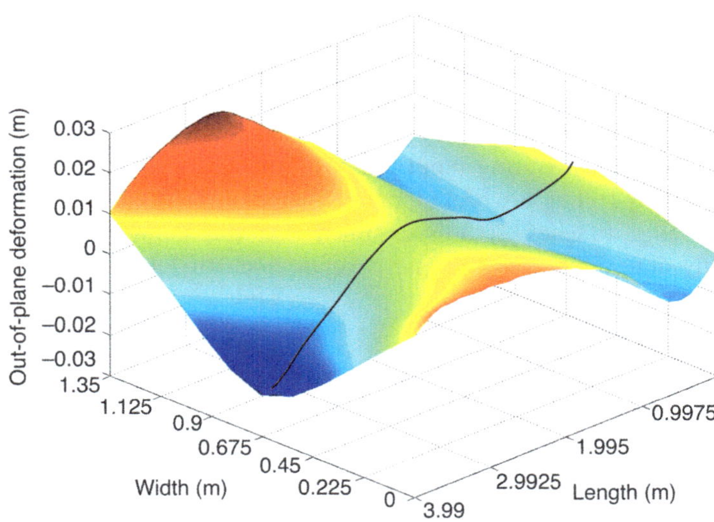

6.13 Net deformation of test plate in Fig. 6.12 due to welding.

6.8.1 Plate flatness standards

There are several standards for the supply of plate, sheet and strip that might be applied to the supply of stock for a welded fabrication or a welded plate (e.g. BS EN 10029:2010[16]). Standard practice for measuring flatness characteristics of steel sheet products is described in ASTM A1030/A1030M.[17] Such standards for plate flatness are in general aimed at the

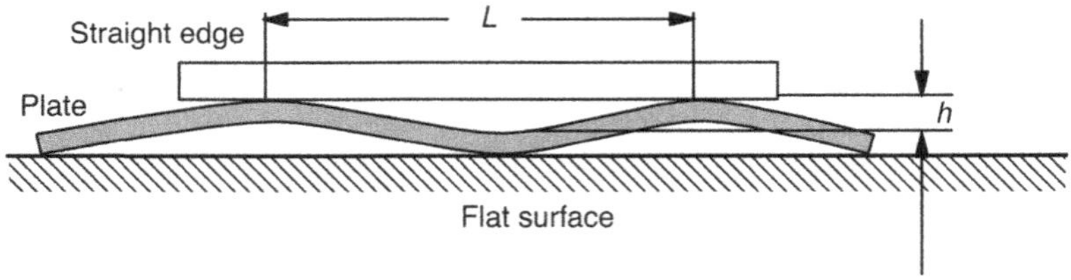

6.14 Standard measurement of plate/sheet out-of-flatness.

assessment of single deviations from flatness, measured over substantial lengths (e.g. 1–2 m (Fig. 6.14)). Gravitational effects are reduced by placing the sheet on a hard, flat surface and the measurement h, over a length L, can be applied longitudinally, transversely or in any other direction. It is often the case with rolled products that such out-of-flat features repeat periodically in a certain alignment, for example, ranged along the longitudinal edges in the rolling direction. Typical standard tolerances on flatness, measured in this way, range from 3 mm to 14 mm when the length L is in the region of 1–2 m and plate thickness is in the range 3 mm to 8 mm. These tolerances reflect what can be readily achieved in production rather than the flatness that may be necessary in a fitness-for-purpose sense.

The ASTM standard[17] includes two further characterisations of out-of-flatness, designated as steepness index, defined by h/L and I. The latter parameter is a measure of the extra length developed by the out-of-flat shape, between measuring points. In principle, this would be the distance traced out by rolling a wheel along the plate between the points where the straight edge touches. If it is assumed that the shape of the feature is perfectly sinusoidal, the extra length can be calculated as $\left(\frac{\pi}{2}\frac{h}{L}\right)^2$. The I factor is then given by

$$I = \left(\frac{\pi}{2}\frac{h}{L}\right)^2 \times 10^5 \qquad [6.9]$$

the 10^5 factor being arbitrary to achieve approximately non-decimal number results for typical out-of-flatness. The values quoted earlier for a 1–2 m span would give a range of I factors from 2.2 to 20. Notice that, for the same out-of-plane deformation, different measuring lengths will give different I factors and this definition of out-of-flatness is not therefore appropriate to characterise the complex distortion profiles generated typically by welding.

6.8.2 Root mean square (RMS) out-of-plane deformation

The out-of-plane displacements from some datum plane, at each point of measurement, can be seen as 'errors' (this statement could also apply to *smoothed* out-of-plane displacements). The datum plane might in the first instance be the zero displacement plane containing the plate support points or it might be the plate profile at a previous stage of fabrication, if the intention is to assess the outcome of applying a welding process. The root mean square (RMS) of displacements is found by calculating the square-root of the mean of the sum of the squares of each displacement. This statistical measure is commonly used to characterise overall errors through a single parameter. It therefore provides a useful parameter to gauge the overall displacement of the plate relative to the datum plane, the initial deformation state or some other intermediate condition. Positive and negative displacements are given equal weight in the formulation.

The strength of the RMS measure is that it gives an approximate indication of typical deviations, positive and negative. The main weakness of the RMS measure is that it is insensitive to the distribution of the varying parameter across the plate. Two deformation patterns with the same RMS value could be visualised, where the first shows relatively moderate levels of displacement over the whole surface and the second shows small displacement levels over most of the surface, but with large deviations in a localised area. This may be undesirable if, for example, the non-flat area in question happens to be along an edge that will subsequently be assembled and welded in turn to another fabrication.

6.8.3 Angular contraction/longitudinal bending partitioning

As noted earlier, out-of-plane deformation after welding is not usually random, but tends to follow a pattern related to two separate actions, namely angular contraction and longitudinal bending. It is therefore useful to process displacement outputs in a manner that can identify these actions. Figures 6.15(a) and 6.15(b) show the point-to-point transverse slope and longitudinal curvature values corresponding to the isometric view in Fig. 6.15(c), as contours. The values given here are local to the points, as provided by the shape smoothing and differentiation process. This offers a clearer understanding of the actions and reveals certain features such as the difference in bending deformation between the start and finish of the weld. This interpretation of the deformation patterns therefore suggests a more insightful use of the RMS approach, by applying this measure to the partitioned parameters.

Table 6.1 shows individual results for the ten small-scale specimens first discussed in section 5.8.1. (N.B. The negative and positive signs are not derived from the RMS calculation but inserted to distinguish sagging from hogging.) As noted earlier, these results show considerable variation when analysed in terms of the three RMS parameters, with two of the specimens showing an opposite curvature to the others and a third being virtually flat.

Table 6.2 shows RMS values for the results in Fig. 6.15. In the case of the

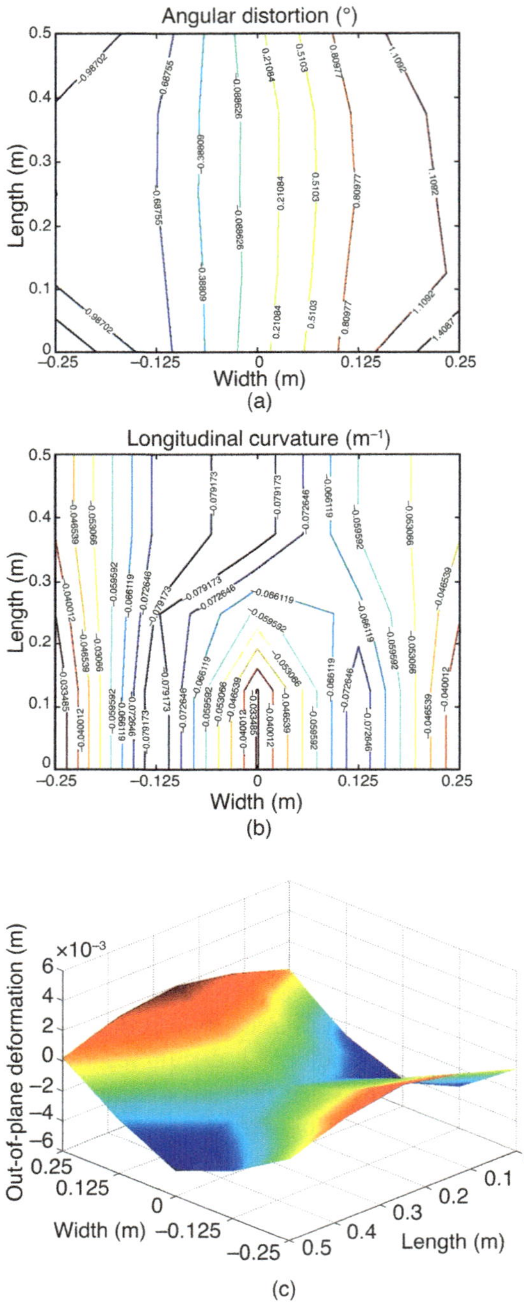

6.15 Deformation of small-scale test specimen: (a) transverse slope contours; (b) longitudinal curvature contours; (c) isometric view.

Table 6.1 Partitioned results for small-scale test specimens

Test specimen	RMS out-of-plane deformation (mm)	RMS angular distortion (degrees)	RMS longitudinal curvature (m^{-1})
1	1.992	0.995	−0.060
2	1.572	0.760	−0.053
3	1.910	1.031	−0.044
4	2.214	1.201	−0.070
5	1.420	0.258	+0.052
6	2.200	1.226	−0.075
7	0.989	0.323	−0.013
8	1.766	0.346	+0.074
9	1.930	1.010	−0.053
10	2.090	0.978	−0.097

Table 6.2 RMS distortion values for welding distortion of small-scale test specimen

Out-of-plane distortion RMS (mm)	Angular distortion RMS (degrees)	Longitudinal curvature RMS (m^{-1})
1.99	0.96	−0.06

transverse slope results, the positive and negative slopes on either side of the weld do not cancel in the RMS measure and therefore the overall RMS result gives a global average of the change in slope values, which can be seen as a global measure of angular contraction. (Some care is needed relative to this interpretation as a straightforward transverse tilt of the plate, without any angular contraction, would also result in a positive RMS slope.) Considering the contours in Fig. 6.15(a) and 6.15(b), there is less variation from the RMS values in the case of the partitioned parameters' transverse slope and longitudinal curvature.

6.9 Conclusion

The value of systematic practical observation and experimental studies in supporting distortion reduction strategies is emphasised in this chapter. It was shown that such observations can point to factors that have the effect of increasing or reducing distortion and are therefore important to address in simulations or other theoretical investigations. It is also evident that several aspects of welding process applications can, at present, be treated only through experimental means. These include measurement of the heat input to the workpiece from the welding process, which is a key driver of distortion mechanisms.

The use of thermocouple arrays to measure the heat input of welding processes was shown to be productive, especially if combined with a matching computational determination to optimise the layout of transducers and index the heat input. A good measure of transducer duplication is desirable to increase confidence in the accuracy of determination. Contexts in which thermography are useful were demonstrated. This method can provide a whole-field transient description, which is useful in various practical situations. However, considerable attention needs to be given to calibration in spatial and thermal respects. Thermocouple arrays are more readily applicable and more likely to give accurate results.

Two strategies for integrated completion of welds and measurement of the consequent out-of-plane deformations were shown. The first of these utilises non-contacting displacement point measuring systems and is designed to determine static displacements before and after each stage of fabrication. The second measurement system is based on four contacting displacement transducers and provides a continuous, dynamic record of the local displacement behaviour of the fabrication, as the weld is completed and during the following cooling down period. The output of this latter system was particularly informative, albeit in a qualitative sense, and a strong case could be made to increase the regions covered by this form of transient measurement in future studies.

Processing the volume of data generated by the deformation measurement systems is not a trivial task and section 6.7 described the application of a specially developed completion and smoothing procedure that is appropriate to the problem of measuring out-of-plane profiles. This combines powerful and flexible smoothing capability with the finite-element concept, which is appropriate to the application. This method completes the deformation profile between and across the measured points and provides strong rejection of measurement noise. The smoothed results are therefore differentiable and make it possible to generate detailed, accurate and informative measures of welded plate profiles. The final section of the chapter identified and discussed several output parameters that can be used to characterise out-of-plane profiles. These form a basis for comparison between computational and experimental determinations and a quantitative index to judge one fabrication procedure against another.

6.10 References

1. Cheng, W., Dydo, J.R., Feng, Z., Chen, Y. and Crompton, J.S., 2000, Finite element modelling of angular distortion in stiffened thin-section panels, *Proc. 9th International Conference on Computer Technology in Welding*, No. 949, pp 120–129.
2. Vasudevan, M., Chandrasekhar, N., Maduraimuthu, V., Bhaduri, A.K. and Raj, B., 2011, Real-time monitoring of weld pool during GTAW using infra-red

thermography and analysis of infra-red thermal images, *Weld. World*, 55, 7–8, pp 83–89.
3. Cobo, A., Mirapeix, J., Conde, O.M., Garcia-Allende, P.B., Madruga, F.J. and Lopez-Higuera, J.M., 2007, Arc welding process control based on back face thermography: application to the manufacturing of nuclear steam generators, *Proc. SPIE, Thermosense XXIX*, 6541, pp 65410F-1–65410F-11.
4. Camilleri, D., Comlekci, T. and Gray, T. G. F., 2004, Use of thermography to calibrate fusion welding procedures in virtual fabrication applications, *Proc. Inframation Conference, Las Vegas, USA*, 5, pp 121–131.
5. Cadden, S., 1987, *Prediction of the in-plane contractions in thin butt-welded plates*, PhD thesis, University of Strathclyde, Glasgow, UK.
6. Michaleris, P. and DeBiccari, A., 1996, A predictive technique for buckling analysis of thin section panels due to welding, *J. Ship Product.*, 12, 4, pp 269–275.
7. Tsai, C.L., Park, S.C. and Cheng, W.T., 1999, Welding distortion of a thin-plate panel structure, *Welding J.*, 78, 5, pp 156s–165s.
8. Huang, T.D., Dong, P., DeCan, L.A. and Harwig, D.D., 2003, Residual stresses and distortions in lightweight ship panel structures, Northrop Grumman, *Technol. Rev. J.*, 11, 1, pp 1–26.
9. Suresh, S. and Dattrajan, H., 1987, *An experimental study of angular distortions in welded structures*, Welding Research Inst., Tiruchirapalls, India, Vol. 8, Pt. 1, pp 7–14.
10. Parmar, R.S. and Arya, S.K., 1989, Studies on angular distortion of butt welds in flux cored arc welding, *J. Inst. Eng. (India), Prod. Eng. Div.*, 69, 3, pp 65–71.
11. Camilleri, D., 2004, *Support tools for the prediction of distortion in the design and manufacture of thin plate welded structures*, PhD thesis, University of Strathclyde, Glasgow, UK.
12. MATLAB, www.mathworks.com
13. Feng, Z. and Rowlands, R.E., 1987, Continuous full-field representation and differentiation of three-dimensional experimental vector data, *Comput. Struct.*, 26, 6, pp 979–990.
14. Comlekci, T., Camilleri, D. and Gray, T.G.F., 2003, Finite element representation of experimental surface deformation data from fusion welded plates, *Proc. 11th Annual Conference of Association for Computational Mechanics in Engineering, Glasgow, UK*, pp 33–36.
15. Cook, R.D., Malkus, D.S., Plesha, M.E. and Witt, R.J., 2007, *Concepts and applications of finite-element analysis*, 4th edn, Wiley, India, New Delhi, India.
16. BSI, 2010, *Hot-rolled steel plates 3 mm thick or above. Tolerance on dimensions and shape*, BS EN 10029:2010, British Standards Institution, London, UK.
17. ASTM, 2011, *Standard practice for measuring flatness characteristics of steel sheet products*, ASTM A1030/A1030M-11, ASTM International, PA, USA.

7
Modelling thermal processes in welding

DOI: 10.1533/9780857099327.126

Abstract: Distortion and residual stress are driven by the thermal gradients generated by the moving heat source. Heat energy enters the workpiece through the fusion zone primarily and procedures are provided to model heat input as a lumped source, including the effects of weld filler deposition and losses due to convection and radiation. Alternative three-dimensional and two-dimensional thermal models are given. Stiffener fillet welding involves more conduction routes and a double-sided fillet stiffener weld is treated. Methods to determine welding thermal efficiency are given, with values for different processes. The characteristics of thermal cutting are discussed.

Key words: convection, radiation, fusion zone, weld pool, heat input model, radial heat source, double-ellipsoidal heat source, weld filler deposition, thermal property non-linearity, fillet welds, welding efficiency, thermal cutting.

7.1 Introduction

This chapter expands on the previous discussion of thermal calculation in Chapters 4 and 5 and aims to show in practical terms how the thermal patterns generated by longitudinal welds may be predicted. Section 7.2 shows how convection and radiation behaviour can be set up in a computational model. Section 7.3 indicates various ways of modelling the heat input arising from the welding process and sections 7.4 and 7.5 present various approaches to features such as simulation of the weld deposition or filling process and the effects of non-linearity in the temperature dependence of material properties. Sections 7.6 and 7.7 deal with alternative three-dimensional (3D) and two-dimensional (2D) computational models. Section 7.8 provides thermal modelling suggestions for fillet-welded construction.

Section 7.9, on welding efficiency, deals with the problems of matching thermal computations to experimental results. If no relevant experimental data on efficiency are to hand, this section also provides some data to assist in choosing an appropriate efficiency. The chapter concludes with a discussion of thermal cutting simulation.

7.2 Convection and radiation

Heat flow to and from a workpiece by convection and radiation is normally very small in comparison to that due to conduction. The conversion of welding process energy to heat energy is in itself difficult to determine and experimental validation is usually required. The effects of convection and radiation on the apparent 'welding efficiency' factor are not easy to distinguish in most applications. However, finite-element programs for thermal modelling will generally include options to describe convection and radiation effects in simple configurations and it is therefore useful to be aware of the basis of these options.

Heat flow from a surface by convection q_c follows Newton's law of cooling, meaning that it is proportional to the difference between the temperature of the heated body T_s and the ambient temperature T_a according to

$$q_c = h_{fc}(T_s - T_a) \tag{7.1}$$

where h_{fc} is a heat transfer or 'film' coefficient. This is evaluated at an average film temperature, assumed to be halfway between the ambient temperature and the temperature of the surface. The coefficient depends on many factors – the thermal and other physical properties of the convecting fluid (usually air), whether the fluid flow is forced or natural, turbulent or laminar, whether the plate workpiece is horizontal or vertical and whether the heated surface in question is on the underside or the upper surface of the workpiece. A general relationship that describes the convection film coefficient as a function of the Nusselt number Nu is given by[1]

$$h_{fc} = \frac{K_{cA} Nu}{L_h} \tag{7.2}$$

where K_{cA} is the thermal conductivity of air calculated at the film temperature $(T_a + T_s)/2$ and L_h is the surface area divided by the perimeter. These effects are described in several textbooks on heat transfer (e.g. reference 1) and some finite-element programs offer suitable coefficients to match common situations.

Heat flow by radiation q_{rad} follows the non-linear Stefan–Boltzmann law

$$\begin{aligned} q_{rad} &= \sigma_{sb}\varepsilon_s(T_s^4 - T_a^4) \\ &= h_{fr}(T_s - T_a) \end{aligned} \quad [7.3]$$

where σ_{sb} is the Stefan–Boltzmann constant and ε_s is the emissivity of the surface. The resulting film coefficient due to radiation h_{fr} (in the same terms as the convection coefficient) can be derived from equation 7.3 as

$$\begin{aligned} h_{fr} &= \frac{\sigma_{sb}\varepsilon_s(T_s^4 - T_a^4)}{(T_s - T_a)} \\ &= \sigma_{sb}\varepsilon_s(T_s^2 + T_a^2)(T_s + T_a) \end{aligned} \quad [7.4]$$

The effects of convection and radiation can be combined in a way that is more convenient for finite-element implementation by expanding equation 7.3 in the form

$$q_{rad} = \sigma_{sb}\varepsilon_s(T_s^2 + T_a^2)(T_s + T_a)(T_s - T_a)$$

and combining this with the convective heat transfer equation to give

$$q_{c+rad} = (h_{fc} + h_{fr})(T_s - T_a) \quad [7.5]$$

always remembering that h_{fr} is a non-linear function of temperature, as implicit through equation 7.4. (Note that it may be necessary to check that there is no inconsistency in the definition of T_s in the software formulation. The convection film temperature is normally based on an intermediate temperature, as stated earlier, and the radiation film temperature is sometimes based on the same value by default, rather than on the actual surface temperature of the component. It is therefore important to check that the film coefficients assigned by the software being used are defined appropriately.)

Assuming that the ambient temperature is known and is not significantly affected by the welding operation, the heat loss at points on the surfaces of the workpiece can then be calculated by the finite-element program through equation 7.5 and thus be taken into account in the determination of internal temperatures in the components. The effects of radiation heat loss will be more marked near to the weld centreline where the temperatures are highest, whereas convective heat loss will be more important further away from the centre.

7.3 Heat input modelling

As indicated in section 5.2 and Fig. 5.1, the heat input due to welding is generally modelled for simplicity as a single ‘lumped’ input, related directly to the welding process energy. In the case of the ‘analytical’ models shown in

Fig. 5.3, the heat input is assumed to be at a point or equally distributed along a line through the thickness of the component, as in the Rosenthal model described in section 4.2. These models imply infinite temperature at the focus point of the heat input. However, the real spatial distribution of heat energy is important and finite-element methods make it possible to simulate that aspect.

Computational software also makes it possible to reflect temperature-dependent thermal properties and other more detailed effects such as local heat generation due to phase transformation. Various heat source models have indeed been devised but, if the region where temperatures need to be known is in the 'far-field' from the zone of intense heat input, it is not always essential to use models showing detailed spatial variation local to the weld.[2] This tends to be the case for distortion, which is driven mainly by far-field mechanics.

The border of the weld pool or fusion zone is an important boundary in thermal input modelling, as this is the major location for the transmission of heat into the component. Nearly all of the heat developed by the arc, together with any other heat-generating features of the welding process, passes through the weld pool and into the body of the component. In theory, the formation of the weld pool could be simulated computationally, so that its shape, size and temperature profile might be predicted from known welding process data such as welding voltage and current, together with travel speed. Unfortunately, this ideal stage has not yet been reached in practical models, i.e. where it is possible to predict the size and shape of the weld pool from basic process settings. In essence, weld pool simulation is a particularly complex multi-physics problem, involving the interaction of electromagnetic phenomena, high-temperature fluid flow in the heat source area and in the weld pool, and metallurgical effects (see section 5.2). In cases where there is a weld edge preparation, defined as part of the design process, it is much simpler to take the 2D profile of the 'filled' edge preparation as an approximation to the transverse size and shape of the fusion zone created.

Three basic approaches are commonly used to model heat input. The first is based on a spatial prescription of heat flux on a surface area, the second on the insertion of elements at a prescribed temperature (perhaps the *liquidus* temperature) and the third is based on the prescription of heat generation within defined elements of the weld configuration (volumetric heat input method). In all these cases, experimental observations of the fusion zone shape and extent are used to a greater or lesser extent to adjust the heat flux distribution, the configuration of elements at liquidus temperature or the heat generation elements. By that means, the estimated weld energy heat input, the observed temperature profiles and the fusion zone shape and size can, as far as possible, be matched.

In one of the earliest analyses, Friedman[3] modelled the heat source in

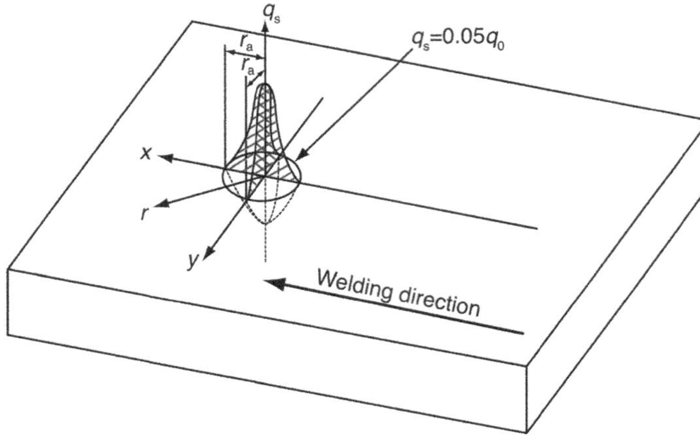

7.1 Radially symmetric normal distribution heat source model.

terms of a surface heat flux, having a radially symmetric, normal distribution as shown in Fig. 7.1. If r is the distance from the centre of the heat source, then the heat flux q_s is given by

$$q_s(r) = q_0 e^{-Cr^2} \tag{7.6}$$

where q_0 and C are constants, determined by the magnitude and distribution of the heat input, hence the parameters q (effective heat input) and r_a (the radial distance in which 95% of the heat flux is deposited) are given by

$$q = 2\pi \int_0^\infty q_s(r)\, r\, dr$$

$$q_s(r_a) = 0.05 q_0 \tag{7.7}$$

Solving the above, the heat input may be defined as

$$q_s(y, x) = \frac{3q}{\pi {r_a}^2} e^{-3(r/r_a)^2} \tag{7.8}$$

The size and shape of the weld pool is determined by the radial parameter r_a. Younan *et al.*[4] used Friedman's heat source model and set the value of r_a to 5 mm.

A more plausible weld pool shape can be established by starting from an ellipsoid with centre at (0,0,0) and semi-axes a, b, c parallel to the co-ordinate axes x, y, z. The Gaussian distribution of power intensity q_{vol} is given by

$$q_{vol}(x, y, z) = q_{m.vol} e^{-Ax^2} e^{-By^2} e^{-Cz^2} \tag{7.9}$$

where $q_{\mathrm{m.vol}}$ is the maximum value of the power density at the centre of the ellipsoid. From the conversion of energy, the heat input is given by

$$q = 4\int_0^\infty \int_0^\infty \int_0^\infty q_{\mathrm{m.vol}} \mathrm{e}^{-Ax^2} \mathrm{e}^{-By^2} \mathrm{e}^{-Cz^2} \, \mathrm{d}x \, \mathrm{d}y \, \mathrm{d}z \qquad [7.10]$$

Equation 7.10 can be expressed in the form

$$q_{\mathrm{m.vol}} = \frac{2q(ABC)^{1/2}}{\pi \pi^{1/2}} \qquad [7.11]$$

To evaluate the constants A, B, and C, the semi-axes values a, b, and c are defined as the distances where $q_{\mathrm{vol}} = 0.05 q_{\mathrm{m.vol}}$ such that

$$q_{\mathrm{vol}}(a, 0, 0) = q_{\mathrm{m.vol}} \mathrm{e}^{-Aa^2} = 0.05 q_{\mathrm{m.vol}}$$

$$A = \frac{\ln 20}{a^2} \cong \frac{3}{a^2}$$

Similarly, $B \cong 3/b^2$ and $C \cong 3/c^2$. Substituting the constants into equation 7.11 yields

$$q_{\mathrm{vol}}(x, y, z) = \frac{6(3)^{1/2} f q}{abc \, \pi \, \pi^{1/2}} \mathrm{e}^{-3x^2/a^2} \mathrm{e}^{-3y^2/b^2} \mathrm{e}^{-3z^2/c^2} \qquad [7.12]$$

where f is the final modification that takes into account the fraction of heat deposited in front of the weld ($f_{\mathrm{f}}, x > 0$) and the fraction of heat deposited at the rear ($f_{\mathrm{r}}, x < 0$), such that $f_{\mathrm{f}} + f_{\mathrm{r}} = 2$.

A better fit to actual weld zone shapes can be obtained through the double-ellipsoidal heat flux model introduced by Goldak *et al.*[5] (Fig. 7.2). Two different values for a are specified, i.e. a_1 and a_2, these being the semi-axis values parallel to the x coordinate at the front and rear of the weld respectively, such that a continuous slope is achieved when $a_1 f_{\mathrm{f}} = a_2 f_{\mathrm{r}}$. Lindgren[6] provides detailed finite-element implementation procedures for this heat input strategy.

Most of the simulations carried out by the authors have used the volumetric heat input method. This is straightforward to implement, in that the element sizes and configurations can be matched to the known weld zone shape and the heat input to the known process parameters, adjusted via an assumed efficiency factor μ. Assuming a triangular fusion zone shape, with a width on the surface of the plate of b_{w} and a penetrating depth of d_{pen}, together with a length of weld formed in a single load step l_{w}, the heat input rate for that volume will be given by

$$q_{\mathrm{vol}} = \frac{2\mu VI}{b_{\mathrm{w}} d_{\mathrm{pen}} l_{\mathrm{w}}} \qquad [7.13]$$

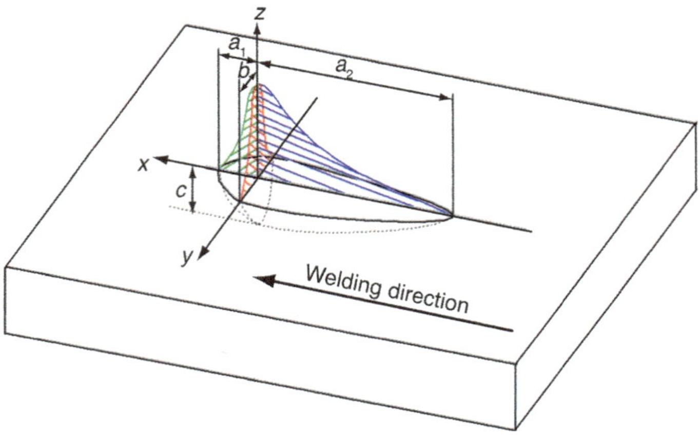

7.2 Double-ellipsoidal heat source model.

in which *VI* is the weld power, calculated from the product of the measured welding voltage and current, and the time over which heat is delivered in the load step is related to the length of weld by the welding speed. Other weld zone shapes can be treated similarly.

7.4 Simulation of weld deposition

The simplest approach to simulating weld deposition is to assume that the weld is present from the beginning (with or without an 'overfill' region). Heat is then generated in elements corresponding to the fusion zone shape and transmitted outwards through other elements. This approach could also be applied to those welding processes described as 'autogenous', where there is no addition of filler material and, often, no weld edge preparation. However, filler material addition may be required for a variety of reasons, as is a weld edge preparation or groove, and there needs to be a way of simulating the transition in thermal capacity between the empty groove on approach of the heat source and the completed, filled weld, behind the heat source. This can be simulated using so-called 'birth-and-death', 'quiet' or 'active/non-active' elements. These are set in place from the outset, but are ascribed initial properties that correspond in a thermal sense more nearly to an empty space. The element property is changed or 'switched on' when the heat source arrives at its location. This approach is not always perfect, depending on the software used, as the thermal capacity of the elements cannot be reduced completely to zero in some finite-element formulations, for reasons of solution instability, but it does provide a way of modelling the weld deposition aspect.

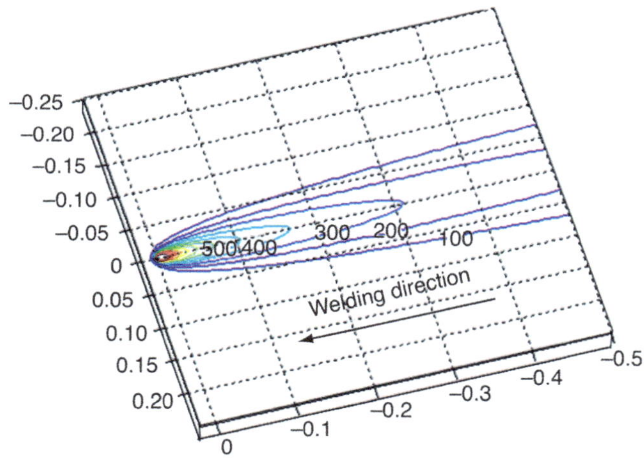

7.3 Predicted thermal distribution in a butt-welded plate using analytical solutions and room-temperature material properties.

7.5 Thermal property non-linearity

The importance of including non-linear properties in the thermal model can be judged to some extent by comparing different determinations. Figure 7.3 shows a thermal contour output, based on Rosenthal's quasi-static 2D theory with *fixed* thermal properties. The governing equation given in Chapter 4 (equation 4.1) is reproduced here for convenience as

$$T - T_0 = \frac{1}{2\pi}\frac{q}{d}\frac{1}{K}e^{-vx'/2\lambda}K_0\left(\frac{vr}{2\lambda}\right) \qquad [7.14]$$

For this example, an average specific heat of 782 J/kg °C was assumed, together with an average conductivity of 70 J/ms °C (linearising the properties given in Fig. 5.4). The energy input rate at the weld was assumed to be 4.74 kW. This figure corresponds to 82.5% of the nominal energy input in a certain trial weld, the process efficiency being determined to give as good a fit as possible between experimental and analytical results.

Figure 7.4 shows a corresponding thermal contour plot, where more realistic variable thermal properties were incorporated.[7] Rosenthal's equations were applied as before, but an iterative solution was used to accommodate the non-linear properties. Comparison of Fig. 7.3 and 7.4 shows considerable difference. In the variable-property version, each of the contours is much longer in the trailing temperature field (i.e. the material stays hotter for longer). The main reason for this is that the non-linear property solution has taken account of the latent heat effect. The resulting greater energy retention in the molten weld pool gives rise to later dissipation of energy in the trailing zone. This could be an important effect.

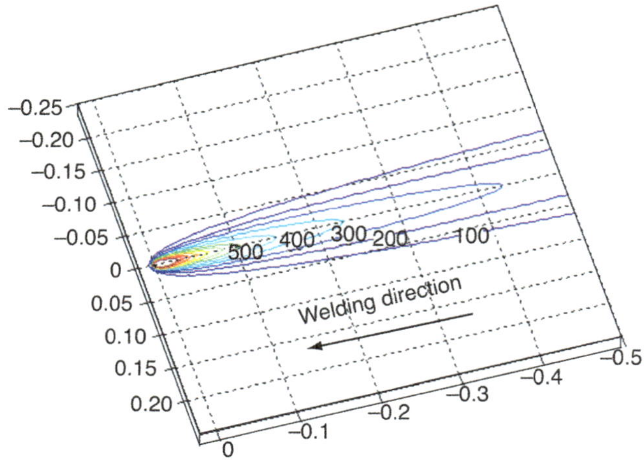

7.4 Predicted thermal distribution in a butt-welded plate using analytical solutions and non-linear material properties.

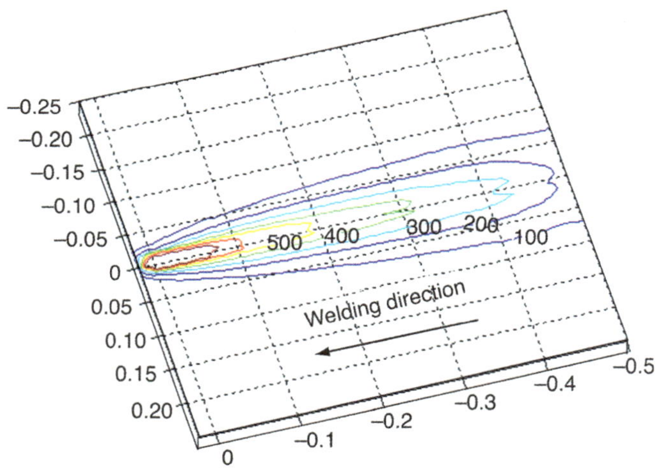

7.5 Thermal transients measured by thermographic camera.

It is useful to compare these theoretical determinations with a measured temperature plot for the same input conditions, as given in Fig. 7.5. These thermal contours were obtained using a thermographic camera.[8] In general, the non-linear property solution reflects the experimental results more closely. However, there is an expansion of the experimental thermal contours near the starting end of the plate (at location –0.5 m) due to the fact that the real boundary is at a finite distance from the heat source and does not correspond to the infinite length assumed in the quasi-static solution. A more precise comparison can be seen in Table 7.1 where

Table 7.1 Comparison of measured temperatures with quasi-static theory using fixed or variable material thermal properties

Position from weld line (mm)	Thermocouple results		Rosenthal – fixed thermal properties		Rosenthal – variable thermal properties	
	Maximum temperature (°C)	Time relative to arc (s)	Maximum temperature (°C)	Time relative to arc (s)	Maximum temperature (°C)	Time relative to arc (s)
15	430	15.7	447	10.4	431	15.6
20	347	30.2	343	17.7	332	27.7
25	291	38.3	280	28.1	272	40.0
35	220	60.8	206	54.2	201	78.1

thermocouple measurements are shown from the same test as that used for the thermographic results in Fig. 7.5. (Note that the indicated temperatures in the thermographic plot are a little higher than those in the thermocouple results, probably due to more reflected radiant heat reaching the thermographic detector.)

Leaving aside the lowest temperatures, which are affected by the finite plate boundaries, the iterative solution using variable thermal properties is more accurate relative to the thermocouple measurements, particularly in the case of the longitudinal or time axis. However, it is useful to note that the maximum *width* of each contour is fairly similar across the two solutions and the measured results, probably because the transverse temperature gradient is steep and the assumed average properties in the simple solution adequately model the heat flow in this direction. This point was emphasised previously in discussing the effect of travel speed in Fig. 4.3 and in relation to the approximate equation 4.4. The comparison in Table 7.1 strengthens the case for using finite-element computations, which can better reflect the effects of non-linear thermal properties.

However, before moving on from analytical to computational methods, it is worth noting a secondary effect of weld travel speed and its interaction with material conductivity. The simplified result for maximum temperature profile across the width, due to Okerblom and shown in equation 4.4, was derived from the Rykalin analysis, but depended on the condition that the heat source would be intense and 'fast moving'. This equation is repeated here in rearranged form as

$$y = 0.242\frac{q}{vd}\frac{1}{c\rho}\frac{1}{T_{\mathrm{m}}(y)} \qquad [7.15]$$

Various approximations are included in this simplification and it is not entirely clear in quantitative terms what the scope of these approximations is. Wells[9] has given a corresponding approximation, based on Rosenthal's

Table 7.2 Comparison of maximum temperatures reached transverse to weld line; energy input to weld identical but travel speed reduced

Weld power	Speed (mm/s)	Maximum temperature reached transverse to weld line (°C)			
		15 mm	20 mm	25 mm	35 mm
Full power input	4.8	443	342	280	208
2/3 power input	3.2	431	335	275	205
Half power input	2.4	419	328	271	203

formulation, but in this case the dependence on travel speed in relation to diffusivity can be identified. The resulting approximation was compared with experimental determinations of the fusion zone width and can be compared with equation 7.14. The approximation is valid to within −6% and +1%, relative to equation 7.14, for $0.1 \geq vy/(2\lambda)$. Wells' formula was applied by Wickramasinghe and Gray[10] to show the dependence on travel speed/diffusivity ratio of the zone width that reaches a given maximum temperature as

$$y = \frac{1}{4}\frac{q}{vd}\frac{1}{c\rho}\frac{1}{T_{\mathrm{m}}(y)} - 0.4\frac{\lambda}{v} \qquad [7.16]$$

where λ is the diffusivity of the material. This equation is readily compared with equation 7.15 and can be rearranged in terms of maximum temperature as

$$T_{\mathrm{m}}(y) = \frac{1}{4}\frac{q}{vd}\frac{1}{c\rho}\frac{1}{\hat{y}} \qquad [7.17]$$

where $\hat{y} = y + 0.4\lambda/v$.The significance of low speed or high diffusivity, in terms of the temperature at a given point, can be judged by comparing the $0.4\lambda/v$ term with y.

The effect of speed can be demonstrated, using this equation, through a comparison of the temperature fields shown in Fig. 4.2 and Fig. 4.3 (Table 7.2). (It should be remembered that these three cases were calculated for identical heat input per unit length of weld travel and, according to equation 7.15, should exhibit identical maximum temperatures at given transverse positions.)

Slower speeds (for identical heat input) have the effect of reducing the temperature at a given width, through the influence of the second term in equation 7.14. The effect is minimal for these parameters but is more significant for slower moving processes such as gas tungsten arc (GTA) welding, especially where material diffusivity is high, such as in aluminium (see Table 4.1). The implication, through the mismatched thermal strain (MTS) algorithm is that, if the correction term has a significant value relative to the primary term, there should be less longitudinal contraction force at low speeds for a given specific heat input.

7.6 Three-dimensional transient thermal computation

As indicated earlier in section 5.6, 3D transient computational analysis should be the most satisfactory approach in terms of modelling all relevant thermal features of the problem. In the analysis illustration that follows, the simulated heat source is moved along the weld line in a series of fine load steps. The finite-element mesh used reflects the thermal gradients represented in the problem and is graded in size from fine mesh near the weld line to a coarser size at some distance from the weld, as shown in Fig. 7.6.

Table 7.3 presents the results of a 3D analysis where a 6 mm thick steel plate was welded at a nominal heat input rate of 1.19 kJ/mm, resulting in a fusion zone taken to be 12.7 mm wide at the top surface and penetrating 5.5 mm. The length of the heated volume was 10 mm, corresponding to one load step. This fusion zone arose from a V-shaped weld preparation 12 mm wide at the surface. A welding efficiency of 82.5% was adopted on the basis of a comparison with previous thermocouple determinations.

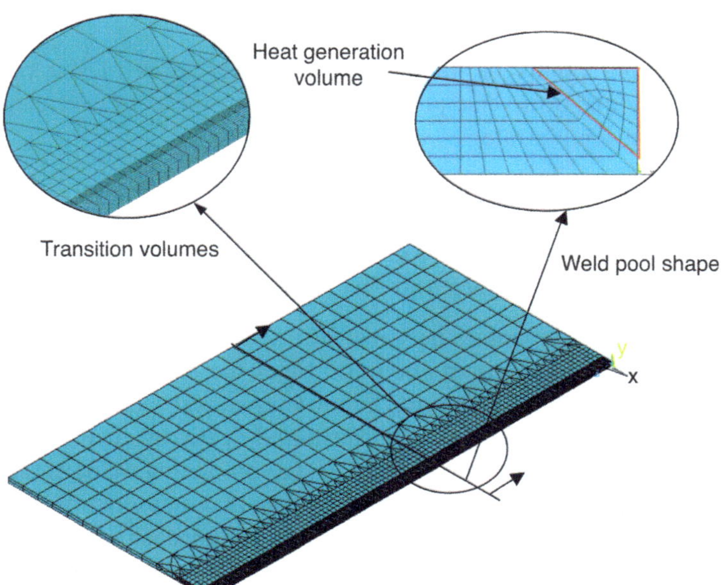

7.6 Mesh design for butt welds.

Table 7.3 Comparison of computed maximum temperatures reached and corresponding peak times for three different heat input strategies

Transverse position relative to weld centreline (mm)		(1) Heat generation elements corresponding to fusion zone shape		(2) Heat generation plus 'birth-and-death' elements		(3) Circular surface heat flux method	
		Maximum temp. (°C)	Time relative to arc position (s)	Maximum temp. (°C)	Time relative to arc position (s)	Maximum temp. (°C)	Time relative to arc position (s)
Upper face	15	427.0	19.1	442.6	20.1	421.4	24.0
Lower face	15	427.2	19.1	442.7	20.1	421.4	24.0
Upper face	20	343.6	31.3	356.6	33.3	346.0	34.4
Lower face	20	343.7	31.3	356.7	33.3	346.0	34.4
Upper face	25	285.8	41.7	297.9	41.7	291.5	46.9
Lower face	25	285.9	41.7	298.0	41.7	291.5	46.9
Upper face	35	212.7	64.0	220.9	64.0	214.2	66.3
Lower face	35	212.0	64.0	221.0	64.0	213.2	66.3

Corresponding points on the upper and lower faces of the plate were identified at four transverse positions relative to the weld centreline.

Three heat input strategies were trialled

1. heat generation elements matched to the measured fusion zone (assumed to be present throughout)
2. heat generation elements with 'birth-and-death' elements
3. 10 mm diameter surface heat flux (Friedman's method).

The differences in outcomes between these strategies are not major, at least not at greater distances than about 1.5 times the half-weld width from the edge of the fusion zone. The maximum temperatures are also very similar for all strategies. The model using 'birth-and-death' elements gives slightly higher maximum temperatures ahead of the completing weld, as there is less material available to absorb heat energy. The maximum temperature peaks at a later point for the circular surface heat flux approach, reflecting the less-embedded nature of the heat source at the surface. It is also worth noting that there is no difference between the upper and lower maximum

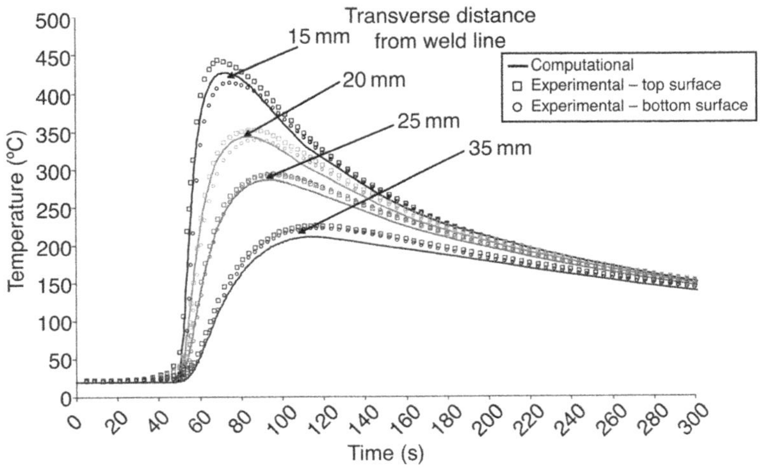

7.7 Comparison of predicted and experimental thermal transients.

temperatures, even at the minimum distance from the weld, despite the asymmetric nature of the heat input through the thickness. This emphasises the role of rapid thermal conduction in steel.

Direct comparison of the heat generation method results with thermocouple trials is shown in Fig. 7.7. Note that overall agreement is maximised in this comparison through the matching of weld efficiency choice. The thermocouple results show some divergence between upper and lower temperatures at points near the arc, but this is very probably due to radiant heat reaching the upper thermocouples directly.

7.7 Transient finite-element model based on two-dimensional cross-section

As indicated in section 5.6.3, considerable simplification and computing economy can be made by adopting a 2D transverse strip model, where the condition of no heat flow from the strip, forward or backward, is assumed. In this transient model, a fixed packet of energy at the specified power is applied to the fusion zone for a period corresponding to the distance travelled by the heat source over the width of the assumed strip. In detail, the heat input is ramped linearly over a period related to the arc diameter. After the packet of energy has been delivered, heat flow is by transverse conduction in the strip, together with convection and radiation as before. The heat generation elements are again matched to the fusion zone shape but the length of the zone in the direction of welding is determined by the product of the welding speed and the duration of heat input.

Figure 5.9 showed a comparison of the 2D and 3D transient results, based

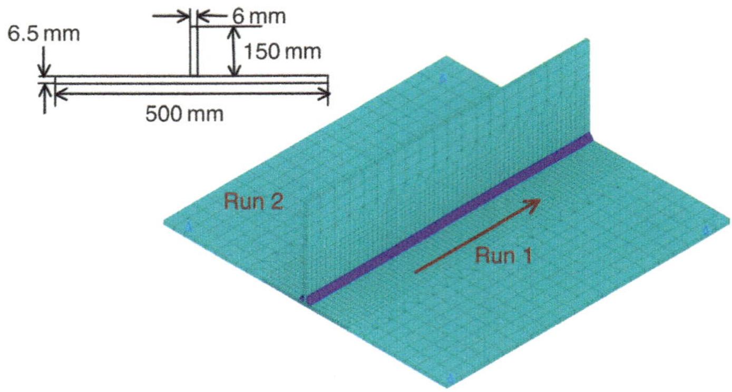

7.8 Schematic illustration and finite-element model of fillet-welded steel plate.

on the same 0.825 efficiency. The assumption of no forward or backward heat flow from the strip results in differences in timing and in maximum temperature values nearer to the weld line, but they are insignificant in practical terms.

Compared with analytical thermal strategies, the 2D finite-element method has much to recommend it, as it is more straightforward to incorporate finite geometry features together with real material properties. The computational time was increased by a factor of eight only, relative to the analytical. The 3D analysis using 'birth-and-death' elements increased the computational time by a further factor of approximately 100.

7.8 Thermal computation in stiffener fillet weld geometries

The considerations discussed in sections 7.2 to 7.7 apply equally to other cross-sectional geometries and in particular to the case where a single, upright stiffener is joined to a flat plate by two sequentially applied fillet welds, as in Fig. 7.8. Here, a 6 mm thick stiffener is welded to a 500 mm square steel plate, 6.5 mm thick. Simulation of the second weld completion has to recognise the presence of the first weld (rather than the un-welded geometry) and the thermal geometry is more complex than for a simple butt weld. Treatment of the interface conditions between the stiffener edge and the plate also raises some questions. Should it be treated as a thermally continuous interface or assumed as a thermal break? Also, heat flow and 'efficiency' may be strongly affected by the exact position of the heat source and weld pool shape, which are under less control in the open fillet geometry than in the case of welding in a grooved edge preparation.

Practical results were also obtained for this model using a metal inert gas

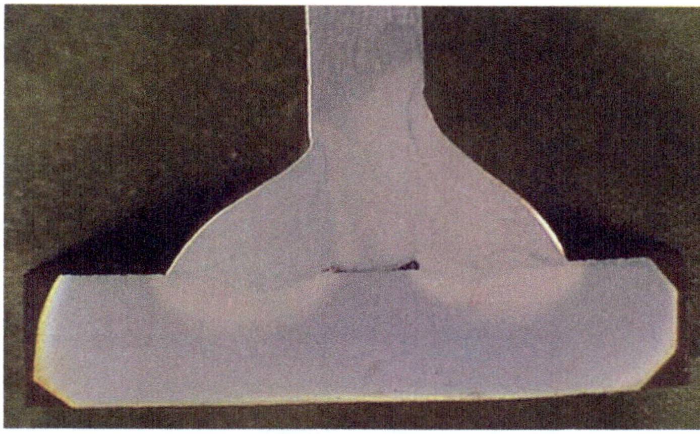

7.9 Macrosection of trial weld – first pass on the right and second pass on the left.

(MIG) shielded process as before, at a nominal process heat input of 1.28–1.41 kJ/mm, corresponding to welding parameters of 4.8 mm/s, 24 V and 256 A for the first pass, with a slight increase in current to 270 A for the second pass. (This increase was not deliberate, as the power source settings were identical for both passes, but it would appear that the electrical impedance in these relatively small specimens was less for the second pass, as the same effect was noted in several trials.) Figure 7.9 shows a sample cross-section of one of the trial welds. Despite the nominally identical welding procedures and parameters, it is obvious that the weld shapes and cross-sectional areas are different, highlighting the uncertainties involved in modelling a process with such variation. The un-welded gap between the base plate and stiffener edge is also apparent.

Given the relative success of the 2D transient approach for the previous butt-weld geometry, this could be seen as a useful starting point for the fillet-weld case. However, the use of 'birth-and-death' elements is particularly important in this case to simulate the *absence* of the second fillet weld when completing the first (right-hand) weld and the *presence* of the first weld when completing the second weld.

Figure 7.10 shows the mesh design used for this model. This accommodates the possibility of incorporating different weld sizes and configurations as necessary and includes fine mesh in the near-weld regions. Heat losses due to convection and radiation were modelled by defining appropriate film coefficients, noting that these should be different for the horizontal plate upper and lower surfaces and for the vertical stiffener surfaces.

The first trial assumed as-designed 45° fillet welds of 8 mm leg length together with full thermal continuity between the stiffener edge and the plate

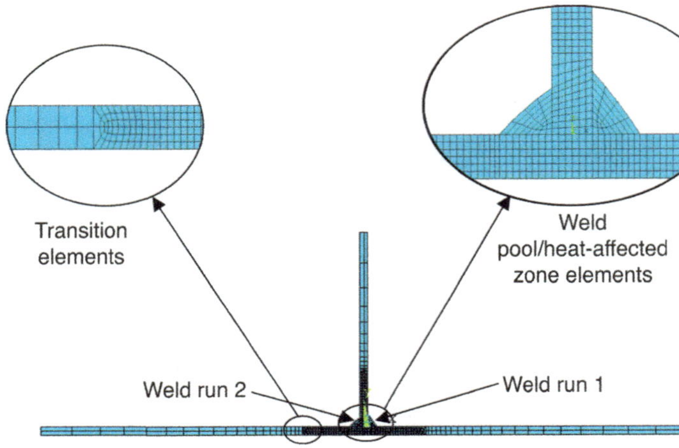

7.10 Mesh design for fillet-weld 2D thermal analysis.

(as if the stiffener was homogeneously bonded to the plate). This simulation certainly reflected extra heat flow from left to right in the plate during the second weld pass, through the already completed first weld and the stiffener/plate interface. However, the temperature profiles did not match the measured patterns.

In the second trial, the nodes at the stiffener/plate interface were separated or 'unmerged', thus preventing any heat flow across the interface. The fusion zones and corresponding volumetric heat inputs were based now on the actual shapes and sizes of the two welds, which, as apparent in Fig. 7.9, were not symmetrical. In a further refinement, each weld volume was split into an inner and an outer region, with different assumed weld efficiencies for each region.[7]

These refinements produced a better agreement with the experimental case, as shown in the corresponding thermocouple results in Fig. 7.11 (agreement being rather better for the first fillet weld than the second). For the second weld pass, adoption of the 'unmerged' interface led to a more accurate time lag of temperature rise at positions on the opposite side of the stiffener from the weld (this arises from the heat energy taking the longer path via the stiffener and the first weld fillet to reach the point of measurement). This suggests that if any interfaces are not intimately bonded, the better modelling assumption is that there is zero transfer of heat across the interface.

However, the agreement between the 2D computations and the experiment also depended on what seemed to be artificially high overall welding 'efficiencies' of 0.935 for the first weld and 0.970 for the second. This suggested that the assumption in the 2D model of zero heat flow in the longitudinal direction is suspect for this geometry, where there are more

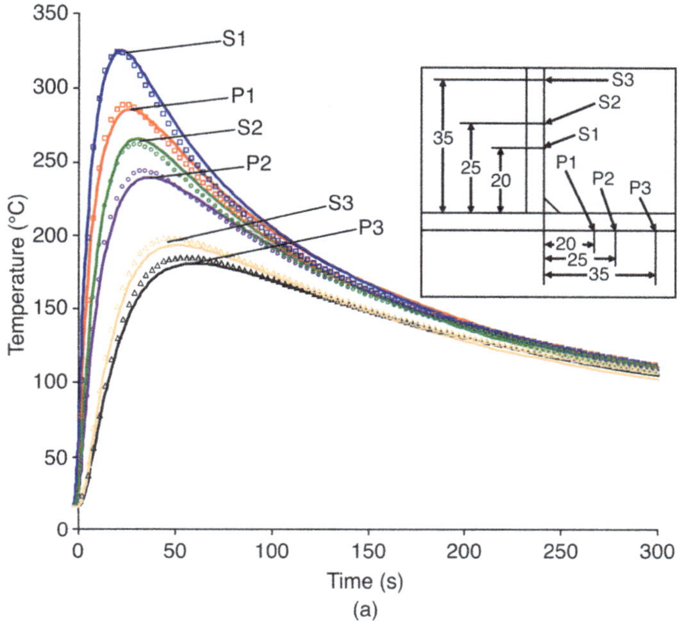

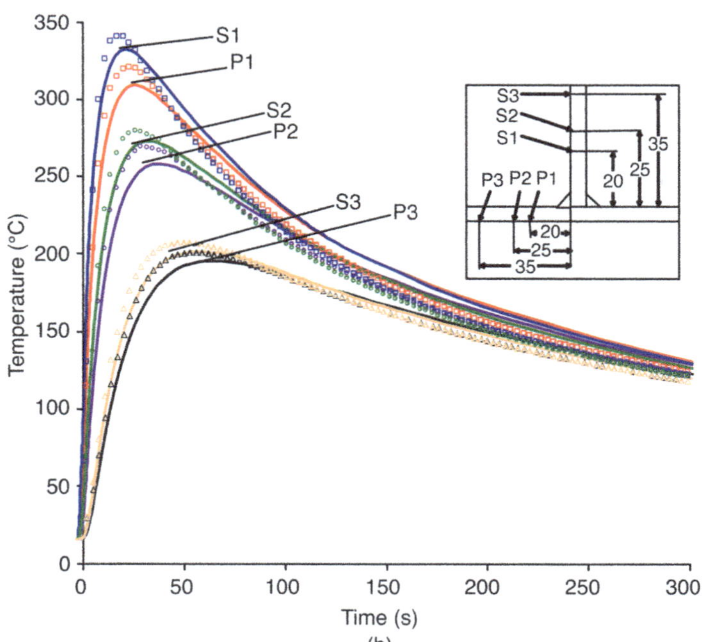

7.11 Comparison of experimental (open symbols) and 2D simulation results (continuous lines) for (a) the first weld pass and (b) the second weld pass.

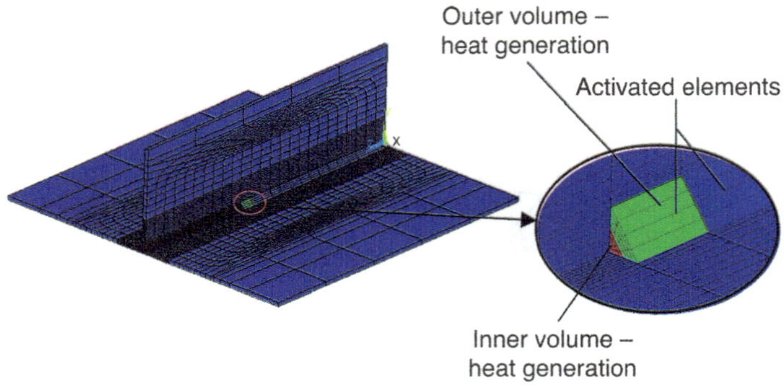

7.12 Three-dimensional finite-element model and heat input positions.

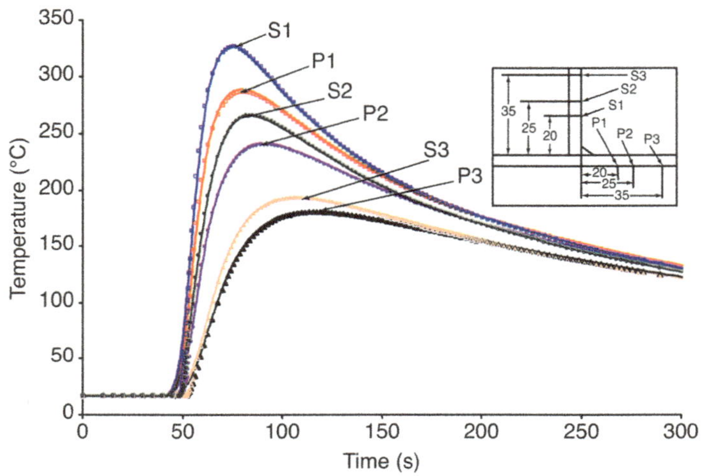

7.13 Comparison of 2D (open symbols) and 3D (continuous lines) thermal models – first weld pass.

thermal paths. The conclusion might be that a 3D thermal model is required for this complex geometry.

Figure 7.12 shows a 3D finite-element model, again including the weld deposition strategy, differing inner and outer heat generation efficiencies as adopted in the best 2D model and unmerged nodes across the stiffener/plate interface. Overall efficiencies could be reduced in the matching process to the more realistic values of 0.845 for the first weld and 0.841 for the second. Figures 7.13 and 7.14 show the thermal histories of the 3D computations for each weld completion. Figure 7.15 gives a flow chart for this analysis.

The differences between the two analyses are fairly small, although more marked in the case of the second weld pass results. However, comparison with Fig. 7.11 suggests that the 3D results approach the actual temperature profiles more closely (especially as more realistic welding efficiencies are

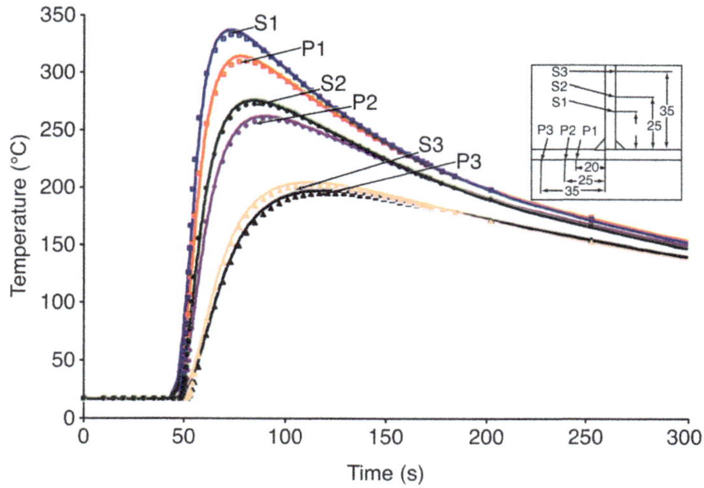

7.14 Comparison of 2D (open symbols) and 3D (continuous lines) thermal models – second weld pass.

used). Hence it would seem that the more accurate analysis may be required where there are more thermal conduction paths.

7.9 Welding efficiency

Various models and refinements that can be used to simulate welding procedures and their thermal outputs have been discussed in the previous sections but the greatest area of uncertainty in relation to prediction of a specific welding outcome lies in the fact that the so-called 'welding or arc efficiency' is normally not known with any accuracy. Table 7.4 shows some published arc efficiency factors for various welding processes.[11–17] The table also includes data obtained by the present authors that cover the cases shown earlier where a gas metal arc (GMA) process was applied with 80% Ar and 20% CO_2 shielding gas and a flux-cored welding consumable.

Most of the published results in the table were not obtained through the matching of experimental and computational data (i.e. 'back-calculation'), as shown earlier in this chapter, but by some form of calorimetry (the results in the table based on back-calculation are distinguished by an asterisk in the 'efficiency' column of Table 7.4). In the calorimetry approach, a sample workpiece is welded and placed within an insulated boundary, and the heat energy retained in the sample is then measured. There is no particular reason to prefer the accuracy of the calorimetric method to the computational matching method, as both approaches are subject to errors.[17] The results in Table 7.4 cover a wide variety of situations and might be expected to be diverse, but some general trends can be seen.

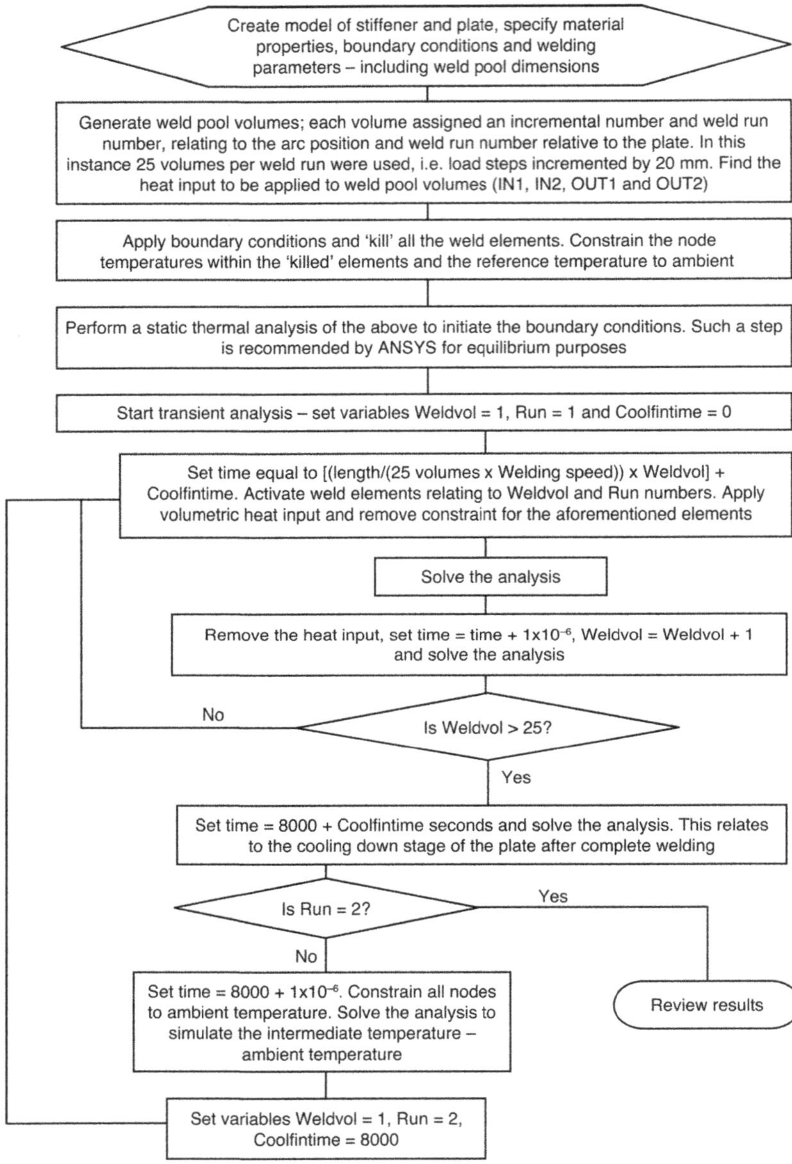

7.15 Algorithm for transient multiple-fillet thermal analysis.

Heat losses from the arc region are relatively high for the GTA process as this process tends to be operated at low power and radiation losses from the arc will therefore be relatively significant.[17] Efficiencies are therefore fairly low, although some differences and trends related to polarity, shielding gas and base material can be seen.

Table 7.4 Published weld efficiencies for various welding processes

Process	Variant	Efficiency	Shielding	Thickness (mm)	Configuration	Material	Power (kVA)	Reference
Gas tungsten arc (GTA)								
	—	0.22–0.46	Argon	—	—	Al	—	Grong[11]
	—	0.55–0.80	Helium	—	—	Al	—	Grong[11]
	DCE +ve	0.57–0.58	Helium	—	Bead-on-plate	Al	2.4–2.8	Cantin and Francis[12]
	DCE +ve	0.60–0.63	He/Ar –75/25	—	Bead-on-plate	Al	2.2–2.8	Cantin and Francis[12]
	DCE +ve	0.52–0.58	Argon	—	Bead-on-plate	Al	1.6–3.8	Cantin and Francis[12]
	DCE –ve	0.85–0.89	Helium	—	Bead-on-plate	Al	1.9–3.6	Cantin and Francis[12]
	DCE –ve	0.85–0.89	He/Ar –75/25	—	Bead-on-plate	Al	1.5–2.9	Cantin and Francis[12]
	DCE–ve	0.76–0.80	Argon	—	Bead-on-plate	Al	1.1–2.4	Cantin and Francis[12]
	DCE –ve	0.44–0.53	Argon	—	Bead-on-plate	Steel	3.8–5.6	DuPont and Marder[13]
	—	0.25–0.75	Argon	—	—	Steel	—	Grong[11]
Gas metal arc (GMA)								
	—	0.70–0.85	—	—	Bead-on-plate	Al	4.5–10.0	Christensen *et al.*[14]
	—	0.66–0.69	—	—	Bead-on-plate	Steel	10.0–12.0	
	—	0.79–0.91	Argon	—	Bead-on-plate	S/steel	6.2–14.0	DuPont and Marder[13]
	—	0.66–0.70	Argon	—	—	Steel	—	Grong[11]
	—	0.75–0.93	CO_2	—	—	Steel	—	Grong[11]
	Dip transfer	0.68–0.72	Ar/CO_2: 90/10	12.7	Bead-on-plate	Steel	—	Joseph *et al.*[15]
	Spray transfer	0.70–0.82	Ar/CO_2: 90/10	12.7	Bead-on-plate	Steel	—	Joseph *et al.*[15]
	—	0.70*	Ar/CO_2: 80/20	4.0	Close-butt	Steel	7.1	Colegrove *et al.*[16]
	Pulsed arc	0.68–0.72	Ar/CO_2: 90/10	12.7	Bead-on-plate	Steel	—	Joseph *et al.*[15]
	Pulsed arc	0.58*	Ar/CO_2: 80/20	4.0	Close-butt	Steel	7.7	Colegrove *et al.*[16]
	Pulsed arc	0.77–0.79	Ar/CO_2: 82/18	5.0	Bead-on-plate	Steel	4.8–5.1	Pépe *et al.*[17]
	Pulsed arc	0.86	Ar/CO_2: 82/18	12.0	Groove	Steel	5.0	Pépe *et al.*[17]
	Cold metal transfer	0.90*	Ar/CO_2: 80/20	4.0	Groove	Steel	3.8	Colegrove *et al.*[16]
	Cold metal transfer	0.80	Ar/CO_2: 82/18	5.0	Bead-on-plate	Steel	3.0	Pépe *et al.*[17]
	Cold metal transfer	0.87	Ar/CO_2: 82/18	12.0	Groove	Steel	3.1	Pépe *et al.*[17]
	Flux-cored wire	0.83–0.85*	Ar/CO_2: 80/20	5.0–6.0	Groove	Steel	5.8	Camilleri[7]
	Flux-cored wire	0.84–0.85*	Ar/CO_2: 80/20	6.5	Fillet	Steel	6.1–6.8	Camilleri[7]
Shielded metal arc (SMA)								
	AC	0.66–0.85	—	—	—	Steel	3.0–7.0	Christensen *et al.*[14]
Submerged arc								
	—	0.81–0.89	—	—	Bead-on-plate	Steel	6.8–11.8	DuPont and Marder[13]
	—	0.91–0.99	—	—	—	Steel	10.0–43.0	Christensen *et al.*[14]

In general, processes and welding parameters that deliver a narrow, concentrated plasma will be more efficient than those with a spread plasma, as less energy is lost in radiation. Christensen *et al.*[14] give data for manual metal arc welding which show that a high-voltage arc (which normally has a greater arc spread) results in 62–72% efficiency, whereas a corresponding low-voltage arc achieves 78–93% efficiency. This work,[14] which covers a large range of processes and experiments, also shows that there is a linear relationship between input power and absorbed power and each process is remarkably consistent over a range of input powers. This suggests that, if other factors are constant, variation in input power generates a *pro rata* increase in absorbed power.

Among the GMA variants, the published efficiencies range from 0.66 to 0.93, taken at face value. Some of the energy in GMA and other processes using consumable electrodes is absorbed in resistance heating of the filler wire and this energy is carried on into the weld pool to contribute to the heat transferred by conduction, thus improving the efficiency. No data for flux-cored wire GMA, comparable to the back-calculated data from the present authors, have been found in the literature, but the apparent efficiencies for this process seem to be high, while remaining comfortably within the 75–93% band summarised by Grong.[11]

Another significant influence apparent in the table arises from the configuration of the workpiece region adjacent to the arc and plasma. Welding on the upper surface, i.e. 'bead-on-plate' or 'close-butt', shows more heat loss than welding in a more confined region such as presented by a prepared groove, where more of the arc radiation is captured by the workpiece. This experience was confirmed by the present authors (note that the groove in the study by Pépe *et al.*[17] took the form of a 7.5 mm deep rectangular slot). By that argument, fillet welding (in effect, welding in a wide 90° groove) should have *lower* efficiency than butt welding in a typical narrow-angle groove, but the authors' results do not bear that out. It may be that the typical multiple conduction paths in fillet-welded configurations increase the absorbed heat energy relative to the losses.

Submerged arc welding consistently shows the highest level of efficiency among the commonly used processes. In this case, the arc is partly insulated from the surroundings, in the radiation sense, by a blanket of flux powder and slag covering, although the latter is also a source of heat loss.

Returning to the computational matching or back-calculation approach, it is instructive to consider the many sources of possible error through the idea of an energy balance. The effective energy delivered to the workpiece is characterised in this method by measuring the actual temperature fields resulting from a given welding process, procedure and configuration and then by finding a computational model and an energy input that gives transient temperature fields that match the experimental case. This back-calculated energy input is then compared with the measured electrical energy input to determine the losses and the apparent efficiency.

Matching of the actual and computed temperature fields is normally carried out on a trial and error basis and can be subjective to an extent. Some compromise will usually have to be made between matching only the peak temperatures at particular thermocouple locations against a more comprehensive global match of overall temperature profiles at several locations. This will be a more sensitive test of the model accuracy and the quality of the match. For example, experimental thermocouple data for the fillet welds described in section 7.8 might be matched more simply in terms of maximum temperatures at a given width y (see Fig. 7.16 and Fig. 7.17).[18]

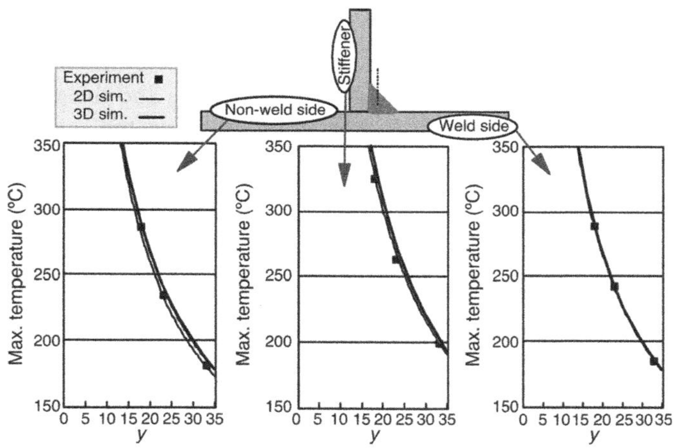

7.16 2D and 3D simulations compared with experimental maximum temperatures at width *y*.

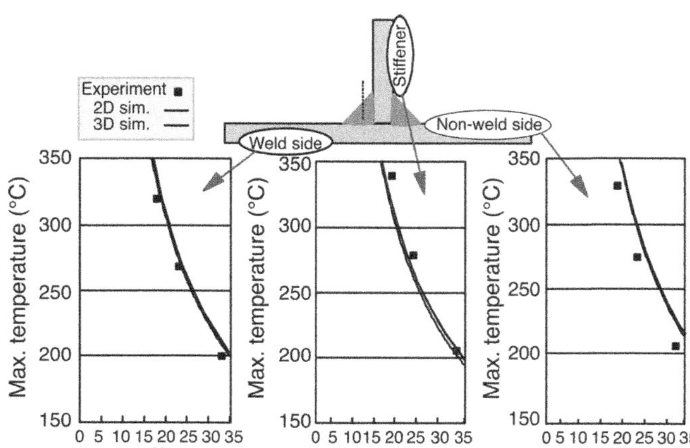

7.17 2D and 3D simulations of second pass.

In this test, three sets of thermocouple arrays were positioned for each weld pass at identical distances from the centre of the respective heat source. 2D and 3D models were simulated successfully using identical efficiencies of 0.70 for the first fillet pass and 0.85 for the second. This compares with values of 0.94 and 0.97 shown previously for 'global' matching of transient temperature profiles according to the 2D model in the first and second fillet passes respectively and 0.85 and 0.84 for 3D model matching. The reason for the apparent success of this simple matching strategy in this case is that the *widths* of temperature contours are insensitive to detailed modelling assumptions although there may be longitudinal differences.

Achievement of exact matching is not helped by the fact that small errors in the positioning of the thermocouples and local radiation heating of the thermocouples can have large effects on the apparent peak temperatures. Duplication of thermocouples in the same relative position to the weld is therefore advisable in order to identify rogue readings and the magnitude of random variations. However, it is worth remembering the simple basis of the 'computationally efficient' model, namely that the peak temperatures at given distances from the centreline of the weld are considered to be significant. It is also a fact that welding process outcomes can vary considerably, even when strenuous efforts have been made to ensure consistent welding parameters and geometry. Welding arcs are essentially flexible conductors and are easily deflected from an intended precise position. A trial involving nine single-fillet welds, carried out using identical process parameters, resulted in back-calculated efficiencies varying from 0.68 to 0.84, without there being large differences in fillet size and configuration.[7]

Such uncertainty concerning the all-important efficiency parameter may be seen as discouraging with respect to the application of computational simulation to practical welding problems. However, in practice, the main benefits of simulation methods do not lie in the prediction of a precise outcome for a specific set of conditions but, more usefully, they provide tools to compare different welding procedures, process conditions, assembly sequences and design configurations. If a fixed thermal efficiency is assumed for all cases, the results will then be relative, rather than definitive, but the information thus provided will nevertheless be invaluable at the design or manufacturing planning stages.[19, 20]

7.10 Thermal cutting

The principles used in this chapter to describe welding thermal transients can also be applied to thermal cutting processes. These are believed to be sources of distortion, albeit to a much smaller degree than welding. The concept of 'efficiency' is again useful to relate the process energy input from an electrical or combustion source to the heat conducted into the plate being cut. However, a substantial amount of heat is 'lost' from the plate because a volume of molten metal is ejected from the workpiece. The other main difference from welding transients is that the cutting process de-couples the material on either side of the cut, whereas in welding it joins them.

Such an approach was studied in the context of laser cutting of 4 mm and 5 mm steel plate under industrial conditions.[21] The cuts using this process were typically parallel-sided and 0.8 mm wide. In the model, it is assumed that heat is conducted and dissipated into each of the cut plates, uniformly through the thickness and along the cut edges. The energy balance can be represented by

$$2H_{\text{flux}} = \eta(P - H_{\text{loss}}) \qquad [7.18]$$

where H_{flux} is the power dissipated in each plate on either side of the cut (in W), η is the conversion efficiency of available power to power dissipated in plates, P is the laser beam power (in W) and H_{loss} is the power lost in removed material (in W).

The power lost due to material removal is a function of the enthalpy used to heat steel from ambient temperature to (at least) the melting temperature (this is dominated by the latent heat of fusion). The cross-sectional area of melted material and the cutting speed are also involved, as in equation 7.19

$$H_{\text{loss}} = \rho(btv)H_{\text{ent}} \qquad [7.19]$$

where ρ is the density (kg/m^3), b is the cross-sectional cut width (measured directly or assumed as laser beam width) (m), t is the thickness of plate (m), v is the cutting speed (m/s) and H_{ent} is the enthalpy of steel (J/kg).

As with welding simulation, the efficiency needs to be established experimentally, by comparing the thermal transients of actual cutting operations with computed simulations. Such a comparison, for a 5 mm thick steel plate is shown in Fig. 7.18.

These results were matched, as shown, using an efficiency of 45%. The nominal applied laser power was 2.34 kW and the power lost due to removal of 0.8 mm cross-sectional width was estimated through equation 7.19 as 0.695 kW. Hence the power dissipated is given by

$$2H_{\text{flux}} = 0.45(2.34 - 0.695) = 0.74\,\text{kW}$$

The overall efficiency, relating power dissipated in the plate to input power, amounts therefore to 0.74/2.34, i.e. 32%.

The problem that arises with this approach in practice is that the input power may be set higher than is strictly necessary to make the cut. Extra power can in that case be dissipated easily in the ejected material and to the

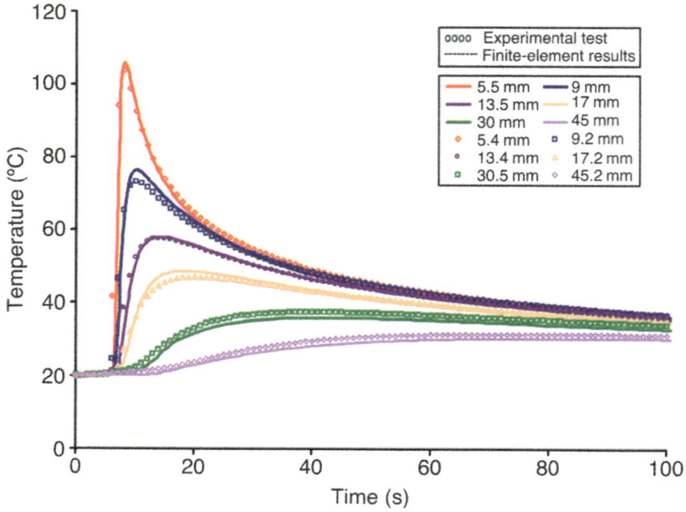

7.18 Experimental and computational temperature transients (5 mm plate).

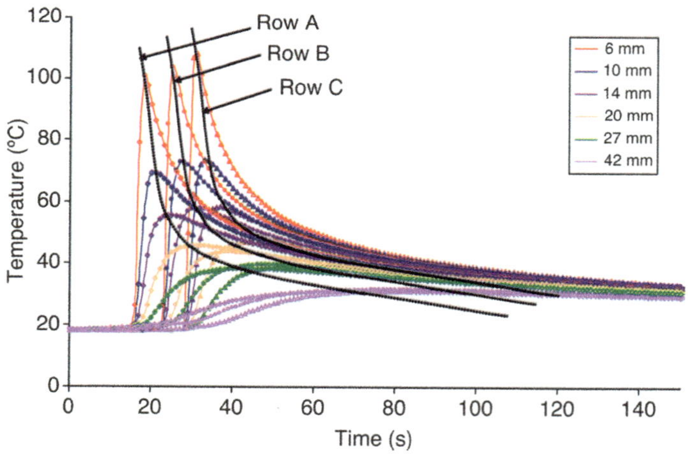

7.19 Experimental and computational temperature transients (4 mm plate).

surroundings, without noticeable effect on the finished cut. This is demonstrated in Fig. 7.19, which shows the thermal transients measured in cutting 4 mm plate using the same laser power travel speed as for a 5 mm cut. There is no significant difference in the thermal transients, from which one must conclude that more heat is being dissipated externally in the 4 mm cut and the same amount of heat is being absorbed by the workpiece through conduction.

It is worth noting that the levels of absorbed energy for cutting are very much smaller than for welding (of the order of 5% or less) and the temperatures measured in this study are therefore substantially less than temperatures generated typically in butt welding a similar thickness. As noted in section 4.5, parts of a steel plate that do not exceed 139°C during welding will experience elastic behaviour only. For the conditions shown in Fig. 7.18, this means that the zone of material that will develop longitudinal residual stress will be less than 5 mm in width. Moreover, as there is no through-thickness temperature variation and the plates are *released* by the cutting operation, the angular contraction phenomenon found in welding will not arise in cutting operations.

7.11 Conclusion

Procedures have been described to model various detailed aspects of welding thermal simulation. These include the effects of convection and radiation (section 7.2), spatial modelling of heat input and fusion zones (section 7.3) and deposition of weld filler material (section 7.4). Incorporation of these modelling options in 3D and 2D thermal analyses was illustrated in sections 7.6 and 7.7.

The influence of thermal property non-linearity was illustrated through experimental and analytical comparisons (section 7.5). It is noted that, although the longitudinal temperature distribution is altered by non-linear properties, the transverse width of isotherms is less affected. A secondary, non-linear effect of material diffusivity at low weld travel speeds was also illustrated.

An initial treatment of thermal simulation in fillet-welded stiffener geometries was introduced in section 7.8. The fillet-welded configuration introduces more conduction paths and makes it more difficult to match simulations reliably to experimental determinations, using a single value of welding efficiency. The results also suggest that 3D thermal analyses may be preferable for these more complex geometries, whereas simpler 2D treatments are adequate for simple butt-welded joins between plates.

The welding efficiency concept was explored (section 7.9) and a table of welding efficiencies for a variety of welding processes was provided, drawn from the literature (Table 7.4). It is difficult to draw definitive conclusions from this diverse database, although there are clear differences between the different welding processes. However, there is a suggestion in the data that the proportion of welding process energy that enters the workpiece through conduction depends on the configuration of the joint relative to the heat source (i.e. the weld preparation). Finally, it was shown that the principles outlined for thermal simulation of welding can also be applied to thermal cutting.

7.12 References

1. Burghardt, M.D. and Harbach, J.A, 1993, *Engineering thermodynamics*, Harper Collins, New York, USA.
2. Radaj, D., 1989, Finite-element analysis of temperature field, residual stresses and distortion in welding, *Weld. Res. Abroad*, 35, 6/7, pp 31–38.
3. Friedman, E., 1975, Thermomechanical analysis of the welding process using the finite element method, *ASME J. Pressure Vessel Technol.*, 97, 3, pp 206–213.
4. Younan, M.Y.A., Wifi, A.S. and Fanous, I.F.Z, 2002, 3D finite element modelling of the welding process using element birth and element movement techniques, *Proc. ASME Pressure Vessel and Piping Conf., Vancouver, Canada*, 422, pp 165–172.
5. Goldak, J., Chakravarti, A. and Bibby, M., 1984, A new finite element model for welding heat sources, *Metall. Trans. B*, 15B, pp 299–305.
6. Lindgren, L.E., 2007, *Computational welding mechanics – thermomechanical and microstructural simulations*, Woodhead Publishing Ltd, Cambridge, UK.
7. Camilleri, D., 2005, *Support tools for the design and manufacture of thin-plate welded structures*, PhD thesis, University of Strathclyde, Glasgow, UK.
8. Camilleri, D., Comlekci, T. and Gray, T.G.F., 2004, Use of thermography to calibrate fusion welding procedures in virtual fabrication applications, *Proc. Inframation 2004 Conference Las Vegas, USA*, 5, pp 121–131.
9. Wells, A.A., 1952, Heat flow in welding, *Welding J., Welding Res. Supp.*, 31, pp 263s–267s.
10. Wickramasinghe, D.M.G. and Gray, T.G.F., 1978, A simple treatment of welding distortion, *Weld. Res. Int.*, 8, 5, pp 409–422.
11. Grong, Ø., 1994, *Metallurgical modelling of welding*, Institute of Materials, London, UK.
12. Cantin, G.M.D. and Francis, J.A., 2005, Arc power and efficiency in gas tungsten arc welding of aluminium, *Sci. Technol. Weld. Join.*, 10, pp 200–210.
13. DuPont, J.N. and Marder, A.R., 1995, Thermal efficiency of arc welding processes, *Welding J.*, 74, 406s–416s.
14. Christensen, N., Davis, V. de L. and Gjermundsen, K., 1965, Distribution of temperatures in arc welding, *British Welding J.*, 12, pp 54–75.
15. Joseph, A., Harwig, D., Farson, D.F. and Richardson, R., 2003, Measurement and calculation of arc power and heat transfer efficiency in pulsed gas metal arc welding, *Sci. Technol. Weld. Join.*, 8, 6, pp 400–406.
16. Colegrove, P.A., Ikeagu, A., Thistlethwaite, S.W., Williams, S.W., Nagy, T., Suder, W., Steuer, S. and Pirling, T., 2009, The welding process impact on residual stress and distortion, *Sci. Technol. Weld. Join.*, 14, pp 717–725.
17. Pépe, N., Egerland, S., Colegrove, P.A., Yapp, D., Leonhartsberger, A. and Scotti, A., 2011, Measuring the process efficiency of controlled gas metal arc welding processes, *Sci. Technol. Weld. Join.*, 16, 5, pp 412–417.
18. Mollicone, P., 2007, *Applied modelling techniques for welding induced distortions and residual stresses*, PhD thesis, University of Strathclyde, Glasgow, UK.
19. Camilleri, D., Gray, T.G.F. and McPherson, N., 2010, Optimizing tack welding fabrication procedures using numerical finite element models, *J. Ship Prod.*, 26, 2, pp 117–134.
20. Camilleri, D. and Gray, T.G.F., 2006, Optimization of welded lightweight

multiple-stiffener plate structures to minimise unwanted shape distortion, *Weld. Cut.*, 5, 6, pp 320–327.

21. Camilleri, D. and Gray, T.G.F., 2006, *Cutting simulation to minimise cutting induced thin plate distortion*, BAE Systems Report, Department of Mechanical Engineering, University of Strathclyde, Glasgow, UK.

8
Computationally efficient methods for modelling welding processes

DOI: 10.1533/9780857099327.156

Abstract: Major simplifications of the corresponding rigorous thermal and thermo-mechanical analyses are applied in this chapter. Transient temperature fields are replaced by fixed temperature profiles. The 'computationally efficient' (CE) methods introduce further simplifications where transverse and longitudinal thermal strains are derived from approximations embodied in the mismatched thermal strain and transverse contraction strain algorithms. In the CE 'analytical' approach, these strains are applied to linear elastic, small strain models, using conventional strength of materials techniques. In the CE 'computational' approach, the strains are applied to linear-elastic, large-strain, finite-element models. Various hybrid stepwise methods are also shown, where a measure of sequential description is included, together with the use of elasto-plastic, finite-element computation.

Key words: reduced solutions, mismatched thermal strain (MTS) algorithm, transverse contraction strain (TCS) algorithm, weld preparation shape, hybrid stepwise approach, transient deformation, sequential model, simultaneous model.

8.1 Introduction

The identifying feature of computationally efficient (CE) solutions is that one or more key simplifying assumptions are introduced, greatly reducing the computational budget. For the simultaneous hybrid (simH) model (section 5.8.1) the main simplifying assumptions, in common with the other CE models, are that the thermal analysis input to the mechanical stage is based on the maximum temperature profile and that the transverse contraction arises entirely from cooling of the fusion zone. However, full thermo-elastic-plastic analyses are applied thereafter in the SimH case. Other CE models are based to a greater or lesser extent on the mismatched thermal strain (MTS) and transverse contraction strain (TCS) algorithms (see sections 4.4–4.6). These algorithms may be applied in an analytical mode through equations 4.4 and 4.5 for longitudinal actions and equations 4.7 and 4.8 for transverse actions. They may also be applied within a computational analysis, as will be shown in this chapter. In all cases, output temperatures from an uncoupled thermal analysis provide the input to the algorithms. The outputs from the algorithms are in the form of mechanical strains, which are then applied to the structure to determine the distortions caused by the welding process. The algorithms are based on simplified and approximate thermo-mechanical analyses and replace otherwise complex and computationally intensive procedures. This chapter therefore has two main sections.

- The first, section 8.2, deals with algorithm-based approaches. An extension to the previously outlined TCS derivation in section 4.6 is given in section 8.2.1, to cover cases where there is already a mechanical bridge between the sections to be joined and the fusion zone is not therefore the sole means of connection. This is followed in section 8.2.2 by a description and an example where algorithms are applied *analytically* to small-scale butt-welded plates. The method whereby algorithms are applied *computationally* to an elastic, finite-element model is shown in section 8.2.3. For various reasons, this is the preferred method for the CE approach.
- The second, section 8.3, describes the various hybrid solution approaches and illustrates these using experimental data from large-scale butt welds.

8.2 Computationally efficient methods based on algorithms

8.2.1 Extension of the transverse contraction strain (TCS) algorithm

Before considering further application of the CE approach, it is useful to show, as in previous work,[1] that the TCS model for transverse strains can be developed to treat a wider range of cases other than that shown in Fig. 4.9 where a simple triangular fusion zone forms the only mechanical connection between the sections being welded. Figure 8.1(a) represents such a case, where a V-shaped weld of surface width b and depth s is placed on a solid connection of depth d. The theory will also apply to fusion zones that are not simply triangular. Figures 8(b) to 8(c) represent, respectively, the

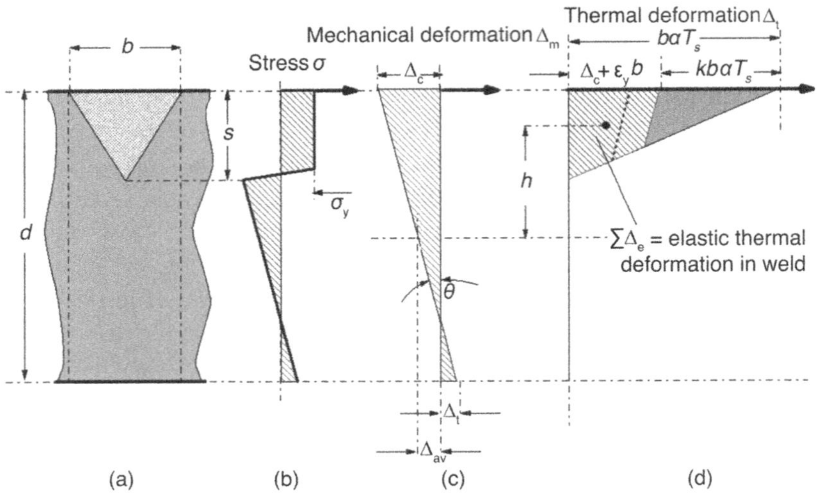

8.1 Transverse contraction model, where full connection is assumed.

transverse residual stress distribution after welding, the mechanical deformations Δ_m generated by compression and bending over the width b and the thickness d, and the thermal deformations of the fused zone as it contracts from the starting temperature T_s. The parameter k in Fig. 8(d) defines the proportion of the surface contraction accommodated by plastic strain. The residual transverse deformations (i.e. the distortions) arise from a summation of the mechanical and thermal deformations, leading to the formulations

$$\Delta_{ave} = \frac{\Sigma\Delta_e}{d} \qquad [8.1]$$

$$\theta = \frac{\Sigma\Delta_e}{I} h \qquad [8.2]$$

where h is the distance between the centroids of the mechanical and thermo-elastic deformations and I is the second moment of the mechanical deformation area.

The total thermo-elastic deformation is expressed from Fig. 8(d) in terms of the parameter k as

$$\Sigma\Delta_e = \frac{1}{2} sb\alpha T_s - \frac{1}{2} sbk^2\alpha T_s = \frac{1}{2} sb\alpha T_s(1 - k^2)$$

leading to

$$\Delta_{ave} = \frac{1}{2}\frac{s}{d} b\alpha T_s(1 - k^2) \qquad [8.3]$$

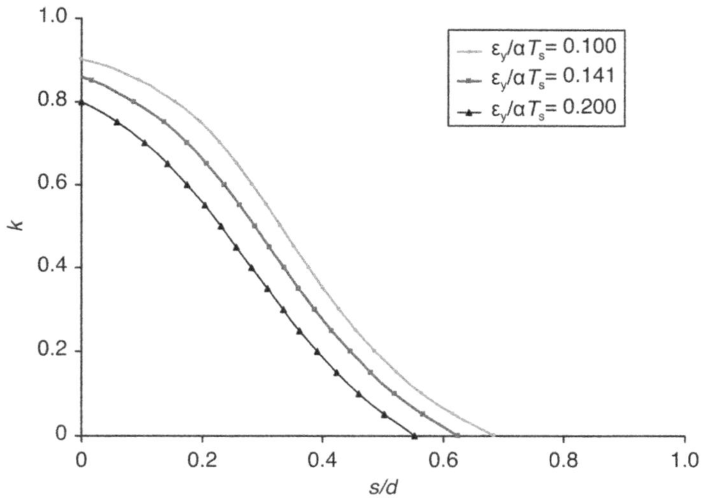

8.2 Relation between *k*, *s*/*d* and $\varepsilon_y/\alpha T_s$ for a triangular fusion zone.

$$\theta = \frac{s}{d}\frac{b}{d}\alpha T_s\left(3(1-k^2) - 2\frac{s}{d}(1-k^3)\right) \qquad [8.4]$$

These equations therefore constitute the equivalents for connected sections of equations 4.7 and 4.8, which apply to joints where the fusion zone is the only connection between the parts. Noting also that $\Delta_t + \varepsilon_y b = (1-k)\alpha T_s b$ and $\Delta_t = \Delta_{ave} + \theta d/2$, these values may be substituted in equations 8.3 and 8.4, to solve for k in terms of s/d and $\varepsilon_y/\alpha T_s$. The result is shown in Fig. 8.2 where a starting temperature of 1000°C was assumed. (The choice of T_s here is arbitrary, but note from Fig. 5.7 that the strength of the sample steel was less than 50 MPa at this temperature. The rationale is that, above this temperature, the material has little or no strength capability to maintain a transverse contraction against the smallest mechanical constraint.)

Note that when the triangular fusion zone penetrates to approximately 60% of the overall section thickness, the value of k drops to zero and the values from equations 8.3 and 8.4 reduce to a case where the fusion zone simply contracts from the starting temperature in proportion to its shape. A similar approach can be applied to other typical fusion zone shapes, such as parabolic and rectangular, with the results shown in Fig. 8.3 (for $\varepsilon_y/\alpha T_s = 0.141$). The relative order of these curves confirms practical experience of the effects of weld preparation and fusion zone shape on angular distortion, where a narrower J-type or U-type preparation usually reduces angular and overall distortion, particularly in the case of multi-pass welds in thick components. The first reason is that, for a given thickness, the volume of weld metal will be less than for a V-preparation and thus the heat

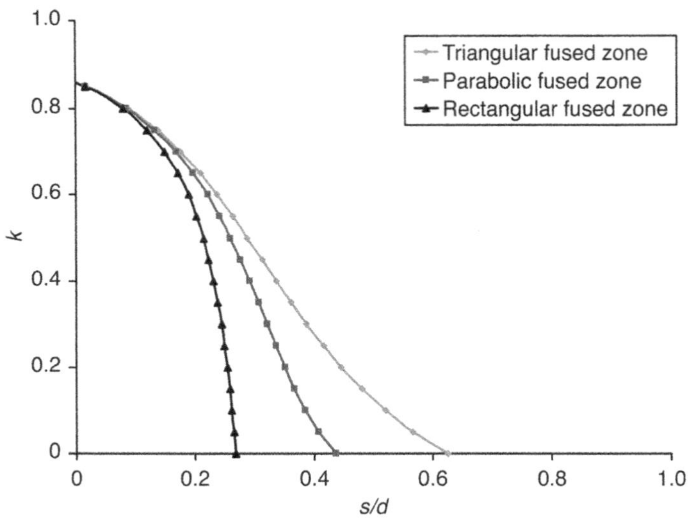

8.3 Relation between *k* and *s*/*d* for various fusion zone shapes and fixed $\varepsilon_y/\alpha T_s$

input will be less. More particularly, in the case of angular distortion, the neutral plane of the fused zone is nearer that of the full section than is the case for the triangular fused zone.

8.2.2 Analytical application of algorithms

The main weakness in applying distortion algorithms to plate-like structures lies in the problem that most simple analytical structural mechanics treatments are unable to provide accurate descriptions of thin-plate deformations when compressive in-plane loads are applied and non-linear out-of-plane movements result. Nevertheless, analytical solutions can provide order-of-magnitude estimates, which are often useful in checking computational determinations.

The small-scale test panels treated as an experimental case study in sections 5.8.1 and 6.8 form a source of real data to cross-check such an analytical treatment. Excluding the atypical results found in test specimens 5, 7 and 8 (see Table 6.1) the average angular distortion of the remaining specimens was 1.02° (0.16° standard deviation (SD)) and the average longitudinal curvature was $0.06\,\text{m}^{-1}$ (SD = $0.02\,\text{m}^{-1}$).

For a width of weld on surface b of 10–12 mm, a weld penetration s of 5–6 mm, $\alpha = 14 \times 10^{-6}$ per °C and $T_s = 1000$°C, from equation 4.7

$$\theta = \tan^{-1}\left(\frac{\alpha T_s b}{s}\right)$$

giving an angular deformation between 1.17° and 1.66°. These results are larger than the experimental values but are nevertheless of the correct order. Also, the plates in the test series were held together at each end by run on/off tabs, and this may have provided angular restraint in the experiments.

The transverse contraction of the plate is given by equation 8.3 as $\Delta_{ave} = \frac{1}{2}\frac{s}{d}b\alpha T_s(1-k^2)$ and, as $s/d = 1$ and $k = 0$, the (assumed uniform) transverse contraction is predicted to be 0.06–0.07 mm (no corresponding experimental result is available).

The average longitudinal contraction of the plate is found by calculating the contraction force through equation 4.5 as $F = 0.335\frac{q}{v}\frac{\alpha}{c\rho}E$ and the data for that evaluation are $q/v = 1.19\text{kJ/mm}$ and $c\rho = 6.12 \times 10^6\ \text{J/m}^3\text{K}$. Calculating the average longitudinal contraction by applying the force uniformly over the 500 mm wide, 6 mm thick cross-section A gives $\Delta_l = \frac{F}{A}\frac{1}{E}l$ where l is the overall length of 500 mm. The result of this calculation is a longitudinal contraction of 0.17 mm (again, there is no corresponding experimental result). In practice, the longitudinal contraction is non-uniform, as shown schematically in Fig. 4.7(c) as the contraction force is concentrated in the centre and the longitudinal sides of the plate will not be fully effective in reacting this force. However, as shall be shown later, a finite-element analysis can model this effect properly.

Such in-plane deformations are hardly problematic in a practical context, but the out-of-plane distortions are more troublesome. The average longitudinal curvature can be estimated by assuming that the longitudinal contraction force F is applied at the centroid of the A + B region in Fig. 4.8, defined through equation 4.4 by the maximum temperatures corresponding to ε_y/α and $2\varepsilon_y/\alpha$. This then generates curvature in the plate (treated as a beam) given by

$$\frac{1}{R} = 0.335\frac{q}{v}\frac{\alpha}{c\rho}\frac{\bar{y}}{I} \qquad [8.5]$$

where $\bar{y}$ is the distance from the centroid of the A + B region to the neutral axis of the V-shaped, angularly distorted cross-section and I is the second moment of area of the V-shaped cross-section. This treatment neglects the 'plate stiffening' effect due to the generation of transverse stresses in the central flat area of the plate, through Poisson's ratio interaction. This has the effect of reducing, by a small percentage, the longitudinal curvature at mid-length, relative to 'beam' theory. Equation 8.5 again shows the major influences of welding and material parameters on out-of-plane plate curvature.

A linear approximation to the stress distribution in region B of Fig. 4.8 simplifies the calculation of the centroid position and the results are shown in Table 8.1 for three different angular distortions, the experimental value

Table 8.1 Distortions calculated through TCS and MTS algorithms

	Experimental value	Analytical value	
		Minimum	Maximum
θ (degrees)	1.02	1.17	1.66
y_{max} (mm)	1.11	1.28	1.81
$\bar{y}$ (mm)	0.90	1.03	1.46
I (mm^4)	10 235	10 625	12 272
$1/R$ (m^{-1})	0.080	0.088	0.108

and the minimum and maximum values found analytically (y_{max} is the peak height of the V-shaped plate from the neutral axis).

The calculated curvatures are larger than the 0.06 value measured in the tests, even when the experimental value of angular contraction is used, but again they are of the correct order. The values of y_{max} are also of the same order as the 1.96 mm root mean square (RMS) out-of-plane deformation measured in the tests. It is interesting to note that the curvature does not change much over the three assumed values of angular deformation, probably because the increased offset of the contraction force is counteracted by an increasing second moment of area as angular deformation increases.

This comparison, based on averaged values, nevertheless conceals significant local variations. The angular distortions at the ends of the plate are greater than in the middle (due to large displacement effects) and the angular distortion at the end of the weld pass tends to be greater than at the start. However, the comparison for these small plates is good – despite the fact that the underlying assumption of longitudinal plane strain throughout is fundamentally incorrect.

8.2.3 Computational application of algorithms

Analytical approaches are useful in terms of illustrating the major influences of the main parameters driving distortion. However, they are bound to be limited relative to computational methods, especially for complex distortions with significant out-of-plane and large displacement influences, which cannot be included in a simple analytical approach. This section uses the basic ideas of the Okerblom idea, but applied instead through finite-element models, as described originally in earlier work.[1]

The further step, beyond Okerblom's original formulation, to implement the MTS algorithm in a finite-element, elastic structural model, is relatively straightforward. The starting point is again a thermal analysis, usually computational, which is used to provide the profile of maximum temperatures $T_m(y)$.

The overall longitudinal contraction can then be generated in the finite-element model by applying an algorithm-derived, temperature–load profile corresponding to artificial temperature reductions $T_L(y)$ at points y across the width of the structural model. This is in no sense a real temperature change. The idea is that the artificial temperature loading and, more correctly, the associated contraction strain forms a convenient way to apply the required analytical strains to the finite-element model. The magnitude of $T_L(y)$ in zone A of Fig. 4.8 is set to develop contraction strain corresponding to $-\varepsilon_y$ and of course zero strain is applied in zone C, which remains elastic throughout. Introducing, for completeness, an ambient temperature T_a, the artificial temperature reductions applied to nodes in the various zones shown in Fig. 4.8 are given by

$$\text{Region A: } T_L - T_a = -\frac{\varepsilon_y}{\alpha} \qquad [8.6]$$

$$\text{Region B: } T_L - T_a = -\left((T_m - T_a) - \frac{\varepsilon_y}{\alpha}\right) \qquad [8.7]$$

$$\text{Region C: } T_L - T_a = 0 \qquad [8.8]$$

The extent of Regions A and B are defined by y coordinates as

$$T_m(y_A) = 2\frac{\varepsilon_y}{\alpha} + T_a \qquad [8.9a]$$

$$T_m(y_B) = \frac{\varepsilon_y}{\alpha} + T_a \qquad [8.9b]$$

It is possible to adopt a further simplification to reduce the element budget and the corresponding computing time, whereby the artificial loading temperatures in zone B, instead of being taken directly from the individual temperatures $T_m(y)$ provided by the thermal analysis, are based on a temperature profile that follows an inverse relationship with respect to transverse position (as indicated in equation 4.4) for the Okerblom analysis. This is given by

$$T_L - T_a = -\left[\left(\frac{A_i}{y} - T_a\right) - \frac{\varepsilon_y}{\alpha}\right] \qquad [8.10]$$

where A_i is a constant derived from the temperatures at the boundary positions of region B.

Implementation of the TCS algorithm is based, as in the analytical case, on the idea that elements in the fusion zone contract from the cut-off temperature T_s down to ambient temperature. The required artificial temperature drop in the transverse direction T_T is found therefore from the required transverse strain such that $\alpha(T_T - T_a) = \alpha(1 - k)(T_s - T_a)$ and therefore

$$T_T - T_a = (1 - k)(T_s - T_a) \qquad [8.11]$$

(This implies that the artificial temperature change is 980°C, if the ambient temperature is 20°C and T_s is 1000°C.) The value of k depends on the shape of the weld and the constraints on angular contraction and is found from Fig. 8.2 and Fig. 8.3.

The remaining feature of the finite-element implementation is that orthotropic coefficients of expansion are used to enable the artificial temperature loads at a given point to generate the required simulated thermo-mechanical strains in the longitudinal and transverse directions. In the longitudinal direction, the actual coefficient of expansion is used with equations 8.6 and 8.7. However, in the transverse and thickness directions, the coefficient is set to zero, apart from in the fusion zone where it is set, using equations 8.11 and 8.6, to a value given by

$$\alpha_T = -\left(\frac{T_T - T_a}{T_a - (\varepsilon_y/\alpha)}\right)\alpha \qquad [8.12]$$

This allows for the fact that the artificial temperature load has already been set in this zone to the value given by equation 8.6.

Application of these contraction strains to the elastic structure will then provide the distorted shape without the need to carry out a complex, non-linear thermo-elastic-plastic analysis. The feature that the transverse and longitudinal actions are solved simultaneously somewhat reflects the observed behaviour, such as shown in Fig. 6.9. It is important, nevertheless, to employ a three-dimensional (3D) large-displacement analysis in order to capture the movements in thin plates subject to compressive loads.[2] The finite-element results that follow were based on models using eight-node, ANSYS Solid 45 elements, together with a non-linear geometric analysis, taking into account large strains and rotations. The plate was assumed to be flat initially.

Figure 8.4 shows in block diagram form how these analyses combine and Fig. 8.5 and Fig. 8.6 provide detail on the deployment of equations 8.6–8.12 in the finite-element scheme. Note that the flow diagram procedures of Fig. 8.5 are optimised to be suitable for large structures. The temperature-linearising approach of equation 8.10 is used and a separate 3D mechanical model is constructed in the thermo-mechanical stage to allow a more efficient mesh with fewer elements. (Extrusion of the 2D thermal model to

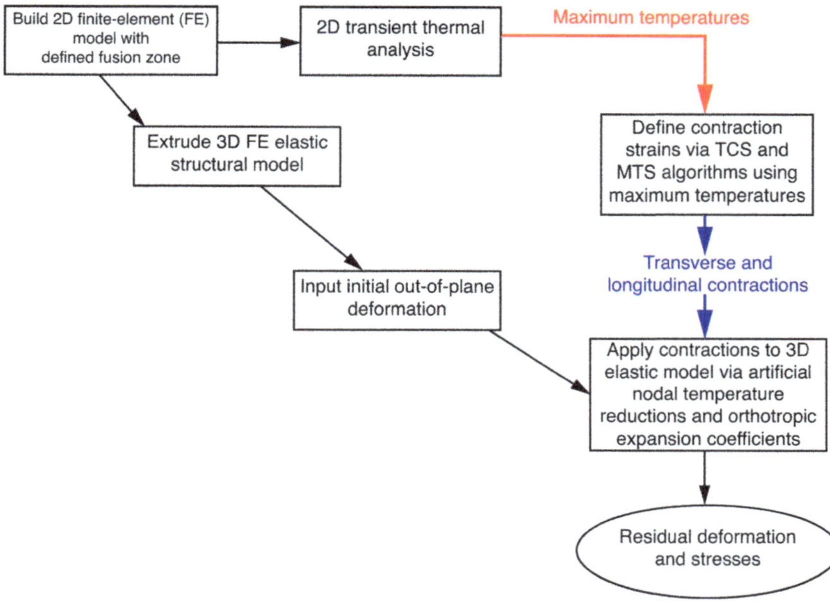

8.4 Computationally efficient simulation strategy.[3]

Table 8.2 Comparison of average theoretical and experimental distortion values

Result	RMS out-of-plane deformation (mm)	RMS angular distortion (degrees)	RMS longitudinal curvature (m^{-1})
CE analytical calculation	1.11–1.81 (y_{max})	1.02–1.66	−0.08 to −0.108
CE computational analysis	2.39	1.067	−0.049
Experimental value[1]	1.96	1.02	−0.06

3D, as shown in Fig. 8.4 is simpler to apply and efficient enough for small structures, but not for large ones.)

If the initial shape of the butt-weld assembly is measured before welding, this can be included in the analysis, using the procedures in Fig. 8.6. This will provide a more accurate description of the final shape. The results of this modelling approach are shown with the corresponding RMS values in Table 8.2.

However, Fig. 8.7 shows that the computational analysis, in particular the use of a 3D structural analysis, provides more detailed pictures of the distortion than can be encapsulated in a single average or RMS value. Figure 8.7 shows that the curvature is more severe along the weld centreline than at the edges and the angular deformation at the ends is slightly greater

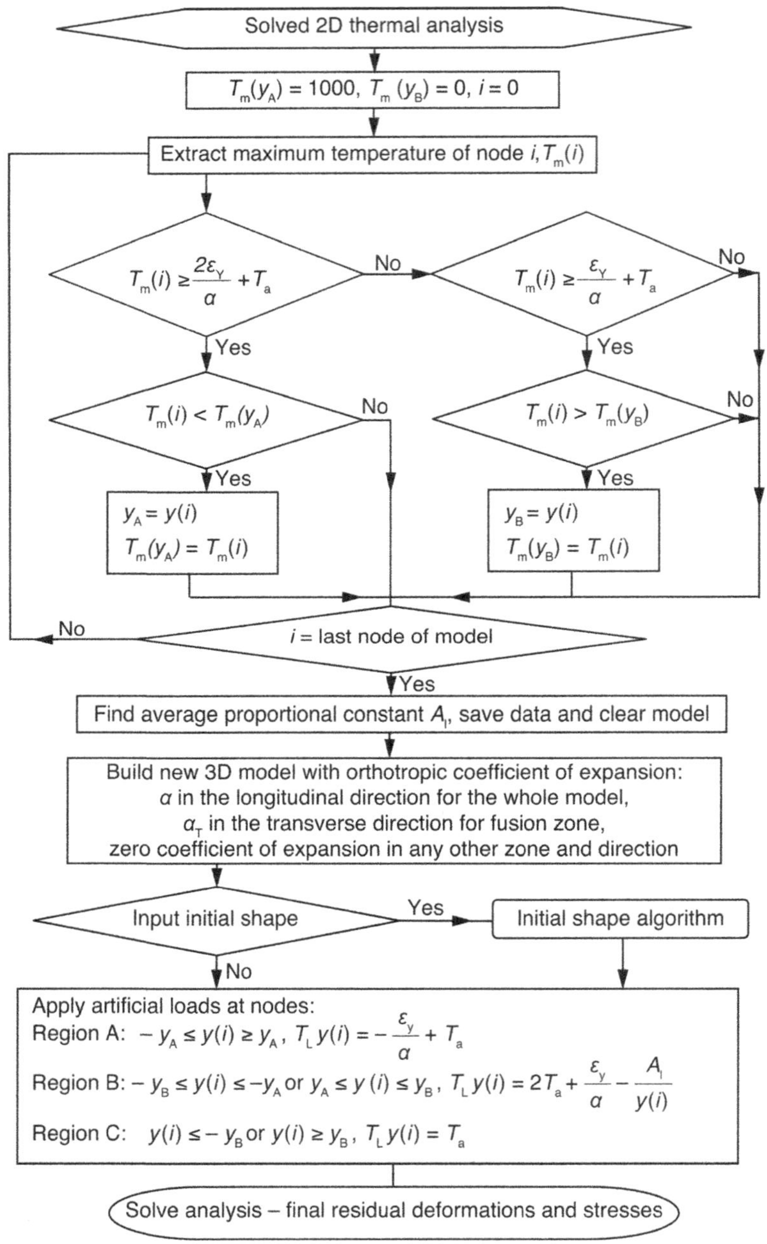

8.5 Finite-element thermo-elastic solution method using MTS and TCS algorithms.[1]

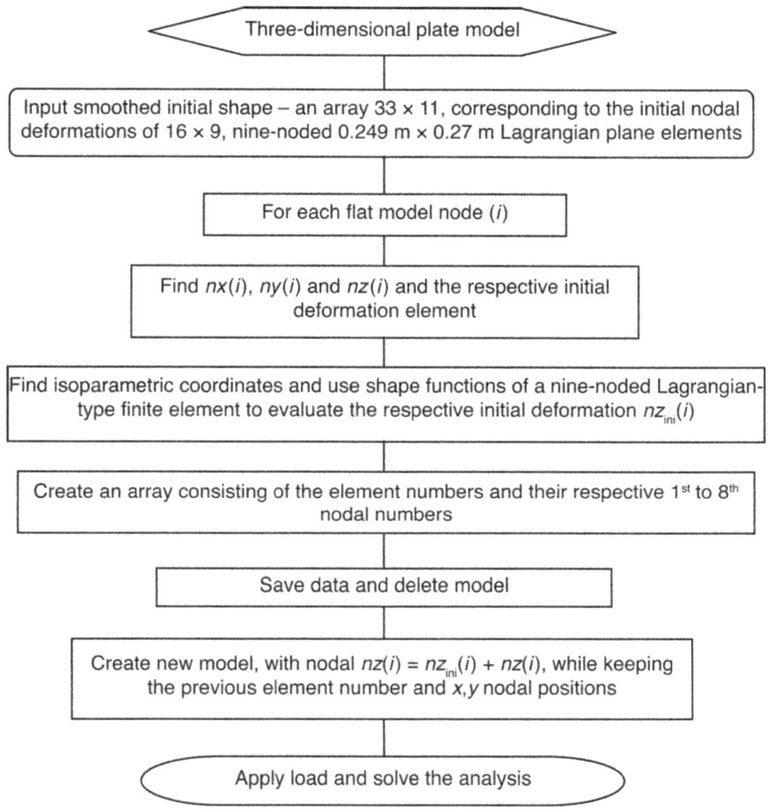

8.6 Initial shape input sub-algorithm.[1]

than at the middle of the plate. (No difference in angular deformation between the two ends is predicted, as the thermal strain algorithm is constant along the length.) As the algorithms generate uniform loading along the length of the plate and the mechanical analysis is elastic, it is clear that the detailed variations along the length arise from 3D structural interactions. Comparison of the predicted out-of-plane profile with the test result, shown in Fig. 6.15(c), shows a good qualitative and numerical match.

In such analyses it is useful to investigate the sensitivity of the analysis to different material property data, as these are often uncertain. The yield strength and coefficient of expansion are two potentially important parameters in this respect and a brief investigation was made into this aspect. Two substantially different but plausible yield strength values were trialled, namely 405 MN/m^2 and 360 MN/m^2. Two different values of expansivity were also trialled, these being deduced as alternatives from experimental data in Fig. 4.5 or Fig. 5.6, depending on the temperature

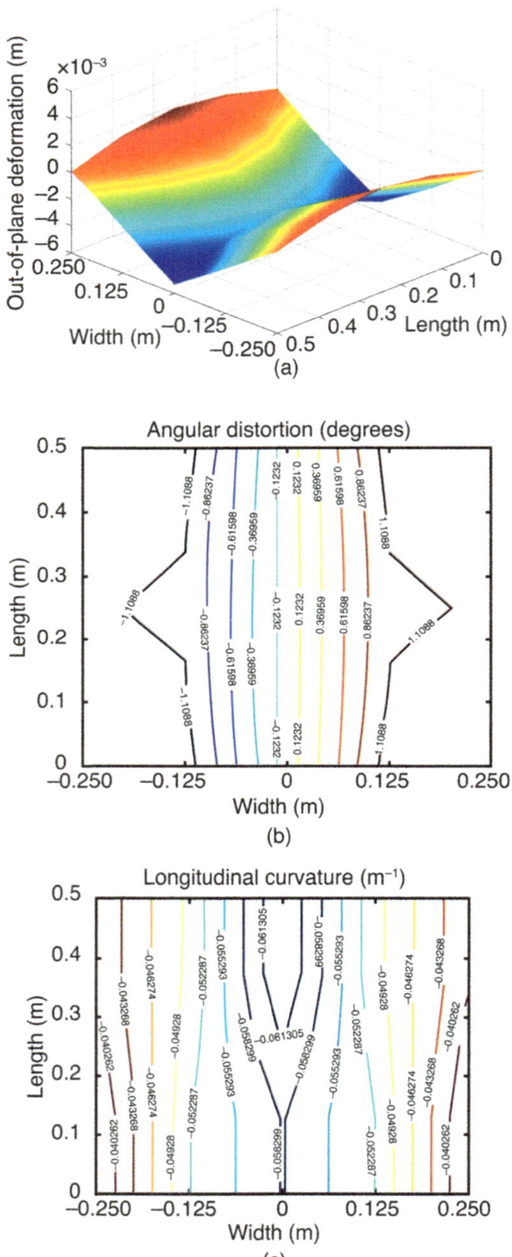

8.7 CE computational analysis for a 6 mm thick, 0.5 m square butt-welded plate: (a) out-of-plane deformation; (b) angular distortion; (c) longitudinal curvature.

Table 8.3 Influence of coefficient of thermal expansion and yield strength on CE results

Coefficient of thermal expansion (degree^{-1})	Yield strength (MN/m^2)	RMS out-of-plane deformation (mm)	RMS angular distortion (degrees)	RMS longitudinal curvature (m^{-1})
14×10^{-6}	360	2.430	1.079	−0.0467
14×10^{-6}	450	2.392	1.067	−0.0491
16×10^{-6}	360	2.845	1.261	−0.0538
16×10^{-6}	450	2.796	1.246	−0.0574

range deemed important, or the material (parent material or filler). The results are shown in Table 8.3.

The interesting effect here is that the yield strength has only a small influence, as suggested in section 4.5. This is due to the tendency towards an inverse relationship between the maximum temperature in the welding cycle at a point and its transverse distance from the weld line. (Note that these results for the small-scale model used the actual output temperatures rather than those generated by the linearisation process described earlier for larger plate models.) The coefficient of expansion has a strong influence, however, generating linear increases of distortion indices with an increase in expansivity, as might be deduced from the previous analytical determinations.

8.3 Hybrid stepwise solution methods

As discussed in the last three paragraphs of section 5.7, the chief deficiencies of the purely algorithmically based method are that there is no capability to describe transient behaviour in the thermo-mechanical stage and the treatment of transverse deformations is too rudimentary in many situations to reflect the 3D interactions that may well be significant. Fillet-welded joints in particular are problematic to treat using the approach described in section 8.2.1, as the transverse bending constraints generated by the stiffened base plates are difficult to define.

Some of the potential modelling deficiencies in the context of transient behaviour can be understood through the experimental observations of transient deformation. Figure 8.8 suggests that, in broad outline, almost half of the main angular contraction occurs during the passage of the heat source and that most of the longitudinal curvature develops during the post-weld cooling period. (This example shows the transient distortion of small-scale specimen 1 in the test series shown in Table 6.1.) This sequential pattern is reflected in the algorithmic treatment. There are minor diversions in the

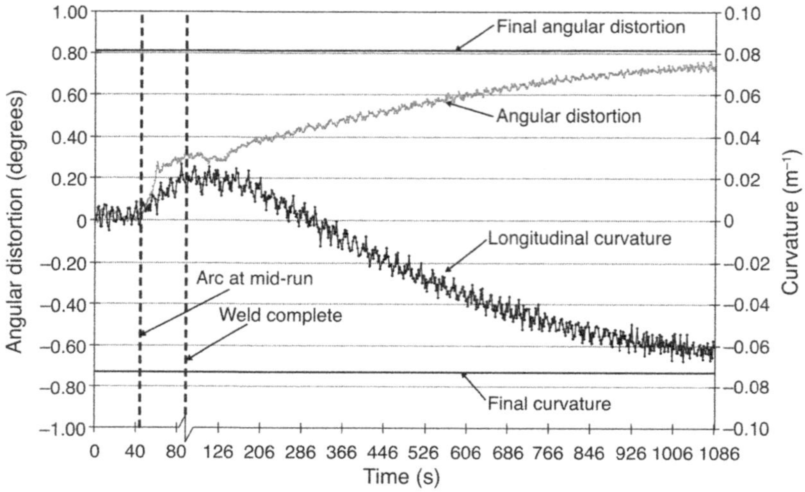

8.8 Transient deformations (small-scale programme) – test specimen 1.[4]

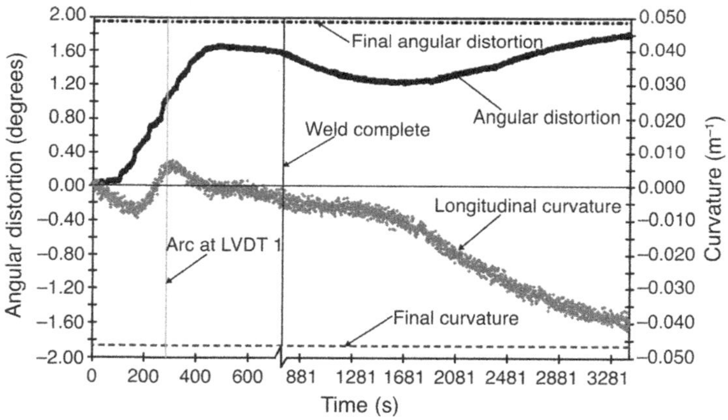

8.9 Transient deformations (full-scale programme) – test specimen 5.[3]

experimental picture, however, as a temporary sagging curvature develops as the weld passes the point of measurement and there are fluctuations in the angular distortion during the post-weld period.

Figure 8.9 shows a similar pattern measured on a full-scale test specimen. In this example, the computational algorithmic method yields a reasonable estimate of the averaged deformations, as shown in Fig. 8.10 and Table 8.4 (specimen 5). (In this case, the initial un-welded profile of the plate was used as the starting point of the computation.)

Figure 8.11 shows the transient deformation behaviour of full-scale specimen 2 (see Table 8.4) where plate deformations during welding and

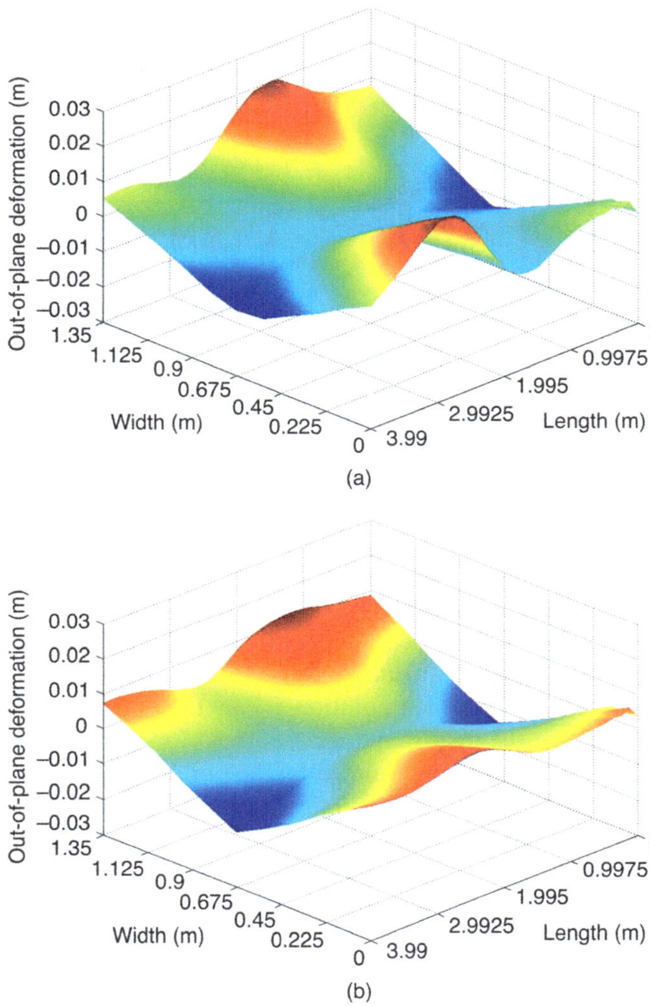

8.10 Experimental (a) and (b) computational deformations (full-scale) – test specimen 5.[3]

Table 8.4 Comparison of experimental and algorithmic computation results (large-scale)

	Specimen 5	Specimen 2
Experimental RMS out-of-plane (mm)	7.11	13.24
Computational algorithmic RMS out-of-plane (mm)	5.55	5.11
Experimental RMS angular deformation (degrees)	1.38	1.59
Computational algorithmic RMS angular deformation (degrees)	1.39	1.28
Experimental RMS longitudinal curvature (m^{-1})	−0.0238	+0.0334
Computational RMS longitudinal curvature (m^{-1})	−0.0101	−0.0099

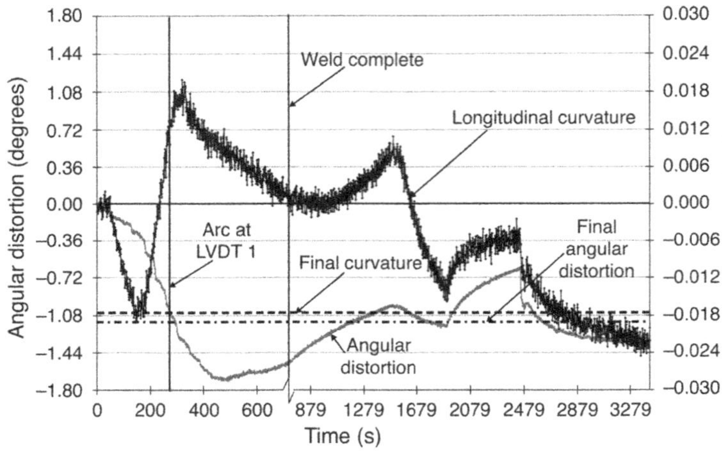

8.11 Transient deformations (full-scale programme) – test specimen 2.[1]

subsequent cooling caused the plate to lift off some supports, leading to curvature reversals. At the final stage, corresponding to the 2479 marker, the plate lifted off the final pair of supports and the cantilevered weight of unsupported plate triggered an unstable reversal of angular deformation and curvature. Algorithmic computation is, in this instance, much less effective in predicting the final shape, as confirmed for specimen 2 in Table 8.4. Such behaviour cannot be simulated by the purely algorithmic method, suggesting that large-scale structures with more complex boundary conditions need a more sophisticated treatment. Hybrid methods can provide this, without too much increase in computational budget.

An outline of the hybrid model approach is given in section 5.8.1 in the context of alternative treatments of butt-welding simulation. The TCS model used in the simple algorithmic analysis is replaced by a non-linear, stepwise, finite-element analysis of transverse strips. Each strip is assumed to be attached at all times to the rest of the plate, in front of and behind the strip, and this allows many improvements to the simulation, including treatment of weld filling, non-linear thermo-mechanical properties, structural restraint and any transient variations in supports or other mechanical boundary conditions.

The basic assumptions of the alternative sequential hybrid (SeqH) and simultaneous hybrid (SimH) solution methods are described in section 5.8.1. The details of the sequential model solution are shown in Fig. 8.12. Transverse contraction of the fusion zone is simulated in a stepwise manner by applying the maximum temperatures $T_m(y)$ to the series of strips in sequence, until the transverse solution is complete. A zero coefficient of expansion is assumed for the parent material in all directions but, in the

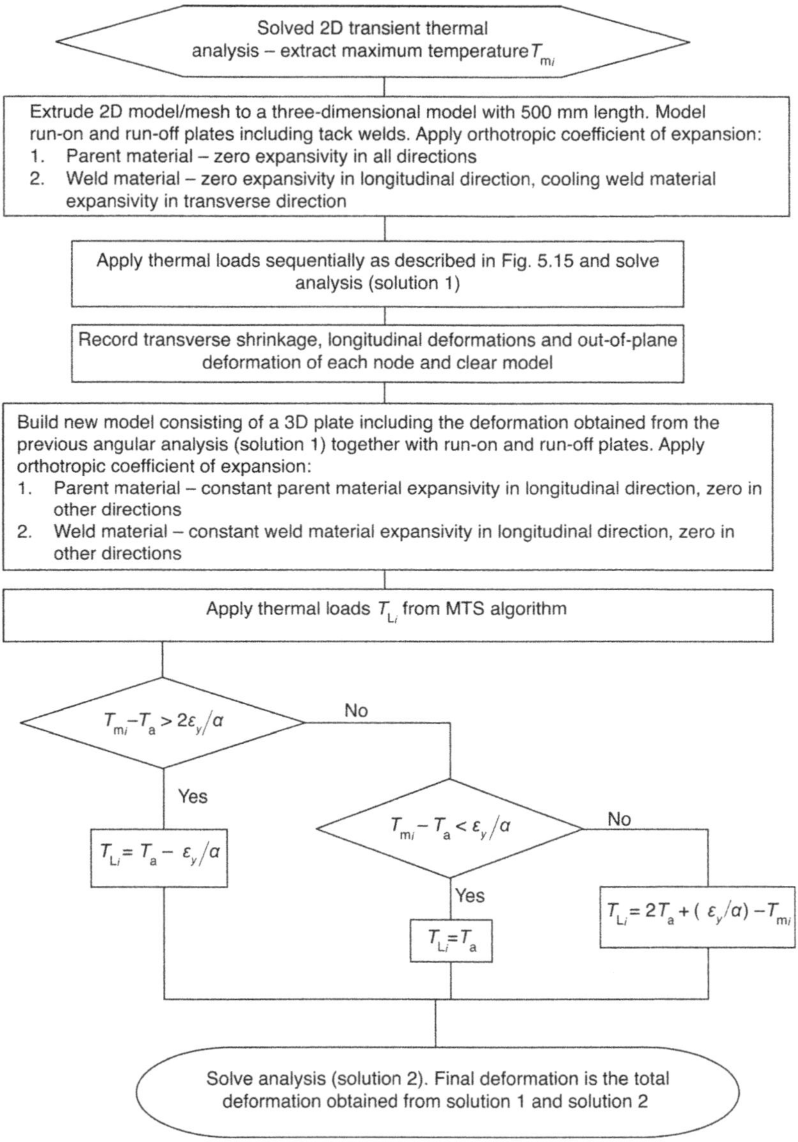

8.12 Sequential hybrid modelling flow diagram.[4]

fusion zone, the cooling contraction coefficient is assumed in the transverse direction and again, zero expansivity in the longitudinal direction. (Note, however, that if transverse stresses are developed this will generate longitudinal deformations through Poisson's ratio.)

Referring to Fig. 8.12, the profile of the angularly deformed plate is then used as the starting point for a computation that is identical to the MTS-

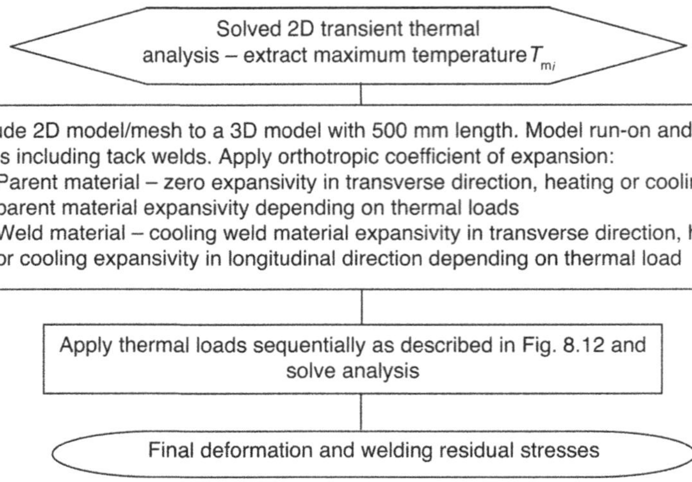

8.13 Simultaneous hybrid modelling flow diagram.

based solution method described in section 8.2.3. Again, the implication is that the angular deformation completes first and the longitudinal contraction and curvature are induced by the longitudinal forces acting on the V-shaped cross-section. This method will therefore be subject to similar problems related to transient shifts of boundary conditions if the actual behaviour is similar to that shown in Fig. 8.11.

The SimH model is much simpler, as shown in Fig. 8.13 although the computational time is much increased. It nevertheless still contains many simplifications relative to the full, transient, thermo-elastic-plastic approach. The treatment is partly 2D and the thermal loading is simplified, as it is based on maximum temperatures rather than the complete field of transient temperatures. Snapshots of the developing transient steps also give something like a realistic picture of the plate behaviour as the weld heat source passes.

Distortion predictions generated by the alternative hybrid methods and the CE algorithmic method are compared with the range of small-scale experimental results in Fig. 5.16. Note, however, that the computations assumed a perfectly flat plate as the starting point.

This comparison brings out several interesting strengths and weaknesses of the various models. Firstly, the single-section hybrid models cannot be expected to be accurate, as they assume a constant temperature profile along the length and no mechanical constraint from adjacent sections. Nevertheless, the single-section seqH and CE models both provide results near to the test outcomes. This is probably due to the fact that they both use the MTS algorithm, which imposes zero strain in the longitudinal direction (plane strain). Increasing the number of sections beyond three improves the performance of the SimH model, with some indications of a plateau for more than ten sections. The larger overall out-of-plane deformation shown by the most refined of the SimH models reflects larger RMS angular deformations, which are boosted by variations along the length, similar to those found in the case of full analyses (see distortion response in Fig. 5.10 (d)). Such variations cannot, of course, be shown by the CE model. The effect on SeqH results is less pronounced and in the opposite direction, as the difference between single and multiple applications of the MTS algorithm will be small.

In summary, there is probably little to choose between these methods in the case of simple butt welds in small-scale configuration. Potentially, they offer a better description of transient behaviour and situations where mechanical constraints and boundary conditions vary along the length of the workpiece. Sequential fabrication of fillet-welded, multiply-stiffened plates forms a particular example of such situations and later chapters will show the application of hybrid methods to such geometries.

8.4 Conclusion

A class of thermo-mechanical analysis, designated 'reduced solutions' has been described. The aim of these solutions is to reduce the computational budget by simplifying the thermal and mechanical models. In the thermal case, the real transient temperature field is replaced by a uniform profile along the length of the workpiece, where the temperatures across the width are the maximum temperatures reached as the heat source passes. In the thermo-mechanical case, algorithm-based, computationally efficient solution methods were described in section 8.2. The algorithms take the temperature profile as input and provide thermal strains as outputs, thus bypassing the more complex elasto-plastic analysis that would otherwise be applied. The thermal strains from the algorithms were then applied to an elastic analysis. Section 8.2.2 described an analytical approach to this procedure and 8.2.3 described a computational approach, where the strains were applied to a 3D elastic, finite-element model. (Section 8.2.1 also provided an extension to the previous description of the TCS algorithm, for situations where there is a mechanical bridge across the sections being joined, before the weld is made.)

Comparison of the 'analytical' and 'computational' approaches supports a conclusion that the finite-element computational approach is to be preferred as it enables large-displacement analysis and can also take account of more complex shapes and boundary conditions.

The main deficiencies of the algorithm-based methods are that modelling of the transverse thermo-mechanical behaviour is often over-simplified and there is no capability to describe transient behaviour as the weld progresses.

These deficiencies can be addressed to an extent through 'hybrid stepwise' thermo-mechanical solution methods, while still retaining a measure of 'reduction' relative to a full elasto-plastic solution method. In these methods, described in section 5.8.1 and section 8.3, the transient aspect is included through a longitudinal division into transverse sections and elasto-plastic analysis is used either for analysis of the transverse behaviour of each section or additionally for analysis of the longitudinal behaviour. Computational time increases for these hybrid analyses, but there is nevertheless a saving over full elasto-plastic analyses because the transverse and longitudinal analyses are uncoupled and a simpler thermal analysis is applied. To an extent, the uncoupled hybrid approach reflects the experimental behaviour of welding distortion as observed.

8.5 References

1. Camilleri, D., Comlekci, T. and Gray, T.G.F., 2005, Computational prediction of out-of-plane welding distortion and experimental investigation, *J. Strain Analysis*, 40, 2, pp 161–176.
2. Berglund, D. and Runnemalm, H., 2002, Comparison of deformation pattern and residual stresses in finite-element models of TIG-welded stainless steel plate, *Proc. 6th Int. Conf. on Trends in Welding Research, Pine Mountain, Georgia, USA*.
3. Camilleri, D., Mollicone, P. and Gray, T.G.F., 2007, Computational methods and experimental validation of welding distortion models, *Proc. I.Mech.E, J. Materials: Design and Applications*, 221, pp 235–249.
4. Camilleri, D. and Gray, T.G.F., 2005, Computationally efficient welding distortion simulation techniques, *Model. Simul. Mater. Sci. Eng.*, 13, pp 1365–1382.

9
Finite-element thermo-mechanical techniques for welding distortion prediction

DOI: 10.1533/9780857099327.177

Abstract: Implementation strategies for thermo-mechanical, elasto-plastic structural finite-element models are given. Meshing strategies are described, together with modelling approaches to incorporate non-linear, temperature-dependent dilatation and strength properties. Case studies on weld tacking assembly are used to illustrate various ways of implementing butt-joint clamping, tacking, weld deposition and tack re-melting in an elasto-plastic model. A further case study on double-sided fillet welding of a stiffener is used to contrast analysis and results between a computationally efficient algorithm-based model and a full elasto-plastic model. Distortion results are compared along with residual stress patterns as computed by the two approaches.

Key words: elasto-plastic structural, expansivity models, meshing, Eulerian/Lagrangian frames of reference, tacking, activation/deactivation, fillet-welded stiffeners.

9.1 Introduction

The simplifying assumptions underlying the reduced solution, thermo-mechanical models described in the previous chapter are relatively sweeping (see sections 4.4–4.6, 5.7, 5.8 and Chapter 8). Transient aspects of the thermal description are usually ignored and the longitudinal and transverse actions are uncoupled. These assumptions depend in turn on the occurrence of highly oriented thermal patterns whereby, at least for steel welded at high speed, the transverse thermal gradient is steep and the longitudinal cooling temperature gradient is shallow. In several of the reduced solution models, the high-temperature mechanical properties of the fusion zone are largely ignored, it being assumed that most of the strain mismatches that give rise to distortion and residual stress occur in a relatively narrow temperature band near ambient temperature, where variations in mechanical properties are small. The hybrid models are an exception to that statement, as material-sensitive, elasto-plastic finite-element analyses are applied partially, either to the angular deformation in the sequential hybrid (SeqH) case or to angular and longitudinal deformations in the simultaneous hybrid (SimH) case.

Given a level of uncertainty as to the validity of these simplifications relative to the solution outcomes, it is necessary to carry out less simplified analyses from time to time, in order to check that the reduced models indeed adequately describe what happens when components are welded in practice and, in certain cases, to evaluate the effects of different thermal and mechanical properties, transient mechanical interactions and welding sequences. This form of analysis was described in section 5.2 (Fig. 5.3) as a 'transient elasto-plastic structural analysis' and the present chapter deals with the formulation and application of such analyses. Of course, this type of model is relatively complex and it is usually necessary to reduce the computing power requirement by limiting the size and complexity of the modelled structure. Note also that we are not here dealing so much with the thermal part of the problem (see Fig. 5.2 and Fig. 5.3), which can be effectively uncoupled from the thermo-mechanical part.

Various topics relevant to elasto-plastic, structural finite-element models (and not covered so far) are treated in section 9.2. This includes meshing schemes and some further implementation details concerning non-linear material thermo-mechanical properties. This is followed by two case studies that exemplify the application of elasto-plastic models. The first (section 9.3) is based on two investigations of weld-tacking assembly and includes the treatment of heat input (see section 7.3), application of active/inactive elements and simulation of weld deposition (see section 7.4). The second case study (section 9.4) shows elasto-plastic modelling (computationally intensive model (CIM)) of a stiffened plate where the stiffener is attached by two fillet welds. The same case is also treated through a computationally efficient model (CEM), where experimental results are available, and the two approaches are contrasted.

9.2 Formulation of thermo-mechanical finite-element model

9.2.1 Geometric aspects

As the thermal and structural parts of the analysis are de-coupled, different meshing schemes and two/three dimensional (2D/3D) choices could be adopted for the two parts of the analysis. However, the thermal gradients and the mechanical strain gradients are likely to be very similar and there are therefore advantages in choosing the same meshing schemes for both parts of the analysis. Likewise, the small computational saving through using a 2D analysis for the thermal part is likely to be outweighed by the complexity and extra work involved in directing the output data from the thermal analysis to matching locations in the necessary 3D structural analysis.

The analyses developed by the authors were based on meshing schemes, which are uniform along the length axis of the structure being welded in the sense that there are no variations near the ends of the workpiece, unless run on/off tabs are being modelled (see Fig. 5.12). There is, of course, a sharp thermal gradient ahead of the heat source along the weld centreline, but it is reasonable to assume that the mechanical effects related to this are obliterated by subsequent melting. An alternative moving-mesh treatment that refines elements ahead of the heat source will be described later, but a uniform cross-sectional pattern of elements greatly simplifies the transient aspect of the modelling process. Various trials showed no suggestion of through-thickness thermal gradient by 10 mm transversely from the weld and so the number of through-thickness elements may be reduced away from the edge of the fusion zone. Large displacement induces out-of-plane bending and, if accurate reading of residual bending stress is required, there could be an argument to maintain more through-thickness elements at greater transverse distances.

Figure 9.1 shows one half of a suitable meshing scheme for a one metre square workpiece, where the effects of alternative lower and upper positions for preliminary tack welds were explored through a full thermo-elastic-plastic analysis.[1] The fusion zones shown in this case relate to a joint where there was no edge preparation but a 4 mm initial gap between the 4 mm thick plates was specified. Note the grading of element sizes – fine in the near-weld areas where the thermal and mechanical gradients are steep, and coarser in the outer regions of the plates. This model is based on eight-noded, Lagrangian brick elements, comprising 25 240 elements and 34 842 nodes. The weld zone is divided into 200 finite volumes, each of 5 mm length.

A corresponding meshing scheme for a fillet-welded structure was shown previously in Fig. 7.9 and Fig. 7.11. This model made use of 22 600 brick elements, comprising 27 915 nodes. ANSYS thermal SOLID70 elements were specified in the thermal case and SOLID45 elements in the structural analysis.

Other meshing approaches have been formulated based on what is called the Eulerian frame of reference, which moves with the heat source (like the Rosenthal/Rykalin quasi-static analyses). (The *fixed* frame of reference is described as Lagrangian.) These will accommodate element size variation in the longitudinal direction relative to the arc. This can be advantageous in

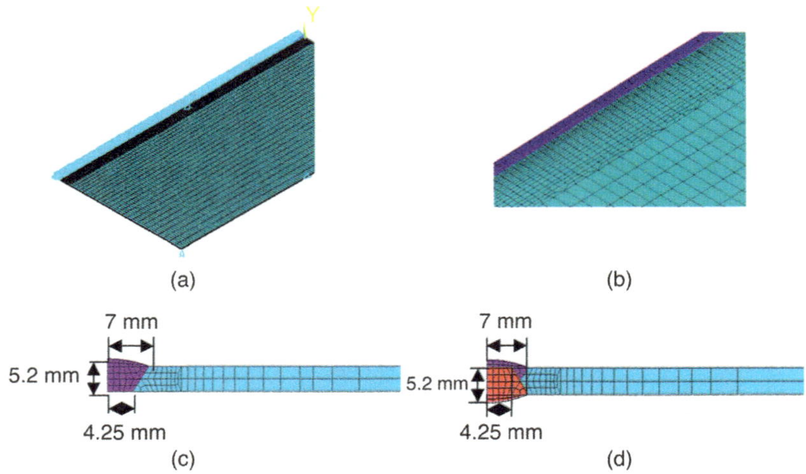

9.1 Meshing scheme for tacked and butt-welded model – 4 mm thick, 1000 mm square. (a) Full model and boundary conditions. (b) Weld zone mesh. (c) Tack and butt weld at same side. (d) Tack and butt weld on opposite sides.

terms of describing phenomena near the heat source, but the steady-state assumption precludes modelling of starting and finishing points, together with any other transient variations along the length of the weld (see Goldak and Akhlagi[2] for further discussion). Thus the recommendation for thermo-mechanical modelling is to use the same transversely graded mesh and the same 3D configuration for thermal and structural analyses alike.

It is also important to model the geometry of the fusion zone as accurately as possible, although this may require some iteration in the thermal analysis and cross-checking against observations of the actual fusion zone shape and size. Modelling methods for the heat input were described in section 7.3 and for the simulation of weld deposition in section 7.4. The 'active/inactive' identification of elements around the fusion zone also defines the thermo-mechanical condition of these elements.

9.2.2 Material thermo-mechanical properties

As noted earlier, the expansivity property relating temperature and thermal strain, particularly during cooling, is an important driving factor in the development of distortion and residual stress. An advantage of full thermo-mechanical analyses is that this property can be included in considerable detail, corresponding to local temperatures and in terms of describing non-linear aspects of the behaviour (see section 5.4). The only limitation is the extent of data available and their relevance to the actual conditions in the weld. In Chapter 5, Fig. 5.13 shows typical dilatation behaviour, as obtained

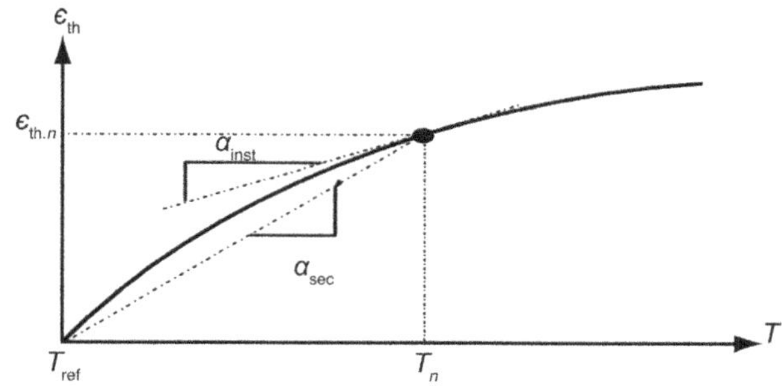

9.2 Schematic thermal dilatation curve.[3]

through dilatometry, Fig. 5.6 shows idealisations of the cooling behaviour and Fig. 5.14 shows idealisations of both heating and cooling, in a form that can be included in a finite-element analysis. The fact that heating and cooling are different causes an immediate complication in that the analysis has to include some means to determine whether a given element is increasing or decreasing in temperature. As indicated earlier, various software packages treat such inputs in different ways and it is important to ensure that there is consistency between the definition of expansivity and the data used. Figure 9.2 shows three common alternative definitions.[3]

In this representation, ε_{th} is the thermal strain developed at an element evaluation temperature T (relative to the thermal strain at T_{ref}) and $\varepsilon_{\text{th}.n}$ is the thermal strain developed by a particular element at temperature T_n. If the data available are finely graded, thermal strains may be input directly from a temperature–strain table for cooling or heating as appropriate. The second possibility is that 'instantaneous' expansivity coefficients α_{inst} are input, appropriate to each temperature of a range of element temperatures. (This was the method used in this work, as indicated in Fig. 5.14.) The third possibility is that 'secant' expansivity coefficients are input, where the value is the mean expansivity between the element temperature and the reference temperature, as shown. The relationships between these alternatives are given by

$$\alpha_{\text{sec}}(T_n) = \frac{\varepsilon_{\text{th}}(T)}{T_n - T_{\text{ref}}} = \frac{\int_{T_{\text{ref}}}^{T_n} \alpha_{\text{inst}}\, \mathrm{d}T}{T_n - T_{\text{ref}}} \qquad [9.1]$$

In some cases, thermal dilatation may be different in different x, y, z directions (or assumed to be so for analytical reasons).

Other temperature-dependent mechanical properties, such as yield strength and elastic and plastic moduli, do not appear to have such a strong influence on the distortion outcome. Yield strength in particular was shown earlier in terms of the algorithmic approach (Table 8.3) to have a surprisingly small influence. However, a full analysis provides the opportunity to include as much property information as is reliably available and that opportunity may usefully be taken.

Figure 5.6 showed how the real non-linear expansivity properties of the parent material and weld metal can be characterised in terms of temperature-dependent linear stages. The corresponding non-linear strength properties were shown in Fig. 5.7. These strength properties can then be readily incorporated in a bi-linear, kinematic hardening model as shown in Fig. 4.6(b).

9.3 Case study: influence of tacking procedures on butt-weld distortion

Weld tacking is often viewed as a trivial operation, but it certainly causes distortion. Significantly high thermal stresses can be caused by tacking, resulting in misalignment of assemblies, alteration of intended gaps between component edges and initial out-of-plane distortion. This topic will be treated in more detail in Chapter 10, but it is useful to discuss key aspects of the necessary thermo-mechanical analysis in the context of the present chapter.

The discontinuous nature of tacking and the scope for different lengths and sequences of tack-laying make a full transient thermo-elastic-plastic analysis very desirable and this case study will outline the measures required to develop such an analysis. Two approaches to the problem are described. The first[4, 5] was concerned only with the tacking operation; there was a corresponding experimental study. The second treatment was entirely computational[1] and was based on the geometric model shown in Fig. 9.1, which includes the tacking assembly stage and the subsequent continuous butt weld. This latter study explored different tacking configurations parametrically and the effects of these on the distortion outcomes of subsequent continuous butt welds, described as 'seam welds'.

The first problem in this case is to determine the thermal input profiles for the tack welds and the seam welds. As always, this has to satisfy two main requirements, firstly that realistic energy inputs are specified, corresponding to actual weld parameters and secondly that the resulting profiles of the fusion zones match the real case. Preliminary thermocouple studies were carried out to cover both the tacking operations and the final seam welds. In the industrial practice studied, a semi-automatic MIG system was used to tack weld and that allows some variation in travel speed through manual application. Hence there was corresponding variation in the thermal profiles along the tacking pass. It was noted that a reduced travel speed was used by the welder at the start and finish of long tacks (40 mm length), thus increasing the heat input in these regions. A reasonable match with the thermocouple determinations was therefore established by adopting 75% weld efficiency at each end of the simulated tack and 48% efficiency in the central 30 mm length. Welding of the smaller 10 mm tacks was adequately simulated using 75% efficiency throughout. Fusion zone sizes were used to gauge efficiency in the seam welding operation, suggesting a fairly low efficiency of around 48% also. The reason for low efficiency may have been related to the large gap between the plates (relative to previous examples where the same process was used in a grooved weld preparation) or to some other feature of the seam welding jig and process conditions.

It is particularly important in this kind of study to take proper account of the effects of discrete weld deposition and fusion zone formation and this is best simulated through the methods described in section 7.3 for heat input and section 7.4 for weld deposition, where elements are present in the model from the beginning, but are switched on and off as necessary. The procedure whereby elements are 'killed' sequentially also simulates re-fusion of tack elements on arrival of the arc during seam welding. The restraining effects of the tack welds are thereby released progressively when re-melting occurs. Such elements therefore underwent two large-strain thermal cycles and convergence problems were encountered as a result. This was controlled by reducing the cut-off or reference temperature from the usual 1000°C to 600°C, so that thermal strains above this temperature were clipped. Various studies suggest that the analysis is not highly sensitive to cut-off temperature. In any case, the boundary condition used in the first study implies some angular restraint, as will be discussed later, and it would therefore be reasonable to assume that this implies a resistance to angular contraction at higher temperatures where the strength of the contracting weld is low. Clamping elements were used in the second study at the tacking stage and these could reasonably be assumed to have a similar restraining effect.

Boundary conditions are also of critical importance in obtaining a realistic model solution. In the first case study of the tacking problem, a transient thermo-elastic-plastic analysis was applied to determine the effects of tacking on plate assemblies, where the results of practical tests on 900 mm square × 4 mm thick plate assemblies were available for comparison. The tacking scheme in this case comprised three 70 mm long tacks, equi-spaced along the weld line and two 35 mm tacks at the ends, so that around 30% of the total length was tacked. A close-butt weld configuration was used (see Fig. 9.3) and the tack-weld fusion zones are seen to be very much on the surface of the plates, rather than fully penetrating, as in the assembly of Fig. 9.1 where a large gap was specified. In the close-butt case, the finite-element model was based on a symmetrical half of the assembly, thus providing transverse constraint at all points along the weld line. However, this boundary condition also generates *angular* constraint on transverse

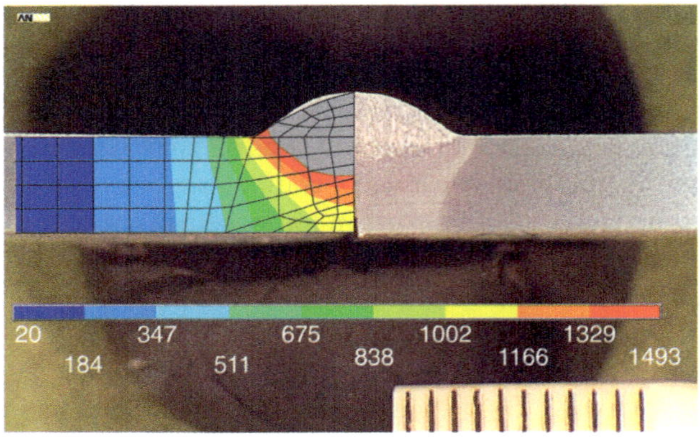

9.3 Simulated and actual fusion zones for tack weld on close-butt geometry (courtesy P. Mollicone).

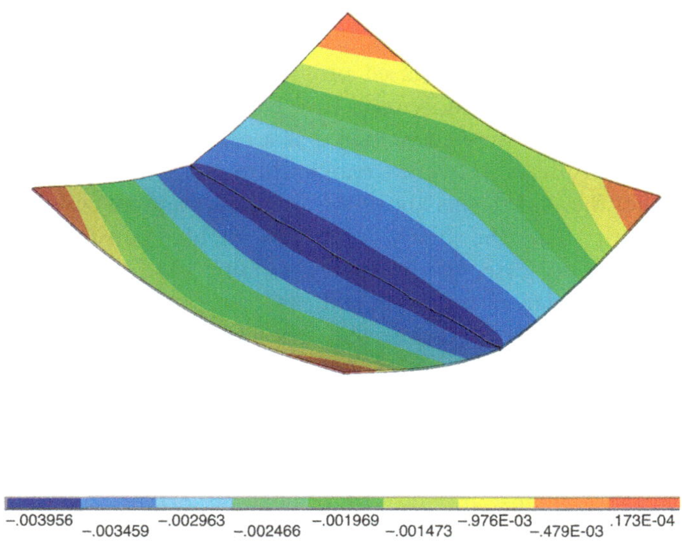

9.4 Computed out-of-plane distortion (m) due to tacking (31% of total seam length) (courtesy P. Mollicone).

deformation along the centreline, which is not apparent in the assemblies as described, where the un-penetrated plate edges should be free to separate. This point will be considered further in Chapter 10, where a comparison with full-scale tests is considered.

Computed out-of-plane deformations caused by the tacking scheme in the first case study are shown in Fig. 9.4.[4] Angular deformation is strong despite the symmetry constraint, as the contracting weld is largely placed on the top surface. Bias of the contracting force to the top surface has also resulted in a sagging curvature, rather than the more common saddle shape seen in continuously welded assemblies of this type.

A substantially different approach to boundary conditions was taken for the second parametric case study of tacking and seam welding, where an initial gap between the plates was specified. Some arrangement is necessary to hold the plates together for the tacking operation in the real case and also in the model. In practice, some form of clamp will be used and this was simulated in the model, as indicated earlier, by several 10 mm long elements, at intervals in the longitudinal gap between the plates, but avoiding the tack weld positions. Zero thermal coefficient of expansion was assigned to these elements and they were deactivated after completion of the tack welds, so that no clamping effect would be present during the final seam weld. Various configurations of simulated clamp spacing were trialled, depending on the associated pattern of tack welds.

Gravitational force was also included in the tacking simulation given that, if supported only at the corners, a 1 m square, 4 mm thick plate assembly will sag appreciably. As the tacks would normally be applied using down-hand welding and the plate would therefore be inverted, *negative* gravitational force was applied to the model for tacking operations on the bottom face of the model. Two versions of the simulated seam welding operation were computed – the first without gravitational force, reflecting a condition where the assembly might be properly supported, and a second version where the tacked plate assembly was supported only at the four corners.

The sequence of activation and deactivation of the various elements (gap element clamps, tack elements and seam welding elements) and with the gravitational boundary condition are critical to this simulation. Figure 9.5 identifies the various elements in the cross-section and the sequence of activation and deactivation is shown in Fig. 9.6. Note that the clamps provide support for the plate during the tacking operation but, as they are removed before seam welding, the three tacks in this case are the only source of support against sagging during seam welding.

The results of this parametric survey will be discussed more fully in Chapter 10 but, for the time being, Fig. 9.7 shows the outcome of applying three 40 mm long tacks on the top surface of the model. In contrast to the earlier study, the deformation is predominantly angular rather than longitudinally bowed. The reason for this is that the tack welds in this second study are more centrally placed within the plate thickness and the overall length of tacking (12% of total seam length) was much less than in the earlier case where the total tack length was 30%. Even at 30%, the overall longitudinal force is much smaller than would be the case for a complete longitudinal weld, where pronounced hogging curvature would normally be generated.

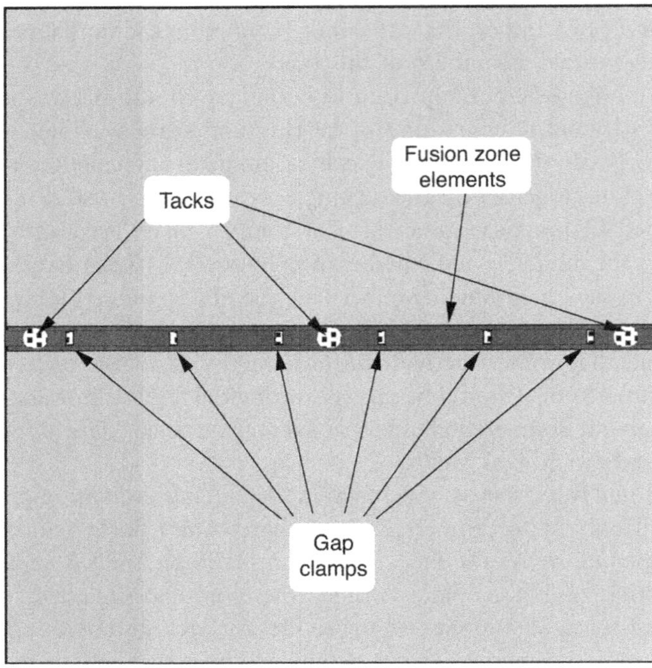

9.5 Layout of active/non-active elements in tacking/welding simulation.

	Tacks	Clamps	Seam weld fusion zone	
Start of tacking	Deactivated	Activated	Deactivated	Tacking ↓
Tack weld formation	Activated in turn during tacking	Activated	Deactivated	
Start of seam welding	Remain activated	Deactivated	Deactivated	Seam welding ↓
Seam welding	Deactivated in turn on heating to reference temperature	Deactivated	Activated on heating to reference temperature	

9.6 Sequence of activation and deactivation of tacks, clamps and fusion zone elements.

Transient thermo-elastic-plastic analyses can thus be developed to provide considerable insight into the effects of different assembly and tacking procedures, covering a programme of configurations that would not be practicable to complete using live tests.

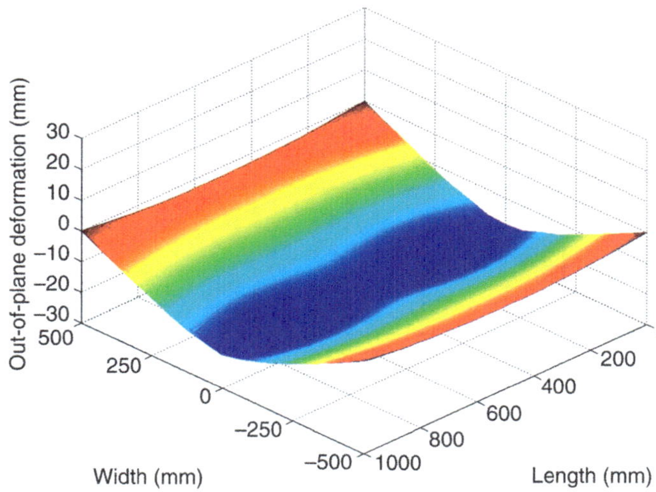

9.7 Computed distortion due to three top-side 40 mm tacks (12% total length).

9.4 Case study: fillet-welded stiffened plate

The thermal part of this case study was treated in section 7.8. The conclusion reached there was that a 2D analysis provides reasonable estimates of maximum temperatures at given locations, but the 3D analysis gives improved estimates of both the temperatures and the times at which given temperatures were reached. Such accuracy is important when the temperature–time outputs provide the input stage of a transient thermo-mechanical analysis. Computationally efficient and full transient thermo-mechanical analyses will be compared in this section, starting from the same thermal analysis output.

9.4.1 Computationally efficient model (CEM)

Transverse and angular contraction caused by each of the fillet welds can be determined in this case through the arguments presented in section 8.2.1, where each fillet-weld fusion zone determines the volume of contracting material and there is a restraint on angular contraction caused by the un-penetrated and un-fused parts of the base plate and stiffener. The weld is treated as parabolic in shape, according to the solution shown in Fig. 8.3. Figure 9.8 illustrates the transverse deformations expected during the first weld pass.

The transverse contraction strain is characterised by partitioning the total fusion zone area A_T into three separate areas, as shown in the top sketch of Fig. 9.8, composed of the fillet contraction in A_{wd}, local bending of the

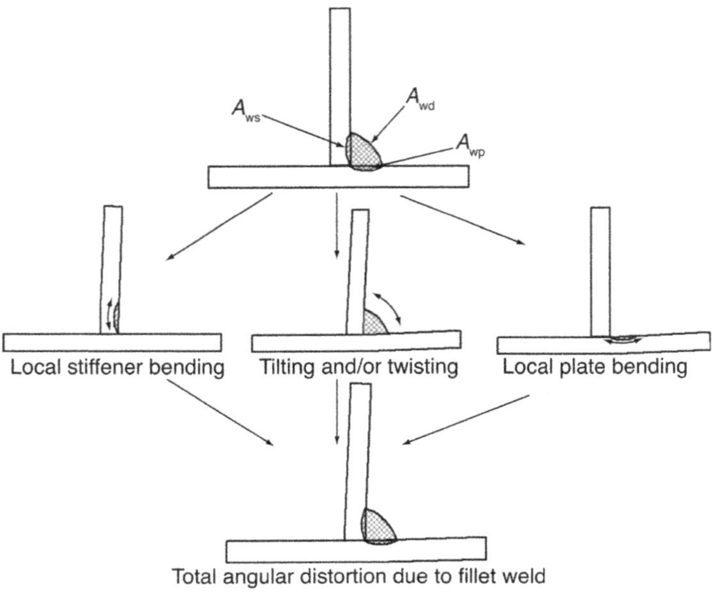

9.8 Transverse deformations caused by fillet welding of a stiffener.

stiffener in A_{ws} and local bending of the plate in A_{wp}. If the stiffener is free to move, contraction of the fillet weld itself may be assumed to be unrestrained (as for butt welds), but angular contractions of the approximately parabolic fused areas on the stiffener and plate are assumed to be restrained by the bending resistance of the underlying material thicknesses, as discussed earlier in section 8.2.1. Various leading dimensions in the transverse contraction strain (TCS) algorithm for fillet welds are identified as shown in Fig. 9.9. The proportions of the 'free' transverse thermal strain generated by cooling from the cut-off temperature T_s in the constrained areas on the stiffener and plate are defined by the parameter k, where the transverse strain is given by $\varepsilon_t = -\alpha(1-k)T_s$.

The parameter k depends on the dimensions in Fig. 9.9 through the relation

$$\frac{p}{t} = \frac{5(1-k^{1.5})}{6(1-k^{2.5})} \pm \left[\left(\frac{5(1-k^{1.5)})}{6(1-k^{2.5})}\right)^2 - \left(\frac{5[(1-k)-(\varepsilon_y/\alpha T_s)]}{8(1-k^{2.5})}\right)\right]^{0.5} \qquad [9.2]$$

As explained earlier, equation 9.2 cannot be solved explicitly for k, but nomograms were provided previously as Fig. 8.2 and Fig. 8.3, where values of k were given for particular values of $\varepsilon_y/\alpha T_s$ and various weld penetration profiles (triangular, parabolic and rectangular). The curve for a parabolic penetration shape and a non-dimensional strain value $\varepsilon_y/\alpha T_s = 0.141$, which corresponds to the material properties in the present case, is repeated here for

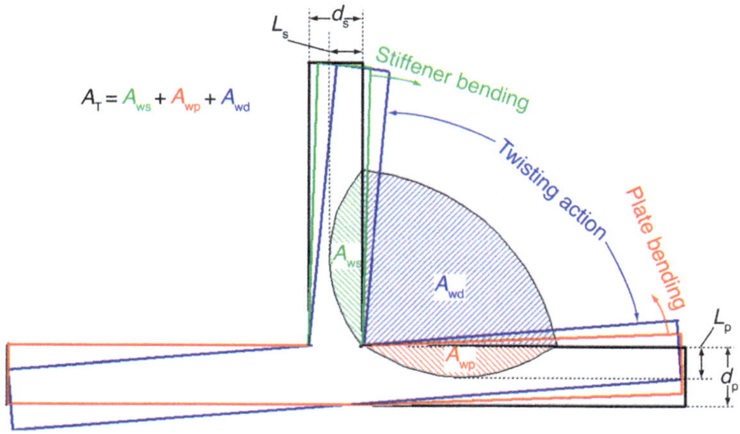

9.9 Transverse thermal contraction of a fillet-welded stiffener.

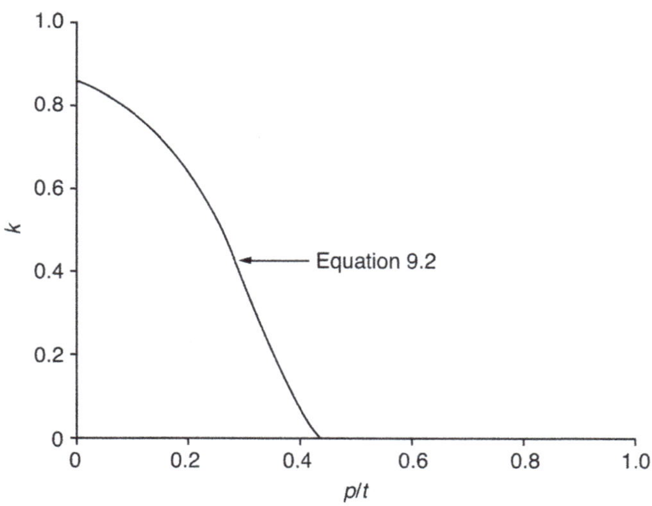

9.10 Relationship between *k* and *p*/*t* for a parabolic fusion zone shape (equation 9.2).

convenience in Fig. 9.10. When penetration into the parent material is small ($p/t \rightarrow 0$) the k value approaches unity and the transverse strain on the surface is very small, whereas penetration to nearly half the plate thickness (unlikely circumstance) would lead to the full transverse angular contraction taking place.

The transverse thermal load to be inserted in the elastic finite-element model is given for all three actions as

$$T_{\mathrm{T}} = T_{\mathrm{ambient}} - T_{\mathrm{s}}\left(\frac{A_{\mathrm{wd}}}{A_{\mathrm{T}}} + \frac{A_{\mathrm{ws}}}{A_{\mathrm{T}}}(1 - k_{\mathrm{s}}) + \frac{A_{\mathrm{wp}}}{A_{\mathrm{T}}}(1 - k_{\mathrm{p}})\right) \qquad [9.3]$$

Restraint of the stiffener against 'tilting' is also something of an open question. If there is no connection between the stiffener edge and the plate before welding (i.e. no tacks or clamps) then the stiffener will tilt freely relative to the plate in the same way as assumed for one side of a butt weld relative to the other (see section 4.6). This will be simulated if the nodes between the two components are unconnected in the finite-element model. If the stiffener is assumed to be more firmly connected to the plate and the nodes are therefore connected, the forces induced by the contraction algorithm (equation 9.3) will tend to bend the underlying plate to a greater extent.

The extent to which contraction of the second fillet weld is restrained by the first also has to be considered. The stiffener is clearly not going to tilt freely, but may bend in the reverse direction to that during the first pass, due to contraction of the A_{ws} and A_{wd} regions. This can also be simulated via equations 9.2 and 9.3.

In this particular case, the test specimens were short and the stiffener was held at each end of the plate by rather robust clamps (run-on and run-off angle tabs welded to the stiffener–plate assembly) and so it was assumed that the stiffener was well connected to the plate and that the stiffener–plate adjacent nodes should therefore be connected.

In closely spaced welds, where there is intermediate cooling to ambient temperature between the two weld passes, the individual MTS (mismatched thermal strain) strain fields will overlap, so that the total MTS due to both welds will correspond merely to a superposition of the two thermal strain patterns, as shown in Fig. 9.11.

An interesting feature of the computationally efficient approach is that the final step, where the MTS-derived loading is applied to an elastic finite-element model, can be carried out in a single load step instead of being applied separately for each weld, as it was here. Due to the elastic formulation of the MTS model, the result is the same as for individual application for each weld. Some consequences of this will be noted later in the comparison of results.

9.4.2 Computationally intensive model (CIM)

In this case, thermal transients are applied as body load forces, in a transient fashion, over 60 ramped load steps during the weld pass and over a further 111 load steps during cooling to ambient temperature. The two weld passes are applied sequentially in opposite directions, as in the actual tests. Weld metal deposition is simulated, as in the thermal analysis, through the element 'birth-and-death' technique. The elements associated with the fusion zones are deactivated before the analysis begins and are then re-activated upon arrival of the welding arc in each pass. Temperature-

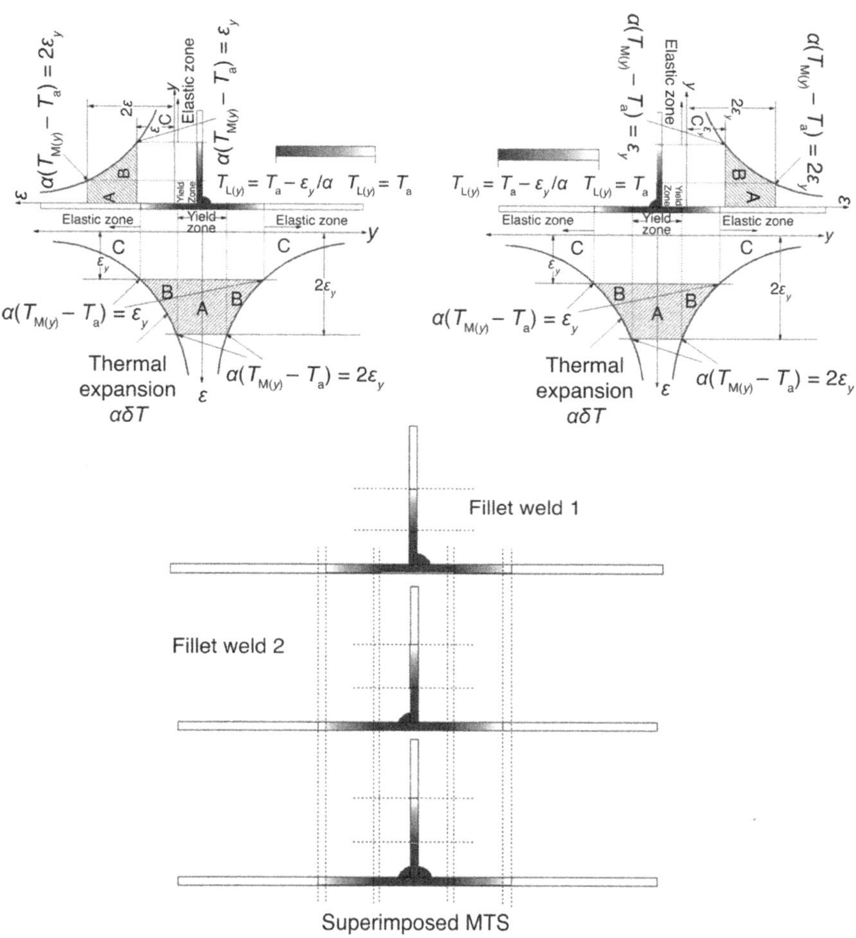

9.11 Mismatched thermal strain in double-sided fillet welding.

dependent expansivity is represented as in Fig. 5.14 and strength properties as in Fig. 5.7. Thus, the model takes into account all the important transient, elasto-plastic behaviour due to each weld sequence, including the corresponding transient behaviour, the direction of the weld passes and some end effects. (The end-tabs themselves were not included in the model.)

Computational instabilities can occur at the highest temperatures and so a cut-off temperature is designated for both parent material and weld metal (1000°C in this case), above which any strains (thermal, elastic or plastic) evolved during heating are disregarded. This is implemented simply by assigning a reference temperature equivalent to the cut-off temperature and a corresponding body load force. This is applied to any elements shown in the thermal analysis to be reaching temperatures greater than 1000°C. These elements are also killed during the heating stage and reactivated during cooling when they reach temperatures below the cut-off temperature. Figure 9.12 shows a flow diagram for this model.[6]

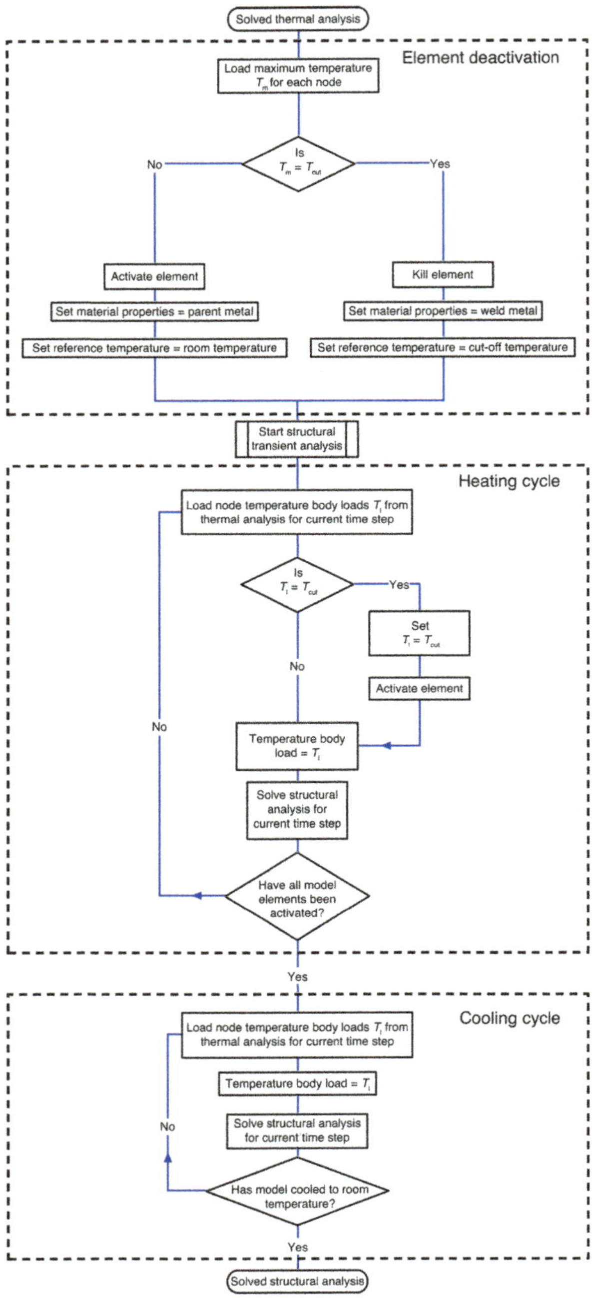

9.12 CIM thermo-elasto-plastic structural model.[6]

9.4.3 Comparison of CEM and CIM results with experimental distortions

The test programme for this study comprised four specimens, assumed to be identical, which were welded according to the basic geometry of Fig. 7.9 together with the process conditions assumed in the computations. Out-of-plane displacements of the base plate were measured before and after each weld pass and distortions for each stage were calculated by subtracting the pre-weld displacements from the post-weld. Transient distortions during each weld pass were recorded using the system described in section 6.6 and are shown in Fig. 9.13.

Several features can be noted. Most of the angular distortion occurs during the weld passes and not in the cooling period. There is slightly more angular movement during the second pass, but that is probably due to the extra heat input for that weld, as described earlier. Longitudinal bowing distortion is rather small relative to typical butt-welded cases, but this is due to the very much greater bending stiffness of the fabrication and the small offset of the contraction force relative to the neutral axis of the cross-section. Nevertheless, the results show initial fluctuation between hogging and sagging deformation as the heat source passes, followed by slow development of sagging during the cooling period.

Figure 9.14 provides a direct comparison of the experimental final distortion profile of the fillet-welded specimen with the CEM and CIM results. (Note that, although all the results are based on the displacements from the initial shape, a flat plate starting point is assumed for the computational cases, whereas the experimental results are based on the actual initial shape. Also, the plate was supported near all four corners, which then represent points of zero vertical displacement.)

These profiles are qualitatively similar, although careful study will reveal that the CEM result is entirely symmetrical, whereas the experimental and CIM results show longitudinal variation. This is an outcome of the CEM application of the load to an elastic model, whereas the CIM loading is applied transiently for each weld to an elasto-plastic structure. The experimental and CIM results also show greater angular contraction at the end of the weld than at the beginning.

A more detailed quantitative comparison is made through Table 9.1 where the averaged RMS values for each form of distortion are compared across the three determinations. In the case of angular and overall out-of-plane deformations, the computational results bracket the minimum, maximum and average experimental results for the four test specimens, the CEM approach giving lower estimates and the CIM giving higher. This outcome might be expected in the CIM results as the angular restraint of the

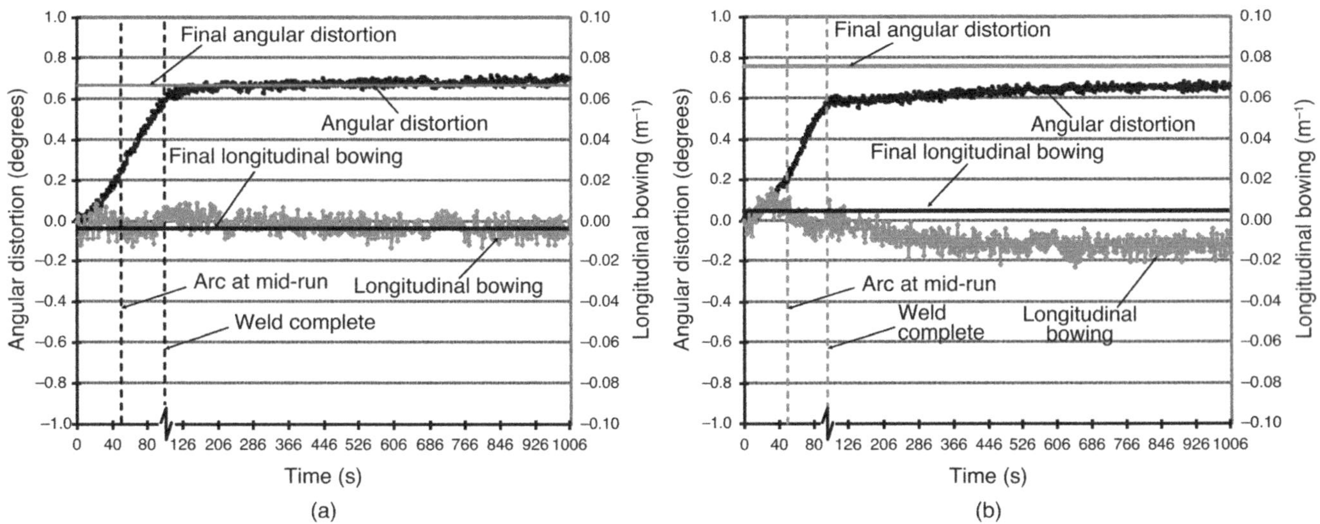

9.13 Transient distortion of base plate in stiffened plate assembly: (a) first weld pass; (b) second weld pass.

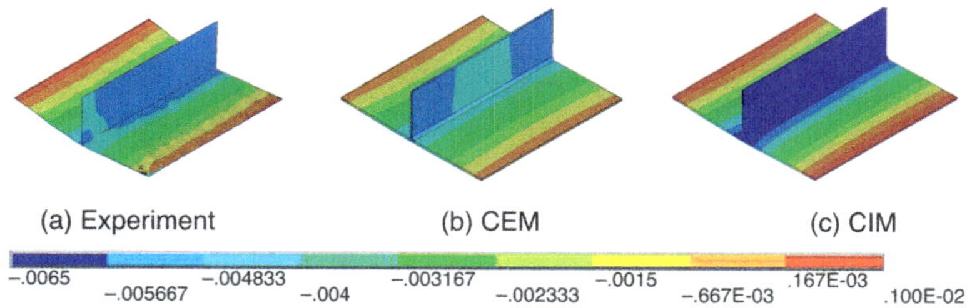

9.14 Comparison of experimental and computational out-of-plane deformations (m): (a) experimental result; (b) CEM result; (c) CIM result.

Table 9.1 Comparison of average experimental and computational out-of-plane deformations

		First weld pass	Second weld pass	Final two-pass total
RMS angular deformation (degrees)	Experimental minimum	0.449	0.572	1.048
	Experimental maximum	0.533	0.730	1.211
	Experimental average	0.499	0.634	1.114
	CEM algorithm-based	0.395	0.494	0.889
	CIM thermo-elastic-plastic	0.714	0.790	1.488
RMS longitudinal bowing (m^{-1})	Experimental minimum	−0.0058	+0.0053	+0.0159
	Experimental maximum	+0.0092	+0.0176	+0.0213
	Experimental average	+0.0038	+0.0124	+0.0189
	CEM algorithm-based	−0.0051	−0.0012	−0.0063
	CIM thermo-elastic-plastic	−0.0111	−0.0002	−0.0127
RMS out-of-plane deformation (mm)	Experimental minimum	1.0129	1.168	2.465
	Experimental maximum	1.2633	1.576	2.880
	Experimental average	1.154	1.360	2.743
	CEM algorithm-based	0.970	1.330	2.300
	CIM thermo-elastic-plastic	1.573	1.793	3.350

end-tabs was not simulated. Moreover, the effects of larger heat input in the second weld pass are seen in the computational results.

The longitudinal bowing deformations are all rather small, for reasons explained earlier, but the test results and the computational results are of opposite sign – the computations show hogging whereas the experiments mostly showed sagging deformations. Little signs of such longitudinal deformation profiles can be seen in the contours shown in Fig. 9.14 and the differences can be considered insignificant.

Another point to note is that there was no difference in the computational outcomes through connecting or disconnecting the nodes at the stiffener–plate interface (which were disconnected in the thermal analysis) as this

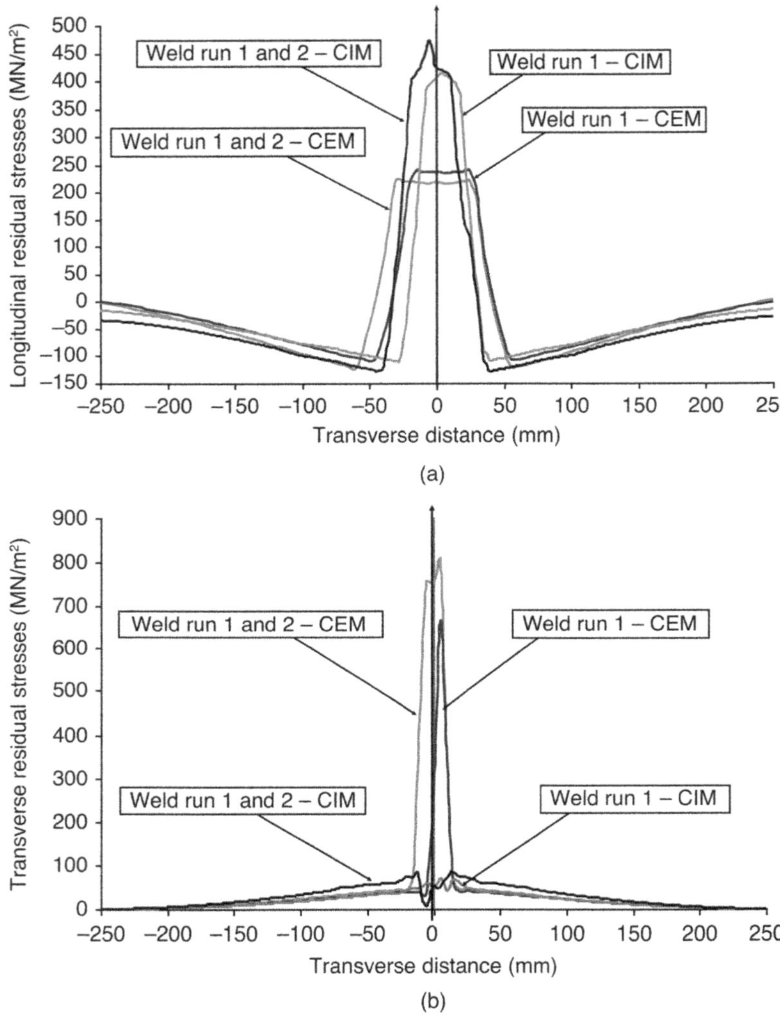

would not affect the longitudinal structural stiffness. Likewise, completion of the first weld did not seem to add significant structural stiffness during the second weld pass.

A further difference between the CEM and CIM results is seen in the description of residual stress fields. Figure 9.15 shows computed longitudinal and transverse residual stress fields on the bottom surface of the stiffened plate, at mid-length.[6] In the longitudinal case, the peak tensile stress according to the CIM rises in the centre to around the uniaxial yield strength of the weld metal, whereas the peak stress is some 30% lower in the

CEM. This is unrealistic and due to the effective release of the plane strain condition in the algorithmic treatment when the thermal load is applied to the 3D elastic model. Likewise, the transverse stress indicated by the CEM is unrealistic and caused by the assumed transverse thermal loading on the elastic model. However, it is interesting to note that the CIM confirms the assumption made in the MTS model that the increase in the tensile stress field due to the second pass is simply related to the offset distance between the two fillet weld passes.

The CIM approach is thus considered to be more accurate in detailed features. However, it has to be borne in mind that the computational budget for the CIM model was around 10^3 times greater than that for the CEM, even in this relatively small-scale fabrication.

9.5 Conclusion

Implementation details have been given for thermo-mechanical, elasto-plastic structural finite-element models. These cover meshing schemes in section 9.2.1 and material property descriptions in section 9.2.2. A case study on the influence of tacking procedures on the distortion of butt-welded plates was given in section 9.3 and section 9.4 considered a case study on the distortion of a plate with a single stiffener attached by two fillet welds.

The tacking case studies illustrate the use of active/inactive elements to simulate clamps, weld deposition and re-melting of tacks by following butt welds. The results show the important effects of weld edge preparation and other geometric features on distortion.

The stiffened-plate, fillet-weld case study contrasts the computationally efficient, algorithm-based model with the full elasto-plastic treatment and both in turn were compared with experimental results. Finally, a comparison of residual stress fields computed by the CEM and CIM was presented and it was shown that the CEM does not give a reliable result for this parameter.

9.6 References

1. Camilleri, D., Gray, T.G.F. and McPherson, N., 2010, Optimising tack welding fabrication procedures using numerical finite element models, *J. Ship Prod. Design*, 26, 2, pp 117–134.
2. Goldak, J. and Akhlagi, M., 2005, *Computational welding mechanics*, Springer, New York, USA.
3. ANSYS, Inc., 2007, *ANSYS 11.0 help documentation*, Chapter 2 General element features, 2.4 Linear material properties, ANSYS, Inc. (Canonsburg, PA), USA.
4. Mollicone, P., Camilleri, D. and Gray, T.G.F., 2008, Procedural influences on

non-linear distortions in welded thin-plate fabrications, *Thin-walled Struct.*, 46, 7–9, pp 1021–1034.

5. Mollicone, P., 2006, *Applied modelling techniques for welding induced distortions and residual stresses*, PhD thesis, University of Strathclyde, Glasgow, UK.
6. Camilleri, D., Mollicone, P. and Gray, T.G.F., 2006, Alternative simulation techniques for distortion of thin plate due to fillet-welded stiffeners, *Model. Sim. Mat. Sci. Eng.*, 14, pp 1307–1327.

10
Simulating welding distortion in butt welding of thin plates

DOI: 10.1533/9780857099327.199

Abstract: Modelling strategies and results from previous chapters are brought together in a comprehensive treatment of butt welding procedures. Various practical issues are treated, including effects on distortion of initial out-of-flatness, disposition of supports, tacking assembly and clamping of components. In the case of clamping, it is shown that excessively constrained clamping strategies can lead to greater levels of distortion than if the components are relatively unconstrained. The outcomes of the various simulation methods for prediction of residual stresses in butt welds are described. Effects in multiple longitudinal butt welds where there are interactions between several welds in an assembly are shown.

Key words: support, out-of-flatness, tacking assembly, clamping, buckling, residual stress, multiple welds.

10.1 Introduction

The aim of this chapter is to bring together various modelling strategies, computational outputs and practical results, in order to inform designers and fabricators involved in the specification of butt-welded structures and their corresponding welding procedures. Much information has been given in earlier chapters through examples and illustrations and it is useful to recall some of these at this stage.

The basic distortions normally found after butt welding two initially flat plates together are illustrated in Fig. 4.7 and these typical movements are present to a greater or lesser extent in most applications. The *transient* development of such distortions is exemplified in the computational results shown in Fig. 5.10. Various modelling strategies for butt welds – transient thermo-elastic-plastic, computationally efficient and hybrid – are outlined in section 5.8.1. Chapter 6 deals with the experimental aspects of measuring and characterising distortion in butt welds. This includes transient observations of plate deformations, while the welding source passes along the seam and then as the plate is allowed to cool back to ambient temperature. Sections 7.6 and 7.7 deal with the computation of thermal transients in a butt-welded plate assembly. Various butt welding example results are given in the context of reduced solution strategies (in Fig. 8.7 to Fig. 8.10). These examples include small-scale and large-scale fabrications and illustrate some effects of initial out-of-plane shape and support conditions. Chapter 9 deals with elasto-plastic structural modelling and includes a finite-element meshing example, detail on modelling non-linear thermal properties and the use of active/non-active elements in simulating changing geometry in fabricating a butt-welded structure.

10.2 Plate support and out-of-flatness influences

Initial out-of-flatness in plates will frequently be present in practical cases and the possible effects of this need to be addressed in any scheme to predict and control distortion. Initial out-of-flatness typically arises from three main sources.

1. Inherent residual stress fields and out-of-flatness in the plate stock supplied. These may be due to manufacturing processes that involve rolling, heating and cooling, together with problems arising in transport and storage.
2. Distortions generated during assembly, clamping and temporary fixing (e.g. tacking) of the adjacent plates.
3. Gravity forces on unsupported areas (if the plate is welded in the flat position).

Whatever the source of initial out-of-flatness, the in-plane compressive forces generated by welding usually magnify the initial pattern, sometimes quite substantially, and it is therefore important to minimise all initial deviations from flat in the assembled-for-welding state.

In the test configurations described in Chapter 6, the smallest practical number of supports was used in order to maintain a statically determinate support condition as far as possible and thereby improve correspondence between the experiments and the computational models. In the small-scale tests (Fig. 6.1), the thickness and in-plane dimensions were such that the sagging displacements due to gravity forces should have been negligible, although there were some problems of initial out-of-flatness due to stock supply and difficulties in aligning assemblies. The large-scale samples, on the other hand, were large enough in area, relative to the support distances, to show small sagging displacements prior to welding. Although there were few supports, the transverse positions of the pairs of point supports on the test bed were chosen to minimise out-of-plane gravitational displacements, on the assumption that the stiffness transverse to the unmade weld seam would be the same as for the rest of the plate (which may not always have been the case). In practice, plates of this size would probably be welded on a bed comprising longitudinal and transverse support bars at closer spacing.

Simulating the supports is in itself problematic, because the support reaction at any point in a statically indeterminate support system will disappear if the plate lifts by even a small amount. Such support conditions can be treated via contact elements in theory, but the pattern of contact can be unstable and may lead to convergence problems in solution. A more robust approach is to specify LINK elements at each desired support position. Such elements can be ordered to exhibit a specified high stiffness under compressive contact loads and to have zero effective stiffness in tension.

The small-scale tests, described in section 5.8.1, were not subject to support/gravity influence as explained above and they therefore provide examples where out-of-flatness influence, due mostly to imperfect stock or pre-welding assembly, can be judged. Figure 10.1 shows a case where the specimen was fairly flat initially and the distortions induced by welding arose largely from the classic pattern of strong positive angular deformation and consequent longitudinal bending in the hogging sense. (Note that the welding-induced displacements in Fig. 10.1(c) were obtained by subtracting the initial displacements in Fig. 10.1(a) from the final distortions in Fig. 10.1 (b).)

The transient development of deformations is shown for this test specimen in Fig. 8.8 and it is worth noting that not all of the angular distortion is attributable to angular contraction of the weld zone as the heat source passes, as more than half is developed during the subsequent cooling period – due, it would seem, to large-displacement structural interactions. Note that this effect would not be accounted for in an 'analytical' application of the algorithm approach but it would be picked up through the final large-displacement, three-dimensional (3D) elastic modelling stage in the 'computational' application used here (see section 8.2.3).

The outcome of the corresponding computational model (assuming a perfectly flat starting point) is shown in Fig. 10.2. This result was obtained using the 'computationally efficient' method, based on the simple algorithms. Numerical comparison of the experimental and computational distortions is given in Table 10.1. The agreement is good, given the simplifications inherent in the algorithm-based computation. In particular, the computation does not include any stiffening effect of the end-tabs against angular distortion, which is present in the experimental result.

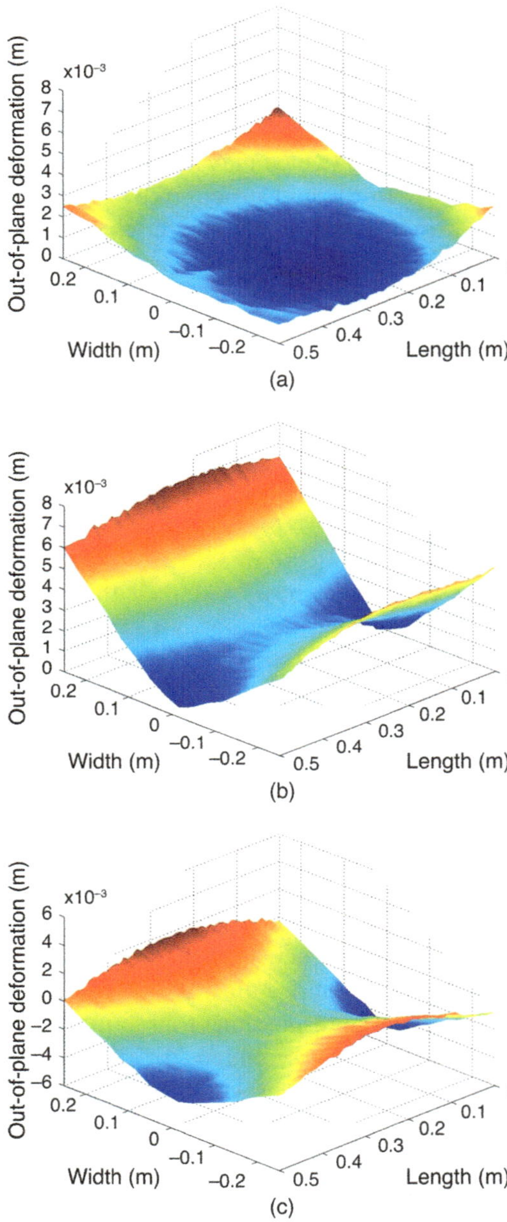

10.1 Small-scale butt-welded specimen (unsmoothed displacements). (a) Initial shape of assembly. (b) Final deformed shape after cooling. (c) Resulting net out-of-plane distortion.

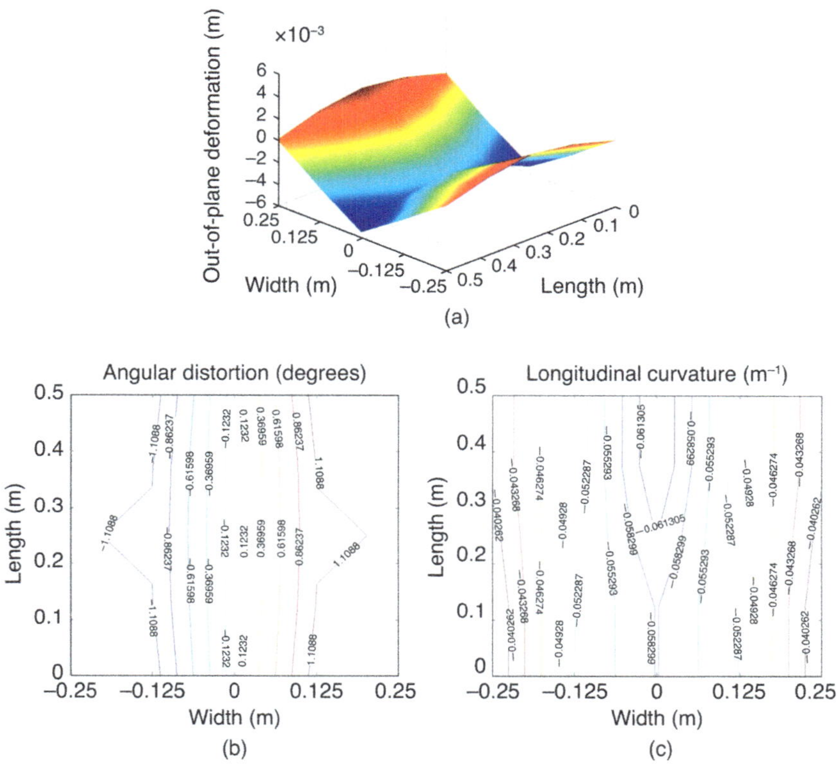

10.2 Predicted distortion using computationally efficient model and assuming initially flat, stress-free plate. (a) Out-of-plane distortion. (b) Angular distortion. (c) Longitudinal curvature.

Table 10.1 Comparison of experimental and computational out-of-plane distortions

	RMS out-of-plane deformation (mm)	RMS angular distortion (degees)	RMS longitudinal curvature (m^{-1})
Experimental	1.99	0.95	−0.06
Computational	2.39	1.07	−0.05

On the other hand, the consequences of starting with a non-flat geometry can be seen in the corresponding results for a different specimen in the small-scale series (Fig. 10.3). The initial out-of-plane state is characterised by a small *negative* transverse deformation (more curved than angular) with a further local deflection in one corner. The net welding-induced distortion (Fig. 10.3(b)) is actually quite small – in fact, the smallest of the series of ten specimens – with root mean square (RMS) values as follows: overall out-of-

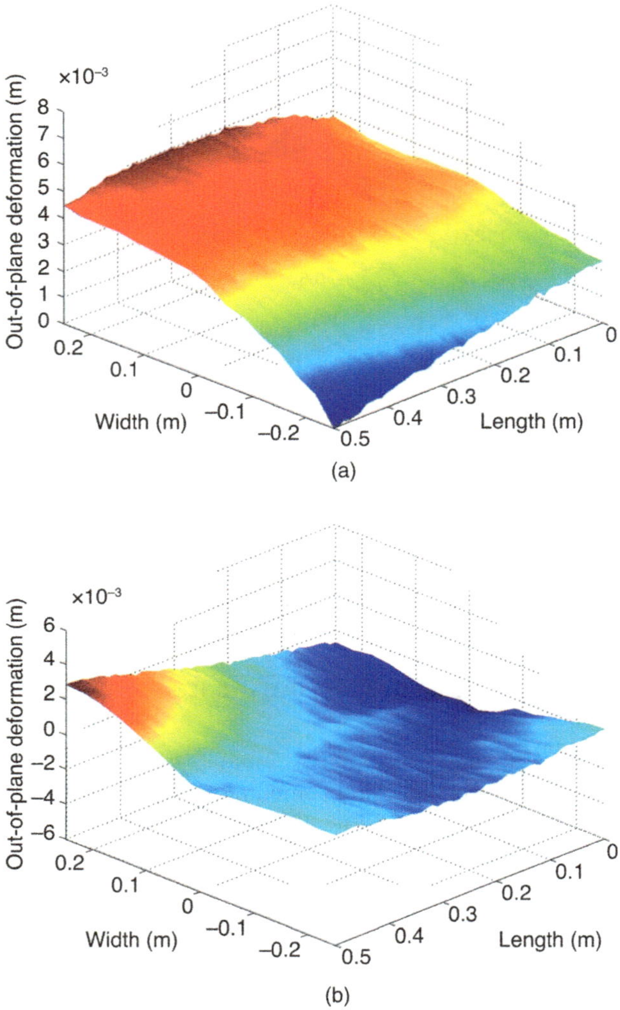

10.3 Butt-welded test specimen with initially 'negative' transverse deformation (unsmoothed displacements). (a) Initial shape. (b) Out-of-plane deformation due to welding.

plane movement 0.989 mm; angular distortion 0.323°; longitudinal curvature $-0.013\,m^{-1}$.

The transient behaviour when welding this specimen was rather complex, as seen in Fig. 10.4, and bears comparison with Fig. 8.8. In this case, a positive angular contraction starts to develop immediately, whereas it is somewhat delayed in the trace of Fig. 8.8. However, due to the initially negative angular deformation, the line of maximum longitudinal contraction is stationed *above* the neutral axis of the cross-section and the plate assembly

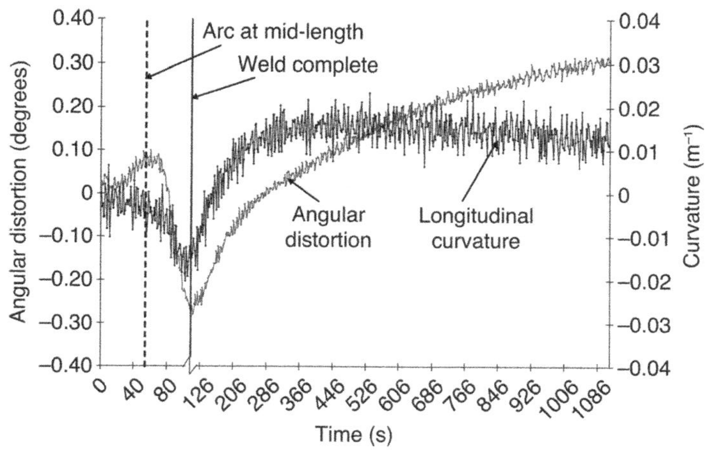

10.4 Transient behaviour from initially 'negative' angular deformation.

was therefore pulled into a sagging curvature. This reverses the angular deformation, due to large-displacement flattening. The plate was then much closer to being flat than the initially flat example (compare the scales on Fig. 10.4 and Fig. 8.8) and more susceptible to deformation instability, due to the compressive load. This caused a reversal of movement during the long cooling period as the heat spread outwards in the plate assembly.

In passing, it is worth noting that it is entirely possible to simulate these complex out-of-plane effects if the initial shape is measured and used as the starting point, although this is perhaps not a strategy that can be readily applied in a production environment. This approach will be discussed later in the context of the large-scale specimen tests.

It is sometimes conjectured that much of the variability found in distortion outcomes is due to initial residual stress fields in the components, random and unknown. These residual stresses may be due to steel mill treatments, thermal cutting and other possible operations prior to assembly. To investigate this point, some of the specimens in the series were thermally stress-relieved before assembly (possibly resulting in some of the initial out-of-plane deformations measured). These specimens exhibited a higher mean out-of-plane distortion after welding and a higher longitudinal curvature. However, a hypothesis test determined that any attribution of this variation to the removal of initial residual stresses had a probability no greater than 5%. Hence, in these test results, there seemed to be little influence of residual stress fields.

The large-scale tests also embodied initial out-of-flatness related to stock supply and assembly imperfections, but the additional effects of gravity were more significant in these tests. In the first place, the self-weight of the plates ensures that the classic single hogging curvature between the first and last of

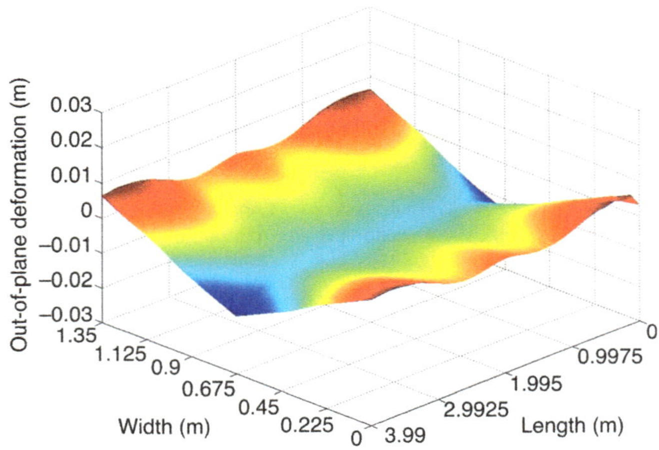

10.5 Predicted out-of-plane deformation of 4 m × 1.35 m × 6 mm thick butt-welded plate. Computationally efficient model used with assumption of initially perfectly flat plate.

the supports is not so likely, although there could be local curvatures in either direction between supports. It is therefore helpful to index these tests against a computational determination for an ideally flat plate and such a case is shown in Fig. 10.5. The locations of the intermediate supports can be seen in the edge ripples of the deformation pattern. The RMS curvature for this case is $-0.0145\,m^{-1}$ (i.e. on average reflecting 'hogging' curvatures) and is small relative to the corresponding curvatures in the small-scale tests. The overall impression of the longitudinal curvature is indeed that the plate remains relatively flat, but the various regions of hogging curvature associated with the support positions have resulted in an overall RMS value that is negative.

However, the deformations of large-scale, butt-welded plates tend to be influenced strongly by the initial shape or out-of-flatness and Fig. 8.6 shows, in flow chart form, how a measured initial shape can be introduced into the 'computational algorithmic' solution method. Figure 8.10 shows an experimental/computational comparison where the initial shape was taken as the starting point. Unfortunately, transient effects arising from shifting support points as the weld is completed will not, in general, be registered by the algorithmic approach. Hybrid methods would therefore be strongly recommended for large fabrications, in order to include a measure of transient analysis through the stepping process while at the same time economising on computing power given the large model size necessary.

10.3 Effects of tacking

Aside from the welding operation itself, a particular practical challenge in fabricating a structure from plate and section components is to assemble the elements accurately and to hold them in place securely during weld completion, in the face of very large thermally developed forces. In many cases, the method used is to apply short tack welds at intervals along the seam or fillet line to be welded. It is also possible that the edges to be welded do not line up conveniently and application of out-of-plane clamping forces may be necessary to bring them into matching alignment before tacking. The tacks are usually placed so that they are re-fused by the weld source during completion of the main weld.

Tack welds may seem like very minor features and are often left to the fabricator's discretion, with no kind of specification or recording of the size or position of tacks used. However, tack placement can have a powerful influence on distortion outcomes. The dynamic angular displacement trace shown previously in Fig. 8.10 shows step changes in angular displacement, which turn out to be precisely at the points where the main weld has re-fused the tacks. The effect is that the tacks restrain angular movement until they are melted, whereupon the plates move immediately to angular positions corresponding to zero local restraint. In the test shown in Fig. 8.10, this has contributed to increased angular deformation and eventually to reversal of the longitudinal bending profile. These deformation steps gradually diminish as the main weld pass encounters more tacks, as more of the main weld has by then been completed and provides greater angular restraint.

The realisation that tacking practice is in fact very important prompted the investigations described in section 9.3. This earlier discussion concentrated on the computational features of the tacking study, particularly the use of transient, elasto-plastic analysis, but here we consider experimental outcomes and practical implications.

In a preliminary study of the tacking operation, steel plates, 4 mm thick and 900 mm × 450 mm in plan, were tacked together to form 900 mm square assemblies, using normal shipyard procedures. An array of tacks was used, comprising three evenly spaced 70 mm long tacks along the length and a 35 mm long tack at each end. In this case, no weld edge preparation was used and no gap left between the plates. Figure 10.6 shows the result of the thermal simulation developed for this case study.

In order to determine a baseline flatness profile of the assembly before tacking, the plates were first held together by three small spot welds and the shape was scanned using the intermediate-size welding rig of the type shown in Fig. 6.7. With the benefit of hindsight, greater scrutiny might have been exercised on the spot-welding operation, which turned out to have a

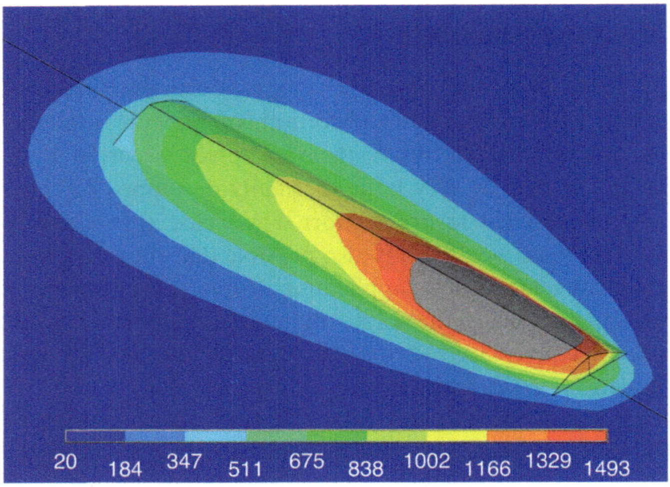

10.6 Contour plot of predicted temperature field for a single tack at point of completion (courtesy P. Mollicone).

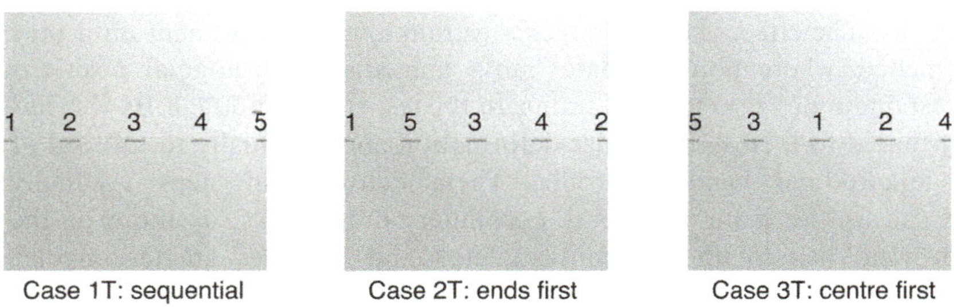

10.7 Tack-laying sequences used (courtesy P. Mollicone).

surprising influence on the distortion outcomes. Three different tack-laying sequences were examined, as this aspect was thought to have a bearing on outcomes. These are described as 'sequential, 'ends first' and 'centre first', as shown in Fig. 10.7.

The results of the tack-weld tests were difficult to understand at first, but two modes of deformation were observed. In a typical example of the first mode, shown in Fig. 10.8, the spot welds had been placed on the top side of the plate and, despite their small size, they generated a surprising level of 'gull wing' deformation, as shown in the initial scan (Fig. 10.8(a)). The five full-size tack welds subsequently placed on the top surface then increased the angular deformation without apparently affecting the longitudinal curvature (see Fig. 10.8(c), which shows the net distortion due to the tack-welding operation only). This outcome indicated that the spot welds had functioned as hinge points for angular deformation in the tack-welding

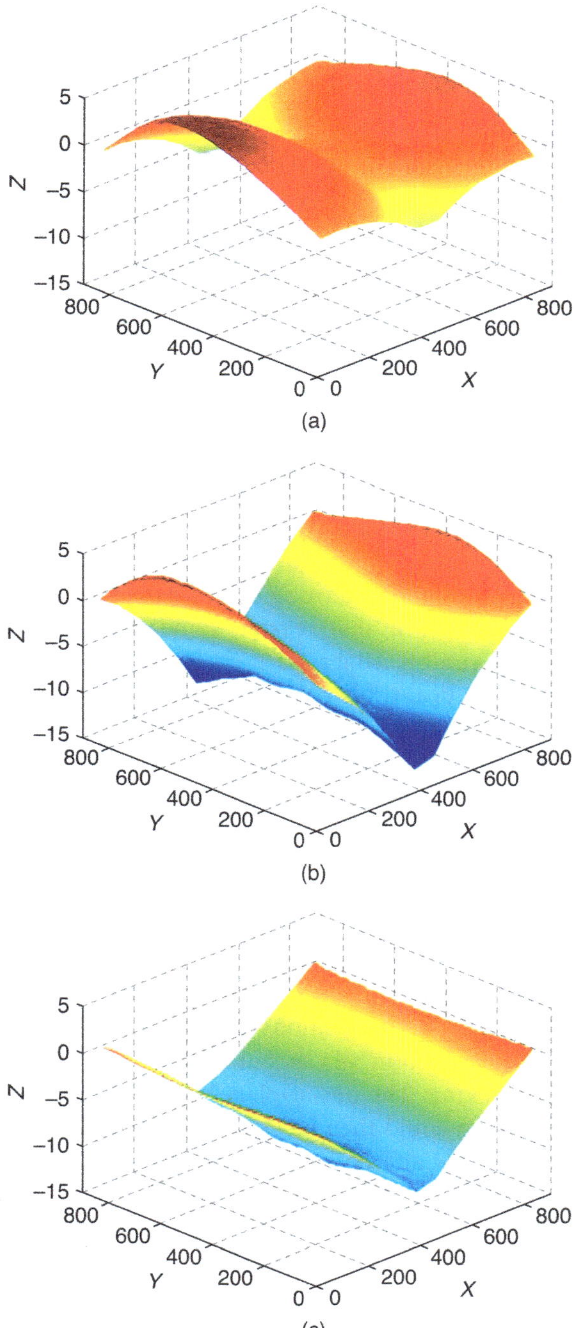

10.8 Out-of-plane distortion – spot welds and tack welds on top side. (a) Initial out-of-plane distortion, post-spot-weld. (b) Distortion after tacking. (c) Net distortion due to tacking (courtesy P. Mollicone).

operation. Comparison of the three sequences of tack-laying showed virtually no effect on the final outcome and this was confirmed by the matching computations.

A typical outcome of the second mode of assembly, where the spot welds had been placed on the underside of the plate (opposite to the later placing of the tack welds on the top side) is shown in Fig. 10.9. The spot welds in fact generated almost the same out-of-plane deformation as for the first mode (inverted) but the subsequent tack-welding operation influenced the angular deformation very little. The implication is that the tack transverse contraction has been resisted by the spot welds. There is a slight change in longitudinal curvature because the line of contraction would have been slightly above the neutral axis of the V-shaped cross-section.

There are several lessons to be learned from this practical trial. The first is that the initial assembly stages are critical in terms of determining the mechanics of response to later stages of heat input, whether during tacking or main seam welding. The small amounts of heat input associated with tack welding (or even spot welding) will drive large angular and longitudinal deformations, which then produce a starting out-of-flatness at the outset of the main welding operation. For that reason, some form of mechanical clamping is preferable, if at all feasible.

As these preliminary practical trials proved to be somewhat inconclusive in terms of identifying a clear strategy for tacking in butt-welded fabrication, a further comprehensive computational study was undertaken, based on transient, thermo-elastic-plastic analyses, as described in section 9.3. The main variables of this study are shown in Fig. 10.10.

Two tack lengths were considered (10 mm and 40 mm) and the number of equally spaced tacks was varied between 5 and 17 for the 10 mm tacks and between 3 and 7 for the 40 mm tacks. In terms of the percentage of total seam length tacked, these numbers correspond to 5–17% of total length for the 10 mm tacks and 12–28% of total length for the 40 mm tacks. Alternative top and underside tack positions were trialled. It was assumed that the assembly was supported in all cases at four corners and subject to gravity loading during tacking. The out-of-plane movements caused by the simulated tacking operation could then be examined as indicated and each shape was then subjected to alternative options of the simulated seam-welding operation, with and without gravity loading. Simulation of the clamping condition was described in section 9.3. The study comprised 32 separate analyses to cover the stated variables.

Figure 10.11 shows the tacking variables at the foot of the diagram and the distortion results in the bar graphs, after the tacking operation and removal of the clamping constraints. The proportions of total length tacked increases linearly from left to right in each table. Note that the 10 mm 17-

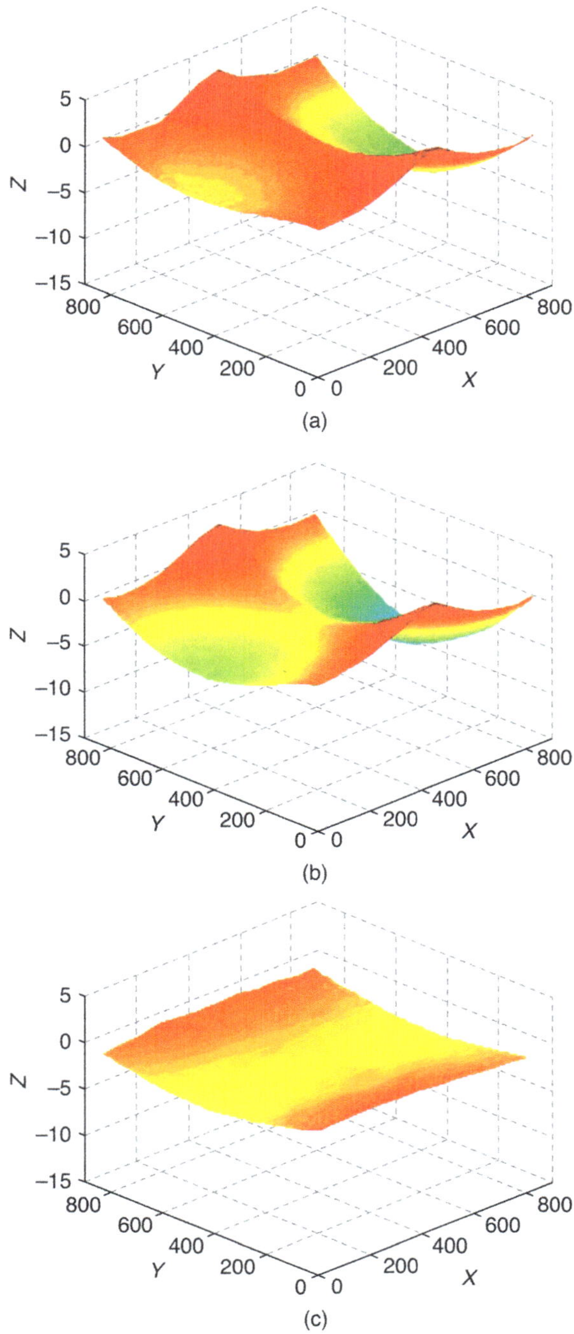

10.9 Out-of-plane distortion – spot welds on underside and tack welds on top side. (a) Initial out-of-plane distortion, post-spot-weld. (b) Distortion after tacking. (c) Net distortion due to tacking.

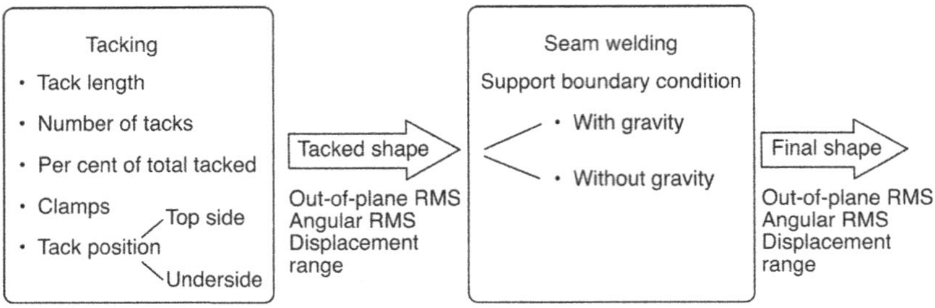

10.10 Variables in thermo-elastic-plastic study of tacking and seam welding.

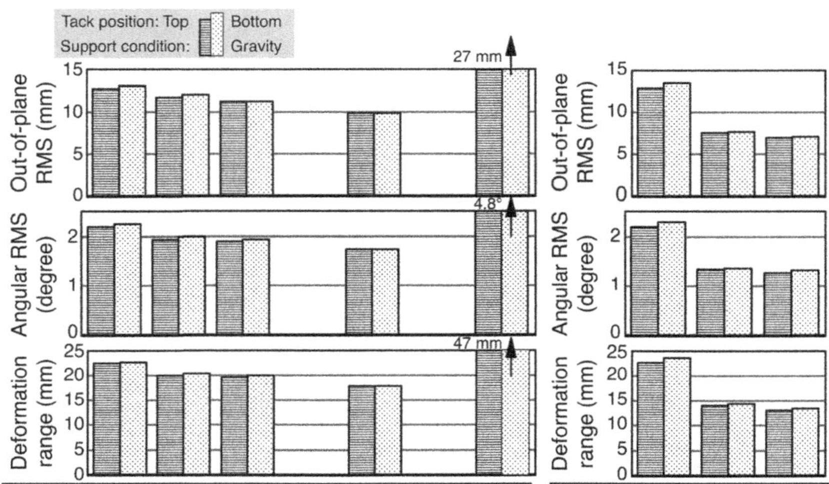

No. of tacks	5	7	9		13		17
% of total	5%	7%	9%		13%		17%
No. of clamps	4	4	4		4		2

10 mm tacks

No. of tacks	3	5	7
% of total	12%	20%	28%
No. of clamps	6	4	6

40 mm tacks

10.11 Variables and results for simulated tacking.

tack configuration included only two clamps, and distortions for this case were very much greater, as shown by the off-scale arrows.

In effect, the loading and boundary conditions for the bottom tack geometry are inverted relative to the top tack geometry and the resulting distorted shapes are therefore also inverted but are otherwise identical in magnitude. The very small differences between top and bottom tacks shown are computational artefacts, due to the lack of rigour in the active/inactive element technique, as noted in section 7.4. (If the supposedly inactive elements are judged to have a very small, but non-negligible stiffness, then the top tack configuration will be slightly stiffer.)

All distortion measures improved a little with an increasing number of tacks, with the notable exception of the 17-tack cases where the two-clamp constraint was insufficient to maintain the shape of the assembly against gravity loading. The reason for improvement with an increasing number of tacks can be seen in a comparison of the 5-tack and 13-tack cases in Fig. 10.12. Angular distortion dominates in both cases, because the longitudinal contraction induced by welding only up to 13% of the length is insufficient to generate the longitudinal hogging shape that is typical of butt welds. However, greater longitudinal contraction is generated in the 13-tack model

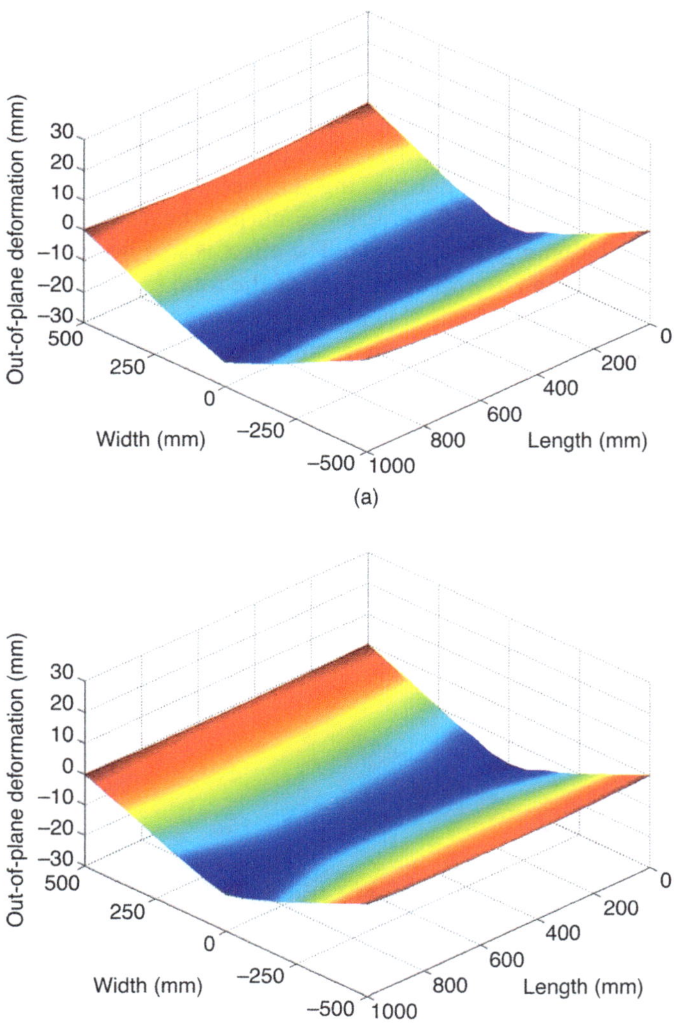

10.12 Effects of increasing the number of 10 mm tacks: (a) 5 top side 10 mm tacks; (b) 13 top side 10 mm tacks.

and this lifted the centreline of the plate slightly, especially at mid-length, and counteracted the slight sagging tendency seen on the long edges in the 5-tack case. This is in fact due to large-displacement flattening and is not so evident at the ends of the plate where the longitudinal bending moment must drop to zero. Hence there is a slight improvement in all distortion measures when the overall length of tacking is increased.

Comparison of the distortions in Fig. 10.12(b) with the results of the 17-tack, two-clamp case in Fig. 10.13 (note the increased scale in the latter figure) highlights the large angular deformations in the latter and confirms the need for sufficient clamps to maintain shape against gravity forces while the tacking operation is being carried out.

Similar behaviour is seen in the 40 mm tacking scheme. The 3-tack configuration shows more distortion than the 5-tack 10 mm case, but again the situation improves with an increasing number of tacks to give a better result than for any of the 10 mm tacking schemes. Comparison of the 3-tack and 7-tack cases in Fig. 10.14 again confirms the role of longitudinal contraction in flattening the assembly. The effect is significant for both tack sizes when something more than 15–20% of the total seam length is tacked.

The clear conclusion from the previous experimental studies and the present computations is that even a small length of tack weld induces angular deformation in the range 1–2.5°, which is in fact fairly similar to typical angular deformations found in fully butt-welded fabrications. In other words, the distortion is determined principally by the cross-sectional shape of the weld tack and not by the number or length of tacks. If initial flatness is the only aim, then the best choice from these trials would be to apply between 5–7 tacks of 40 mm length. However, although initial flatness

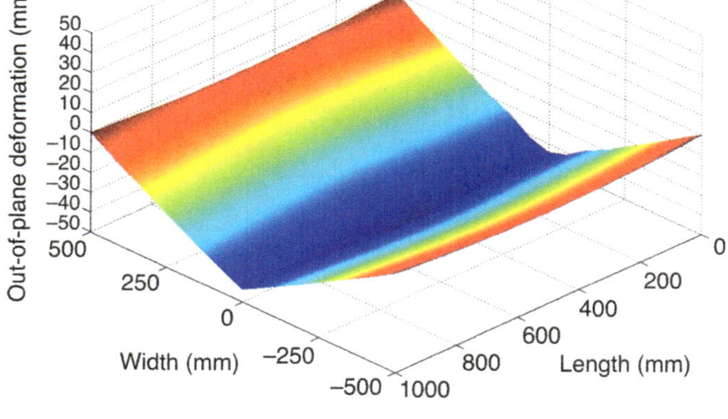

10.13 Effects of insufficient clamping.

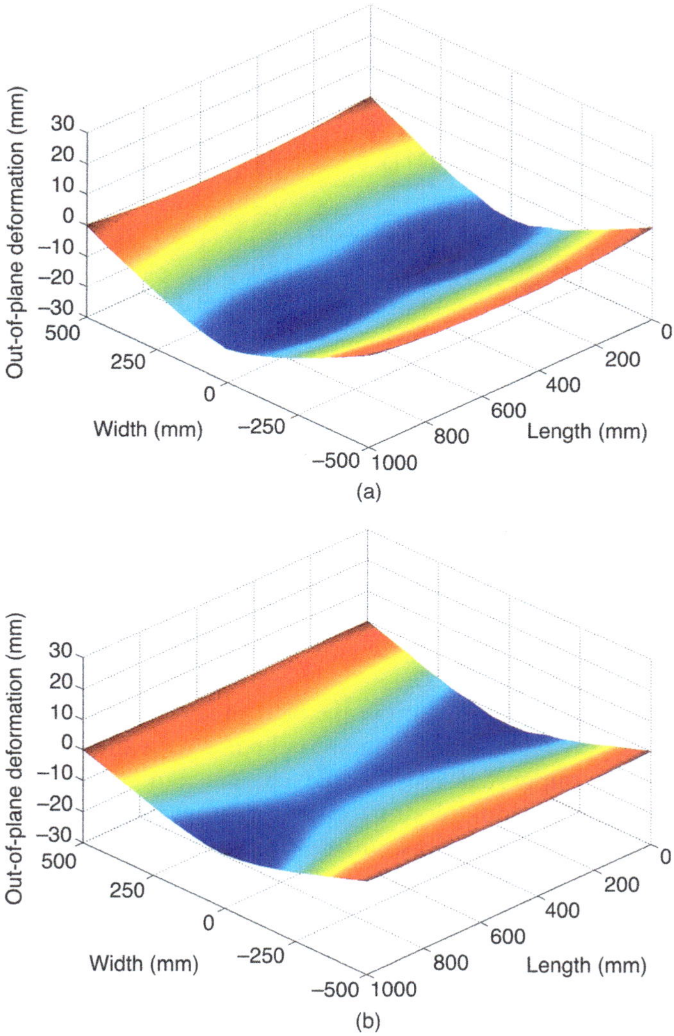

10.14 Effects of increasing the number of 40 mm tacks: (a) 3 top side 40 mm tacks; (b) 7 top side 40 mm tacks.

is always desirable, the influence of tack pattern on the final seam weld has still to be considered and, as we shall see, that influence is not as obvious as it might seem.

In the case of the seam-welding operation, the simulated clamps are removed after the tacking operation and have no further influence. The further distortions generated by seam welding the variously tacked geometries are shown in Fig. 10.15. Four different results are given for each number and size of tack. The first two results bars in each series of four

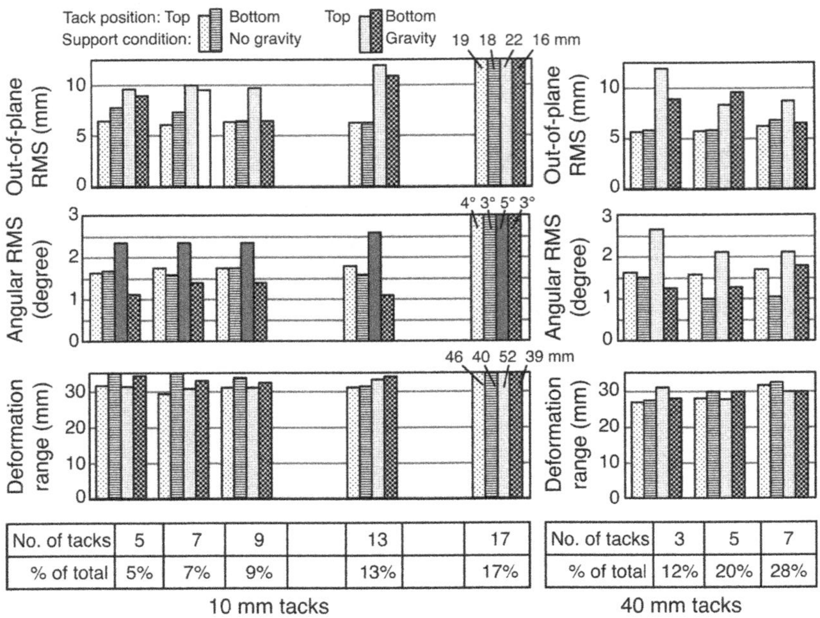

10.15 Effects of seam welding the previously tacked cases.

cover assemblies tacked at the top or the bottom and seam welded in a fully supported, no-gravity condition. The second pair covers the situation where the assembly is supported only at the corners and gravity is applied. Although the initial geometries of top- and bottom-tacked assemblies are the same, the bottom-tacked cases present an inverted deformation profile relative to the seam weld and the tack constraints are also on the opposite side of the plate, relative to the seam weld.

Application of the seam weld mostly reduces the distortion produced by tacking, as measured in terms of RMS values, although the *range* of displacements generally increases. The reason for this latter finding is that seam weld contraction tends to lift the centre of the plate, thereby flattening the angular contraction in that region.

The other initially surprising feature of these results is that the relative distortions of the different cases are not what might be expected from the outcome of the tacking stage. That is, the tacked plates that exhibit better flatness in the results of Fig. 10.11 (generally more tacks and greater total length) do not maintain this better position in the results of Fig. 10.15. In other words, a better initial shape (after tacking) does not necessarily yield a better final shape after seam welding. Another general finding is that, in the cases where gravity loading is applied, seam-welding distortions are usually greater, especially so in the case of top-tacked assemblies.

Examination of the transient deformations during the seam-welding

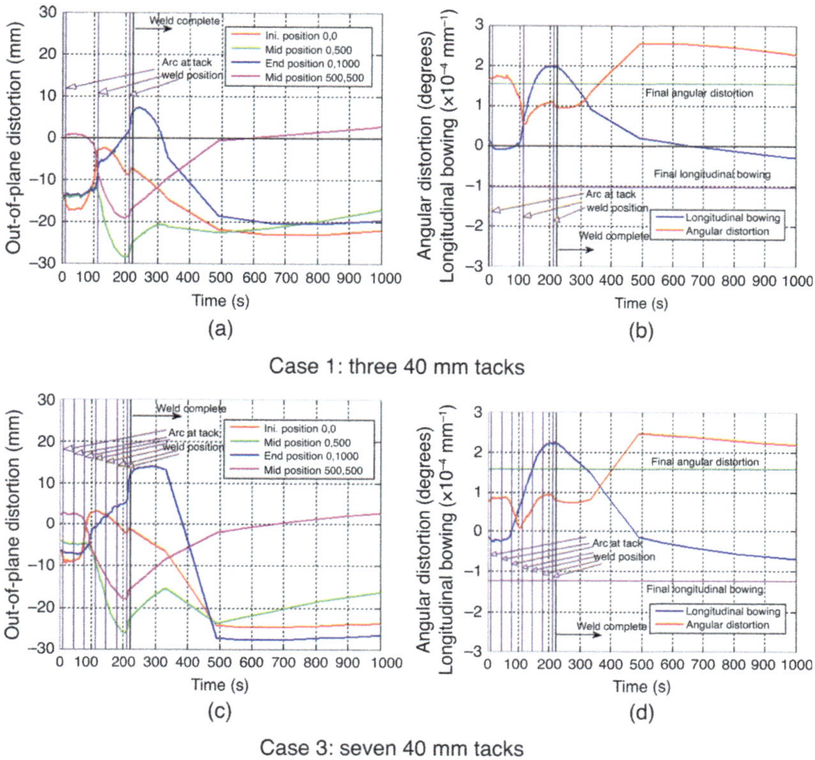

10.16 Transient displacements of 40 mm tacked plates during no-gravity seam welding: (a), (c) transient displacement; (b), (d) transient deformation.

operation provides some clues to the reasons for this slightly contradictory outcome. Figure 10.16 contrasts the transient deformations of the 40 mm 3-tack and 7-tack configurations. Figures 10.16(a) and 10.16(c) track the out-of-plane displacements of four points on the plate – three along the centreline (at the weld) (0,0; 0,500; 0,1000) and the fourth point (500,500) on the edge at mid-length. These points are used to construct global angular and longitudinal distortions. The corner points set the baseline zero displacement. Figures 10.16(b) and 10.16(d) track the angular distortion at mid-length and the longitudinal bowing along the centreline (note that these distortions are not the same as the RMS values that take account of variations all over the plate).

Rapid changes of displacement can be seen at points where the seam weld encounters and re-melts tack welds. The basic patterns of displacement and distortion are similar for both tacking schemes, but the range of global movement in the 7-tack scheme is much greater than is the case for the 3-tack scheme. Essentially, the 3-tack scheme is more flexible and can

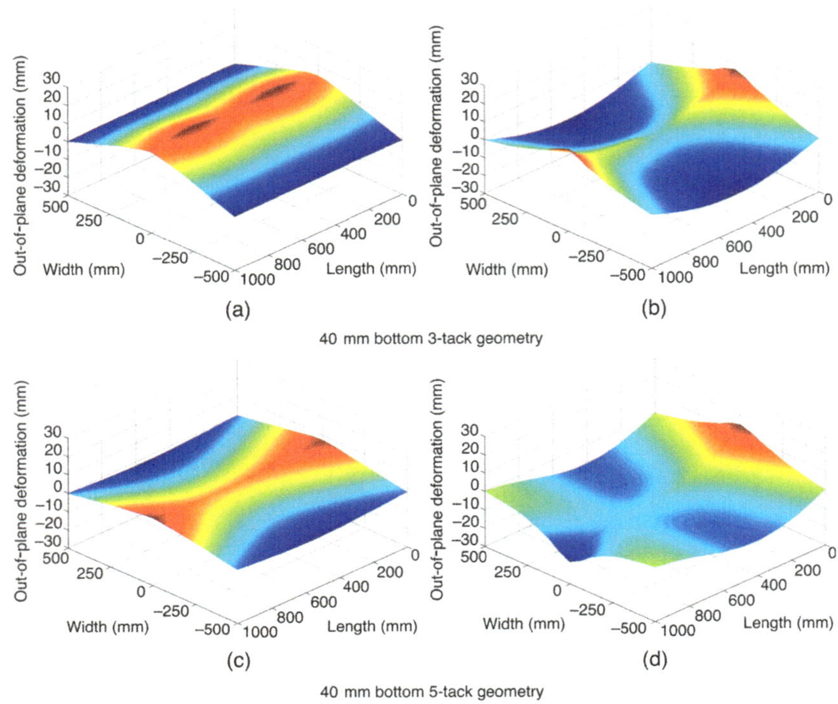

10.17 Profiles of 40 mm bottom-tacked assemblies. 40 mm bottom 3-tack geometry: (a) initial shape post-tacking; (b) after seam welding. 40 mm bottom 5-tack geometry: (c) initial shape post-tacking; (d) after seam welding.

accommodate local expansions and contractions as the heat source passes along, whereas such movements cause larger overall movements in the more firmly attached 7-tack example. Hence the advantage of better initial shape in the 7-tack assembly is lost in the seam-welding operation. Note also the rapid changes in direction of movement, due to large-displacement interactions between the transverse profile changes along the length and the longitudinal bending curvature.

The behaviour of bottom-tacked cases exhibits further complex behaviour. Figure 10.17 shows the initial and final profiles of the 3-tack and 5-tack cases. In these cases, the weld line is *above* the neutral axis of the inverted V-shape and longitudinal contraction pulls the plate towards a sagging geometry. However, interaction between longitudinal and transverse bending and flattening actions has caused buckling in the 5-tack case, due to the overall stiffer behaviour. The transient out-of-plane displacement and deformation profiles for the 5-tack geometry in Fig. 10.18 show that the buckle occurs just at the point of weld completion.

It is worth noting in passing that the usual recommendation of 'balanced'

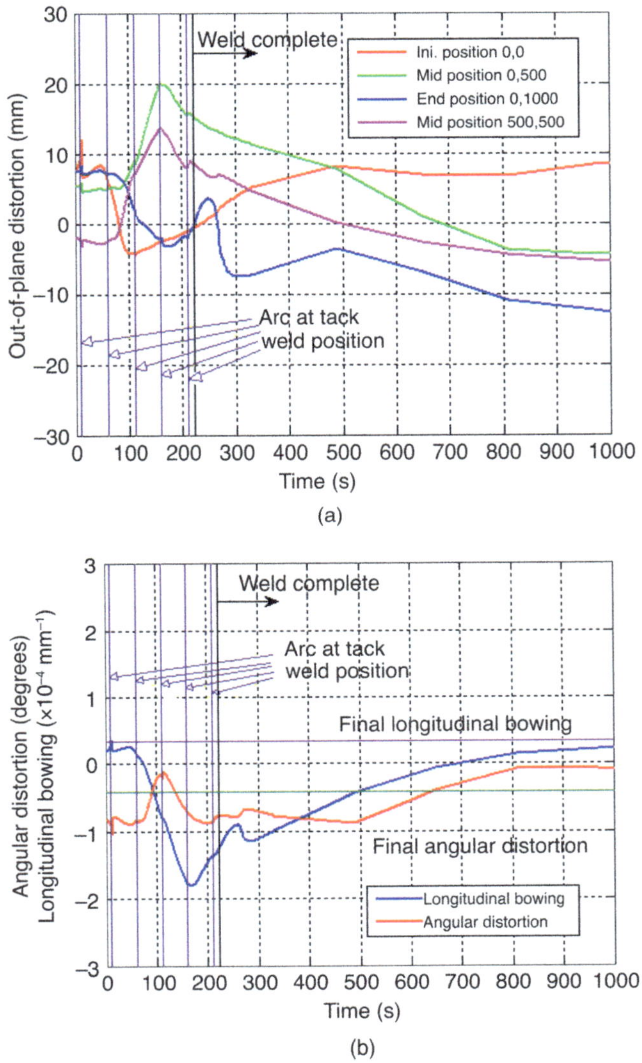

10.18 Buckling development in constrained 40 mm bottom 5-tack geometry. (a) Transient displacement. (b) Transient deformation.

welding – in this case placing the tack welds on the opposite side from the seam weld – has not had the desired effect of reducing the distortion. The angular distortion is indeed reduced a little in some cases, but the RMS and ranges of displacement values are often increased. The reason is that the distortion mechanics are more heavily influenced in this geometry by longitudinal bending (and buckling in the case of the inverted, bottom tack profiles).

The results for seam welding with gravity loading included (see Fig. 10.15) mostly show increased angular and longitudinal deformation (see Fig. 10.19). Gravity in this case has increased the angular deformation, as the angular contraction is working in the same sense as it was during tacking and most of the hogging longitudinal deformation in the no-gravity case has been negated by longitudinal sagging between the supports when gravity is applied. The gravity effect is emphasised in the some of the cases that have fewer tacks and are more flexible.

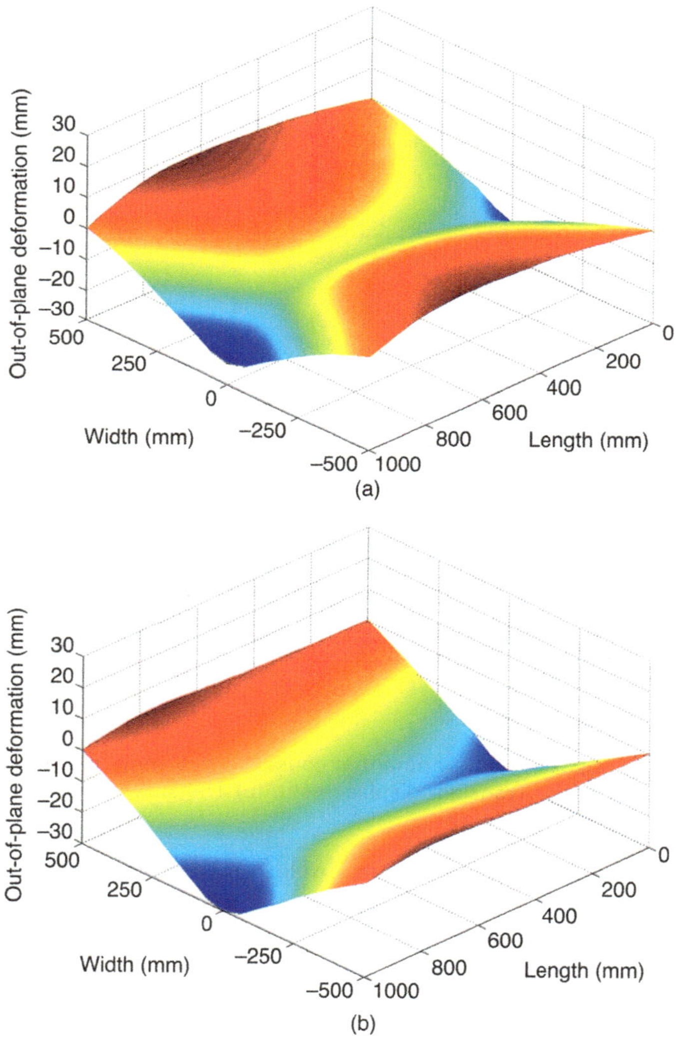

10.19 Effect of gravity loading when seam welding 13-tack assemblies. (a) 10 mm 13-tack, no-gravity. (b) 10 mm 13-tack, gravity included.

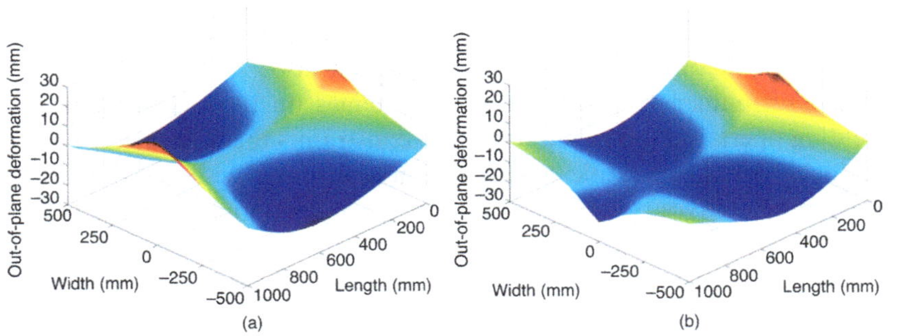

10.20 Effect of gravity loading on 5-tack assemblies. (a) 10 mm 5-tack, no gravity. (b) 10 mm 5-tack, gravity included.

Gravity can also influence the intervention of buckling. Figure 10.20 compares the 'inverted' 10 mm, 5-tack geometry with and without gravity loading. The combination of a relatively flexible tacking scheme and gravity loading has tipped the balance towards buckling. Paradoxically, the RMS angular deformation of the buckled shape is one of the smallest obtained, but the deformation range and RMS out-of-plane displacements are poor.

These studies, covering three different tack lengths and several tacking frequencies, show that if a tacking scheme is used to hold components in place, considerable thought has to be invested in planning the layout and execution of the scheme because many of the possible choices have contradictory influences. The following provides general guidance from the specific set of examples investigated here.

- The distortion caused by tacking is dominated by angular contraction and depends on the size and shape of the tack weld and consequently the specific heat input (J/mm). To a first approximation, the angular distortion generated does not depend on the number or frequency of tacks and is very similar to the levels that would be generated by a continuous weld using the same welding process parameters. This seems to apply despite the fact that the absolute heat input of the tacks over the seam length will be very much less. However, longer tacks (occupying more than about 15% in these examples) begin to generate some longitudinal hogging distortion, which flattens the plate to an extent.
- The obvious function of tacks is to maintain the relative in-plane and out-of-plane edge positions of the components to be welded during the following continuous weld. However, they also provide bending stiffness along the centreline and this is important if the plate is subject to out-of-plane gravity force during the seam-welding operation, when clamps have been removed. Large angular distortions will also be generated in

the tacking operation if gravity force is present and there are too few mechanical clamps to maintain centreline stiffness.

- The main influence of different tack schemes on the following seam-welding operation lies in the structural constraints that the tacks impose on local thermo-mechanical deformations associated with passage of the heat source. Layouts with many closely spaced tacks are not as effective in this respect because the constraints are higher. However, using fewer tacks contradicts the stiffness requirement of the previous point and the essential need is to find a compromise between these conflicting requirements. For the case considered, fewer but longer tacks provided a more effective solution.
- Placing tacks on the opposite side from the main weld to give a 'balanced' welding condition was not particularly effective in the cases considered. Angular distortion was sometimes reduced, relative to 'same-side' tacking, but buckling tended to negate any benefits, particularly in the stiffly tacked cases where buckling was more likely.

These points are for a specific case, but they demonstrate the value of simulation in understanding the effects of different procedures without having to carry out extensive trials.

10.4 Clamping effects

Large, flexible plate structures can be awkward to handle in a fabrication shop and the methods used to support and/or restrain deformations during manufacture have been found to be critical in terms of distortion control.[1] If some kind of stiff frame or support is used, it might be assumed that clamping the fabrication to the frame should help to maintain the required shape during welding operations. Practical experience does not encourage this view, but it is useful to be able to understand why clamping is usually not as beneficial as might be imagined. Computational simulation provides a tool to make such an examination. Accordingly, the simple butt-welded plate geometry considered in the first of the tacking studies was used as a model to explore support and clamping. Three different forms of boundary

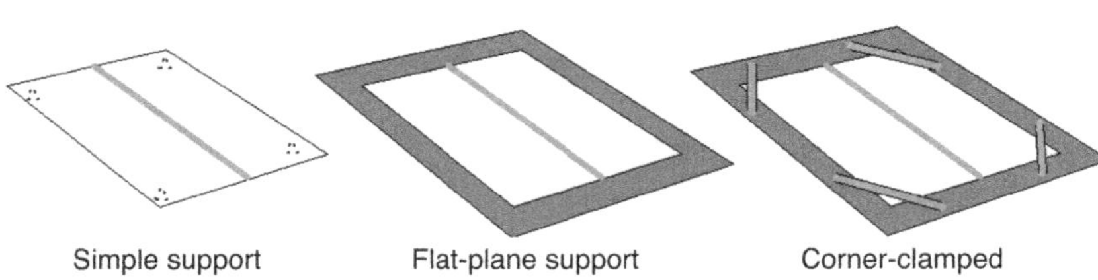

10.21 Alternative support conditions for butt welding.

condition were applied, as shown in Fig. 10.21. These layouts represent common industrial practice.

All models were subject to gravitational loading. The simple-support condition provides the least constraint, although note that there is a symmetry constraint along the centreline. The flat-plane support condition should provide better support against gravity loading and the plates are free to rise from the flat surface in this instance. This condition was simulated by attaching the plates to a large number of LINK elements that had a high stiffness in compression and a low stiffness in tension. The corner-clamped case restrained vertical displacement along the lines of the simulated clamps. Identical thermal inputs were used for the three cases, these being the same well-researched conditions used in earlier studies. No differences in heat conduction to the support fixtures were assumed. Given the nature of the likely mechanical behaviour, full, transient, thermo-elastic-plastic analyses were applied.

The outcomes of the analysis are seen in Fig. 10.22. The corner-clamped case is also shown before release of the clamps, showing that clamping has been quite effective at this stage, except near the ends of the weld. (Note that the near end in the isometric view corresponds to the end of the weld pass and that the transient analysis properly reflects the difference in thermal conditions between the start and finish of the weld.) It is immediately obvious that the corner-clamped case has deformed in the opposite sense from the unrestrained cases, both in terms of angular deformation and longitudinal bending. Understanding the reasons for this marked difference required a frame-by-frame comparison of the deformations taking place in the unconstrained cases and the corner-clamped case. As noted in previous chapters, transverse angular contraction as the heat source passes is normally accompanied by downward local bulging in the longitudinal direction as the parent material surrounding the weld zone expands. In time, this expansion is recovered and the material contracts longitudinally, leading to hogging deformation, due to the V-shaped cross-section. However, in the clamped case, initial downward displacement is impossible and the hot material can only bulge *upwards* longitudinally, thus forcing the cross-section into an inverted V-shape. As a consequence, this bends the joint transversely and permanently. (This can be seen in the end regions of the plate, even before release of the clamps.) As the axis of post-weld contraction is now slightly above the neutral axis of the cross-section, an elastic stress pattern corresponding to a sagging bending distribution is generated and evidences itself when the clamps are removed. It is also worth noting that unstable deformation patterns could be seen in the transient animations at the point where the weld was around 75% complete.

A more direct comparison of outcomes is given in Fig. 10.23. This shows that the corner-clamped case does not give a better result than the free cases,

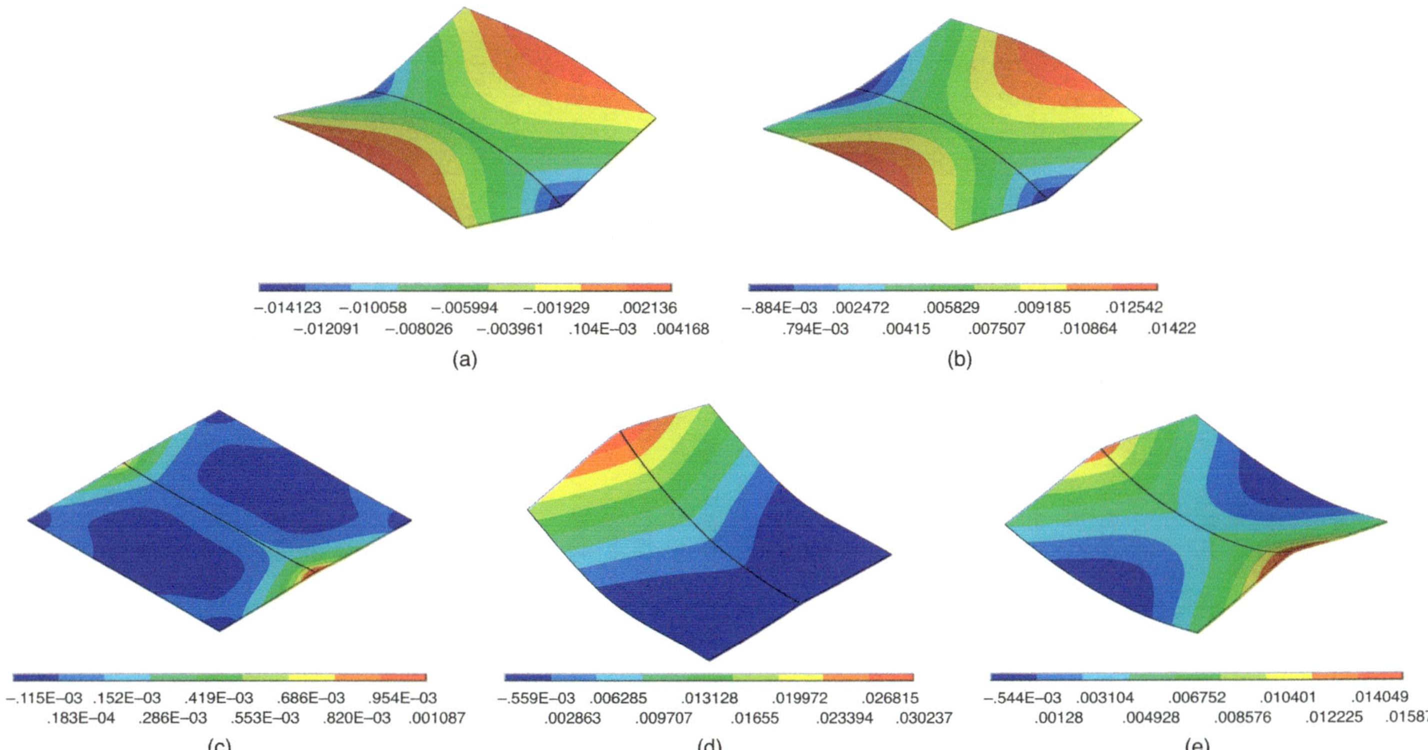

10.22 Simulated out-of-plane deformation. (a) Simple support. (b) Flat-plane support. (c) Corner-clamped support before clamp release. (d) Corner-clamped one end released. (e) Corner-clamped support after clamp release (courtesy P. Mollicone).

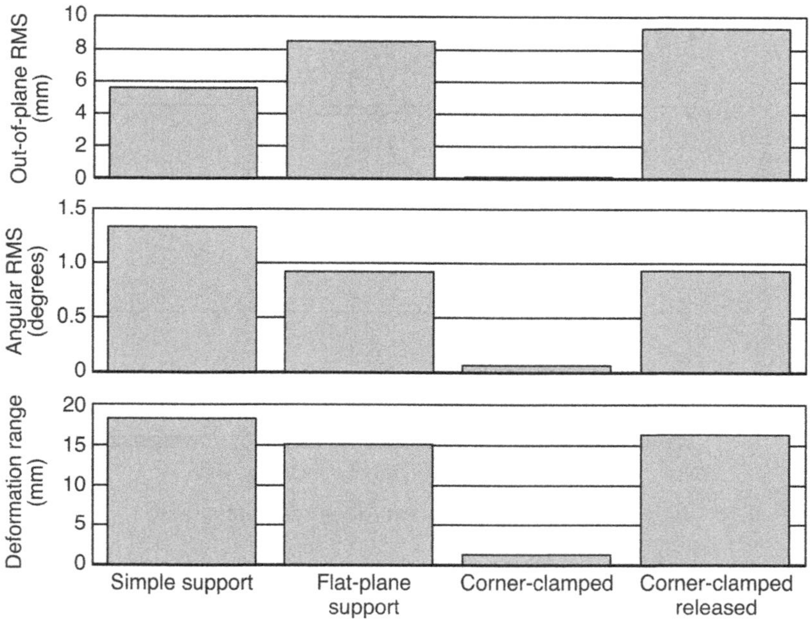

10.23 Distortion levels for different forms of support.

notwithstanding the out-of-plane influence of gravity force in the simple-support condition. Indeed, the intervention of buckling seen in the clamped case is a feature to be avoided. As with the tacking review, it appears that too much restraint can have negative consequences.

10.5 Residual stress in butt welds

An introduction to the effects of welding on residual stress was given previously in section 4.7. Figures 4.11 and 4.12 showed approximate estimates of the longitudinal residual stress pattern, based on applications of the mismatched thermal strain (MTS) algorithm. However, it was noted that the accuracy of these estimates is poor and that a full, transient, thermo-elastic-plastic analysis is necessary if there is a need to predict accurate local and maximum values of residual stress in a given case.

Lindgren[2, 3] concluded that, if residual stresses are to be determined, simulations must at least account properly for elastic strains, thermal strains and an inelastic strain component (normally plastic strain). Radaj[4] identified thermal stress as the main driving factor for welding residual stresses. However, it is also agreed, not least by the present authors, that it is necessary to include metallurgical transformation strains, if these are relevant (see section 4.3). This is especially the case if any of the material transforms at a relatively low temperature, where it also develops significant

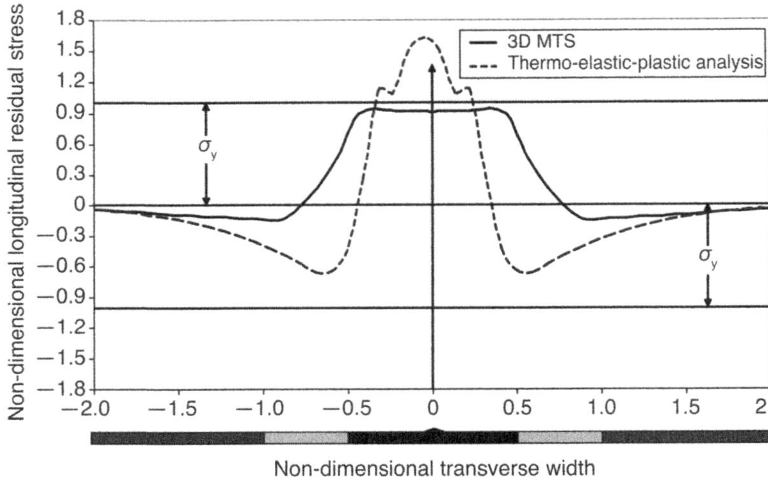

10.24 Various evaluations of longitudinal residual stress.

mechanical strength. Other studies on low-carbon steel by Free and Goff[5] and on aluminium alloy by Zhu and Chao[6] contradict this conclusion and suggest that it is not essential to include such non-linear, temperature-dependent properties in the formulation. However, it should be necessary a priori, unless proved otherwise for a specific case.

Figure 10.24, repeated from Chapter 4, contrasts the previous algorithm-based, longitudinal residual stress patterns with just such a thermo-elastic-plastic analysis, where the detailed thermal properties used in the analyses of Chapter 9 were applied. The '3D MTS' result derives from application of the MTS algorithm fictitious cooling outputs to a 3D elastic structure. The stress in this figure is computed at mid-length and mid-thickness of a butt-welded plate.

The differences between the algorithmic and thermo-elastic-plastic determinations are substantial. The more accurate version shows a higher maximum stress, due to inclusion of a work-hardening material property in the formulation and there is also a larger compressive residual stress in the regions flanking the weld zone. However, it is not always clear that maximum accuracy is invariably required to determine questions such as the effects of residual stress on the fatigue performance of a given structure or on its tendency to buckling failure. It may therefore be of interest to explore residual stress predictions based on simpler modelling approaches, such as discussed in Chapter 8. The residual stress results of the two reduced solution 'hybrid' analyses – the sequential hybrid (SeqH) and the simultaneous hybrid (SimH) methods – will therefore be compared with a transient, thermo-elastic-plastic analysis.

In both the SeqH and SimH methods, the true transient temperature inputs are replaced by the simple cross-sectional maximum temperature profile, derived from a 2D thermal analysis. It is also assumed in both cases that the transverse contraction depends entirely on the cooling behaviour of the fusion zone and the transverse expansivity of parent material is therefore set to zero. Element birth-and-death options are employed to simulate weld filling and the fusion zone elements are cooled from a reference temperature.

In the SeqH analysis, the mechanical behaviour of transverse strips is computed using thermo-elastic-plastic analysis, with non-linear material properties as outlined in section 5.5 for strength characteristics and in section 5.8.1 for expansivity and phase transition strains. Stepwise analysis can be used with this method to reflect a certain degree of transient behaviour but the results shown later are based on a single step. In the second stage of the analysis sequence, the MTS algorithm is applied to the angularly distorted plate, via fictitious cooling loads (see section 8.2.3) and using elastic material properties and linear coefficients of expansion (transverse expansivities in the weld and parent material are set to zero). In the SimH analysis, thermo-elastic-plastic analysis is applied to *both* transverse and longitudinal actions. A cut-off temperature of 1000°C was used, although a separate study showed that a 600°C cut-off temperature yielded similar results. In the thermo-elastic-plastic case, a cut-off temperature of 1493°C was used, together with isotropic coefficients of expansion.

Longitudinal and transverse residual stresses at mid-length and mid-thickness are shown in Fig. 10.25 for all three types of analysis. The transient, thermo-elastic-plastic result is taken from a study by Mollicone *et al.*[7]

The results show the SeqH model to be unsatisfactory, as the maximum indicated stress is some 30% less than the yield strength of the material and the width of the tensile zone is substantially greater than for the more advanced analyses, as in Fig. 10.24. This is due to the release of load in the 3D model when the plane strain condition is relaxed. Both the SimH and thermo-elastic-plastic solutions capture the elevation of the fusion zone longitudinal stress by work-hardening and the results for these two methods are fairly similar, despite the difference between the use of maximum temperature profile as thermal input in the one case and full transient thermal profile in the other. The transverse stresses are rather small, but again the SimH model agrees better with the transient thermo-elastic-plastic case. This nevertheless suggests that the simple assumption whereby transverse contraction is attributed solely to the fusion zone is effective.

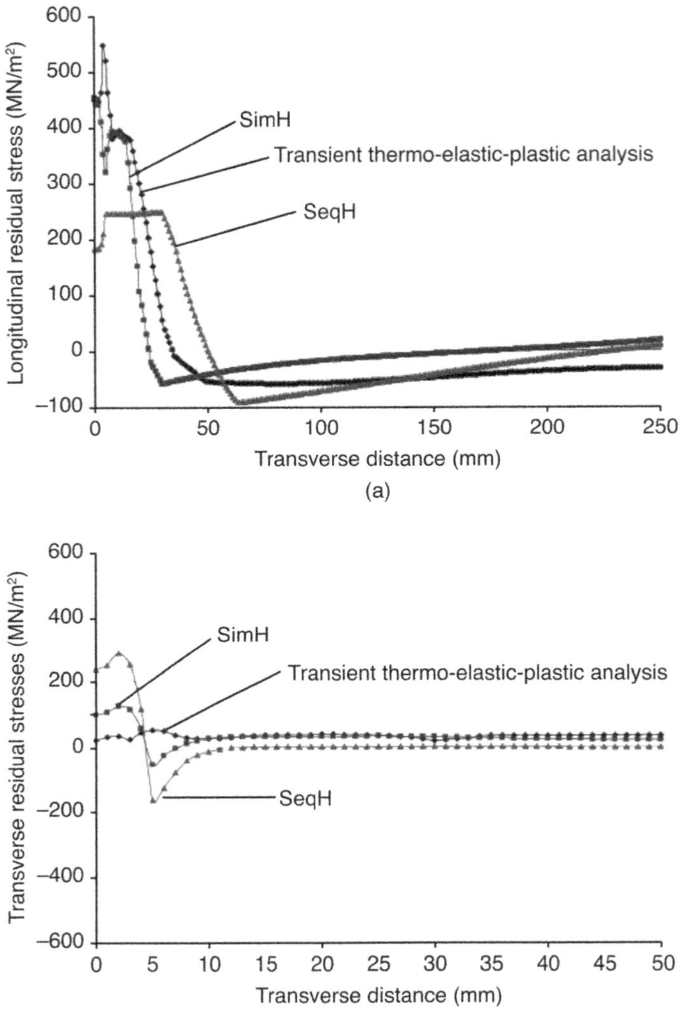

10.25 Residual stress comparison – hybrid models and transient thermo-elastic-plastic analysis. (a) Longitudinal stress at mid-length and mid-thickness. (b) Transverse stress at mid-length and mid-thickness.

10.6 Multiple butt welds

Most fabricated structures will involve more than one weld and, as noted in the study of tacking followed by butt welding, the distortion outcome of each weld will be influenced by its initial conditions, which in turn will depend on the outcomes of previous weld passes. Hence, the addition of further welds should be seen as a potentially non-linear process in effect. The two main factors that influence this are the initial shape of the fabrication, being the result of previous operations, and the initial stress present where the weld is to be placed, including any residual stress from previous welds. A further factor that has a bearing is the distance between welds. If the new weld under consideration is sufficiently far removed from the preceding welds, so that there are no significant residual stresses within the field of the new weld, then it can be treated as a weld in isolation. At the opposite extreme, if a weld is placed exactly coincident with an earlier weld (equivalent to re-fusing the original and with no new filler addition) then the various zones should go through exactly the same cycle as for the original and no *new* distortions should be generated.

If the build-up of butt welds in the fabrication has been tracked computationally, then the finite-element analysis should take care of the changing shape of the structure and the final residual state should also be available, albeit at varying levels of accuracy, depending on the type of analysis used. These data will then establish appropriate initial conditions for further operations. This feature is in itself a strong argument for adopting computational simulation to plan the fabrication procedure. However, the simplified algorithmic methods can also provide some insight into what might happen if a weld is added to a fabrication with one or more nearby previous welds.

Figure 10.26 recalls the original explanation in section 4.5 of the MTS algorithm for longitudinal stress and also uses the same width non-dimensionalisation as in Fig. 4.8 and Fig. 10.24. The stress–strain graph in the top-left oval (a) tracks the history of a point on the width where the heating and cooling strain is given by $\alpha T_m(y) = 2\varepsilon_y$. Starting from zero stress and strain, free expansion would take the strain to the point A and the plane strain condition then brings the strain level back to zero at point B. Release of plane strain allows elastic recovery to point C and free

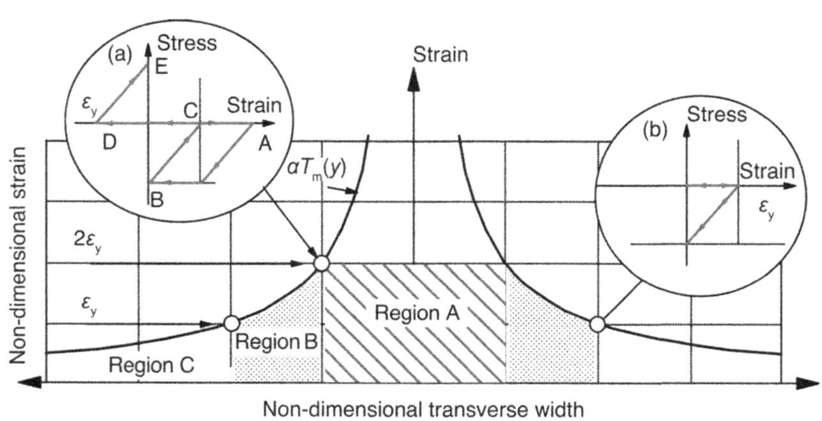

10.26 Basis of MTS algorithm – no initial stress.

contraction during cooling then brings the strain level to point D, with the final imposition of plane strain raising the stress level to the limiting level of tensile yield. Similar arguments (b) define the wider limit of the MTS contraction force area.

If it is now assumed that a butt weld is placed in a region where the global longitudinal stress, due to residual stress or any other cause, is tensile with a value σ_i, the contraction force developed according to the MTS algorithm is reduced from that previously, as shown in Fig. 10.27. Here, the argument for the position of the upper point is unchanged and therefore the width experiencing tensile yield stress is unchanged. However, material at the lower point on the width can now expand and contract by an amount given by $\alpha T_m(y) = \varepsilon_y + \varepsilon_i$ while still returning to the initial global tensile stress value and material outside this zone can expand and contract elastically.

Figure 10.28 completes the argument in a case where a weld is placed in an area where the global stress is compressive. In this case, the contraction force is increased (rather substantially) and there should be greater

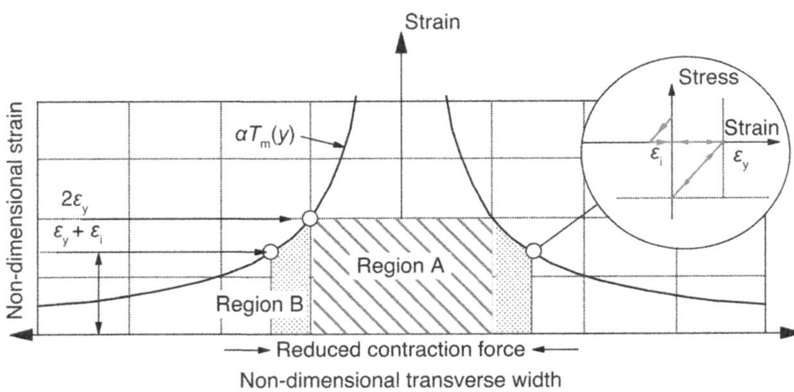

10.27 MTS algorithm – initial tensile stress.

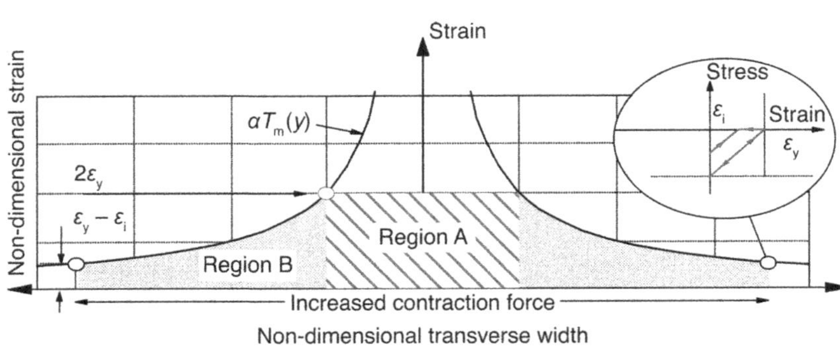

10.28 MTS algorithm – initial compressive stress.

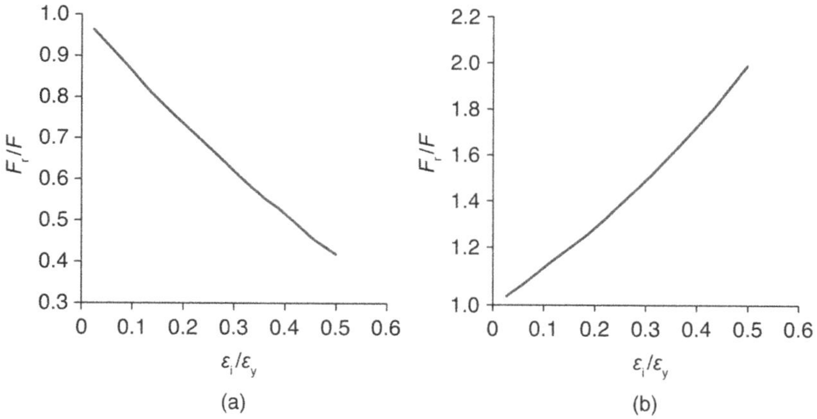

10.29 Effect of initial stress on contraction force in MTS algorithm. (a) Tensile initial stress. (b) Compressive initial stress.

distortion. The altered contraction force F_r has been shown[8, 9] to be given by

$$\frac{F_r}{F} = \frac{\ln[2\varepsilon_y/(\varepsilon_y + \varepsilon_i)]}{\ln 2} \qquad [10.1]$$

where F is the contraction force in the absence of an initial stress. The sign of the initial strain is changed in the case of initial compressive stress. This simple result is of course based on a specific temperature distribution (Rosenthal/Rykalin quasi-static analysis), but nevertheless gives a useful indication of the likely effect on distortion of an initial stress. Figure 10.29 provides the relationship in graphical form for alternative initial tensile and compressive stress fields.

10.7 Conclusion

Various computational treatments given in previous chapters have been applied here to butt-welding operations. Several practical aspects of fabrication were investigated and results given. These include out-of-flatness effects and the consequences of insufficient plate support (section 10.2), the use of tacking in pre-welding assembly and non-linear effects on following butt welds (section 10.3) and the outcomes of applying clamps to constrain thermal strains during welding (section 10.4). Approximate treatments of residual stress computation were given in section 10.5 and the strengths and weaknesses of different computational approaches relative to residual stress determination were discussed. Finally, simple treatments for interacting multiple butt welds were given in section 10.6.

10.8 References

1. Spicknall, M.H., Kumar, R. and Huang, T.D., 2005, Dimensional management in shipbuilding: a case study from the Northrop Grumman Ship Systems Lightweight Structures Project, *J. Ship Prod.*, 21, 4, pp 209–218.
2. Lindgren, L.E., 2001, Finite element modelling and simulation of welding, part 2: improved material modelling, *J. Thermal Stresses*, 24, pp 195–231.
3. Lindgren, L.E., 2001, Modelling for residual stresses and deformations due to welding – 'knowing what isn't necessary to know', *Math. Model. Weld. Phenom.*, 6, pp 491–518.
4. Radaj, D., 1989, Finite element analysis of temperature field, residual stresses and distortion in welding, *Weld. Res. Abroad*, 35, pp 31–38.
5. Free, J.A. and Goff, R.F.D.P., 1989, Predicting residual stresses in multi-pass weldments with the finite element method, *Comput. Struct.*, 32, pp 365–378.
6. Zhu, X.K. and Chao, Y.J., 2002, Effects of temperature dependent material properties on welding simulation, *Comput. Struct.*, 80, pp 967–976.
7. Mollicone, P.G., Camilleri, D., Gray, T.G.F. and Comlekci, T., 2006, Simple thermo-elastic plastic models for welding distortion simulation, *J. Mat. Process Technol.*, 176, pp 77–86.
8. Camilleri, D., 2005, *Support tools for the design and manufacture of thin-plate welded structures*, PhD thesis, Department of Mechanical Engineering, University of Strathclyde, Glasgow, UK.
9. Camilleri, D., Comlekci, T. and Gray, T.G.F., 2005, Design support tool for prediction of welding distortion in multiply stiffened plate structures: experimental and computational investigation, *J. Ship Prod.*, 21, 4, pp 219–234.

11
Simulating welding distortion in fillet welding of stiffened plate structures

DOI: 10.1533/9780857099327.233

Abstract: Results from weld simulations and experiments for fillet-weld procedures are brought together to provide information on distortion in stiffened plate fabrications. The results of thermocouple and thermographic studies are used to expand on thermal aspects, described initially for small-scale fabrications in Chapter 7. Computationally efficient solution methods are applied to large-scale fabrication cases and the results compared with experiments. Special attention is given to application of the transverse contraction strain (TCS) algorithm and ways in which it can be fine-tuned to match different models of transverse constraint. Modelling methods for multiple parallel stiffeners are given and the results of full-scale experiments are compared with the results of a computational case study in which different sequences of stiffener attachment were applied.

Key words: double-sided fillet welds, stiffeners, thermocouples, thermography, computationally efficient treatment, transverse constraint, transverse contraction strain (TCS) algorithm, multiple stiffening, weld sequence.

11.1 Introduction

Various issues treated under the heading of butt welds arise also in the context of fillet-welding construction – namely, support and flatness problems, assembly and the use of weld tacking, residual stress interaction and the effects of carrying out multiple welds, which are an inevitable part of stiffened plate construction. However, as fillet welds are mostly being used to attach stiffeners, the resulting stiffened fabrications will normally exhibit significantly greater out-of-plane bending resistance than the corresponding flat plates. This should reduce some of the gravity-induced and buckling distortion problems typical of flat plate, butt-welded construction, particularly if the stiffeners are pre-tacked before completion of the main attachment welds. That, of course, depends on the assumption that the assembled fabrications offered up for welding behave integrally, in the structural sense.

The thermal simulation of fillet welds is often more problematic than for butt welds. There is usually no weld edge preparation in fillet welding of thin materials and the cross-sectional configuration of the fillet tends therefore to be less consistent and less predictable than for butt welds. A limited description of the thermal complexities involved in analysing a two-pass pair of fillet welds was given in section 7.8 but more experimentally based detail will now be given in this chapter (section 11.2). Computationally efficient (CE) models applied to small-scale fabrications were also contrasted in section 9.4 with computationally intensive thermo-mechanical models for this single-stiffener geometry. This comparison will now be extended to large-scale fabrications.

Reduced solutions are of particular value for large-scale stiffened panels involving several welds and such treatments are explored in sections 11.3 and 11.4. The transverse distortion behaviour of stiffener fillet attachments is particularly uncertain, and section 11.3.1 considers different CE models in the context of the transverse contraction strain (TCS) algorithm. In multiple-stiffener designs, the order of welding stiffeners is a source of potential variation and a case study is used to show how this can be modelled and how the outcomes may differ. These results are compared with experiments.

11.2 Plates with double-sided continuous fillet-welded single stiffeners: thermal aspects

Figure 11.1 shows experimental temperature transients at various cross-sections along the weld length, during the first-pass weld on a stiffener in the small-scale test series. The welding process parameters were held constant. This shows substantial but consistent variations in temperature magnitude between the first, second and third rows (A, B and C) of transversely positioned thermocouples along the weld length on the stiffener. At 20 mm height on the stiffener, there is a $-13^{\circ}C$, $+14.4^{\circ}C$ variation in temperature between the different thermocouple rows, with smaller but corresponding variations at greater distances from the weld. Some of this variation may be related to end-boundary conditions, but it is more likely that minor fluctuations in arc position and thermal gap between the stiffener and plate have an influence. Figure 11.2 shows that at points equidistant from the heat

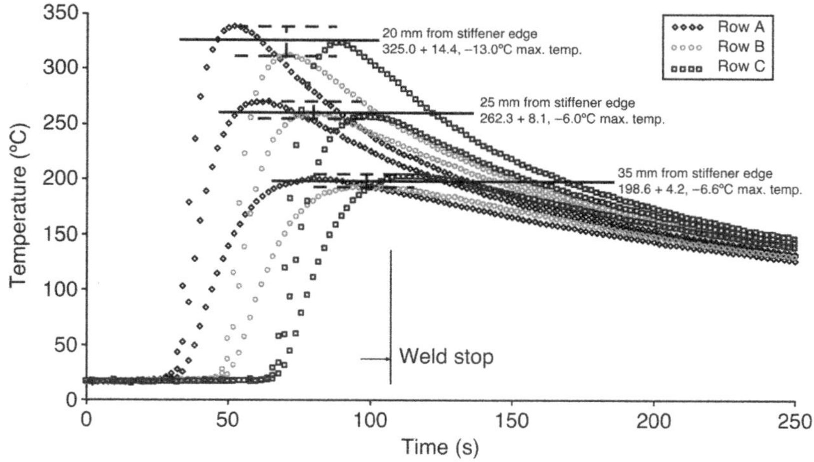

11.1 Temperatures on stiffener at three transverse cross-sections.

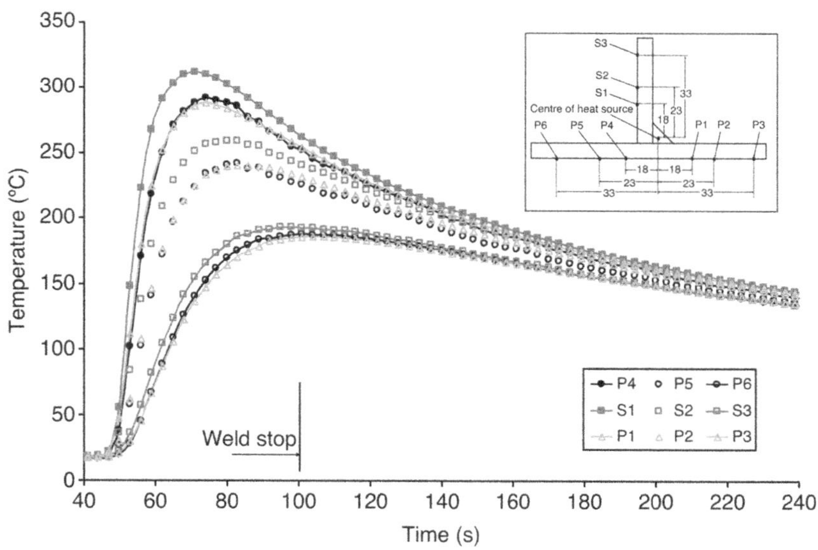

11.2 Temperatures at mid-length – equidistant transverse points from heat source.

source on the plate, temperatures are similar, but the stiffener temperatures at corresponding distances are significantly higher.[1]

This is confirmed by the thermographic study of Fig. 11.3, which shows higher temperatures over larger areas in the stiffener than in the base plate (see Fig. 11.3(a)).[2] This is to be expected for several reasons. Assuming that the heat from the arc and weld fusion zone is equally directed to the stiffener and the plate, heat can flow from the weld in one transverse direction only in

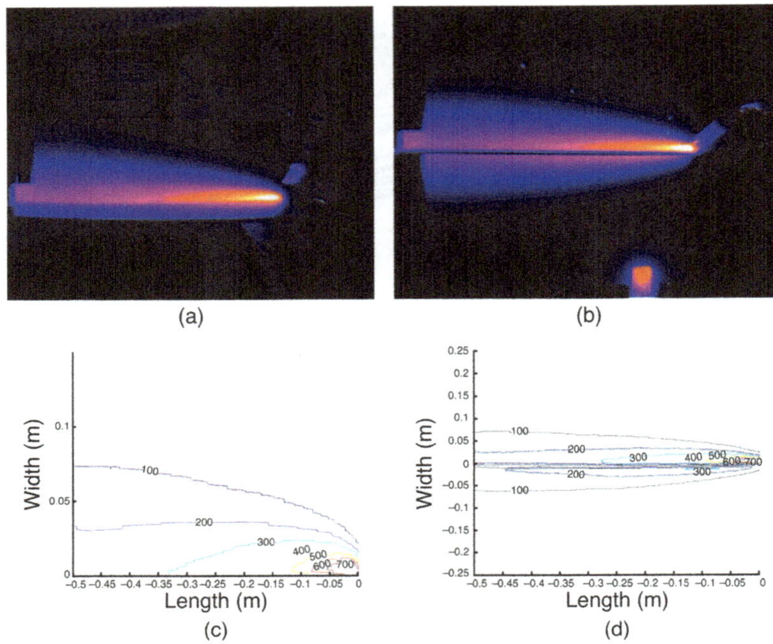

11.3 Thermographic image of (a) elevation view of stiffener and (b) plan view of plate. Converted temperature contour plots of (c) elevation view of stiffener and (d) plan view of plate.

the stiffener but in two directions in the plate. Moreover, the stiffener is slightly thinner than the base plate and the convection environment is different.

Further complications arise in modelling the second-pass fillet weld, discussed in part previously in section 7.8. The thermocouple readings in Fig. 11.4 again show higher readings on the stiffener but also reveal a time-lag in reaching maximum temperature in the plate at equal distances on the opposite side of the stiffener from the weld. This is readily attributed to a secondary flow of heat through the stiffener and the already completed first weld. This suggests also that there was little heat flow across the discontinuity between the edge of the stiffener and the base plate during either pass. The temperatures developed in equivalent positions in the second weld pass were all higher than in the first pass and this is attributed mostly to the increase in input energy, mentioned earlier in section 7.8.

It can be appreciated therefore that considerable care is needed to establish a representative fusion zone cross-section and corresponding thermal inputs for the two fillet-weld passes, along the lines of Fig. 11.5 and Table 11.1. These are large welds in relation to the plate and stiffener thicknesses. Note that, although the energy input was greater for the second

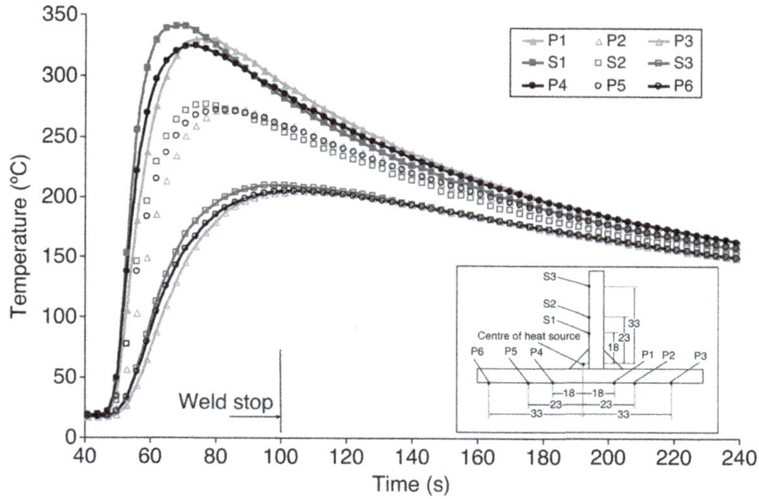

11.4 Fillet weld – second-pass temperatures on stiffener and plate.

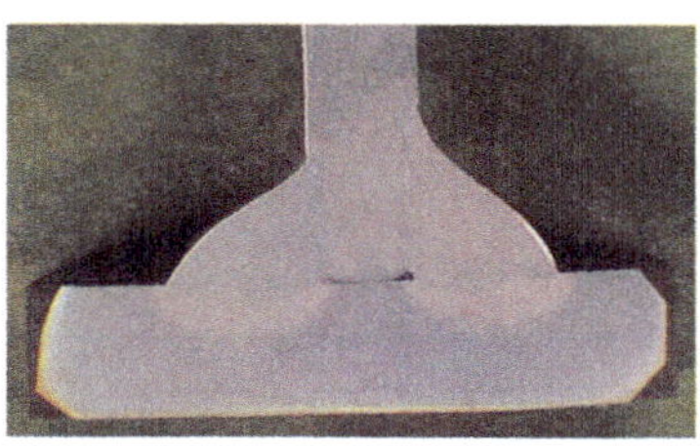

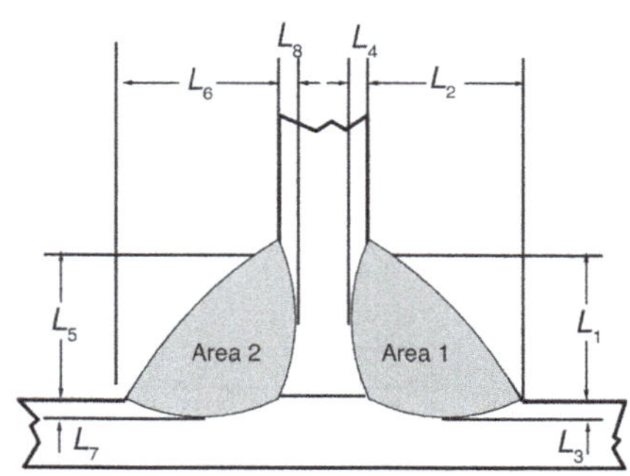

11.5 Fillet welds – first pass on right (area 1) and second pass on left (area 2).

Table 11.1 Fillet-weld configuration dimensions – small-scale tests

	Second pass Area 2	First pass Area 1
Area (mm^2)	38.2	43.9
Vertical weld size L_5 and L_1 (mm)	6.9	9.4
Horizontal weld size L_6 and L_2 (mm)	7.8	7.1
Penetration on plate L_7 and L_3 (mm)	0.9	0.7
Penetration on stiffener L_8 and L_4 (mm)	0.9	0.3

weld pass, the cross-sectional area of the fillet-weld fusion zone was smaller at this particular cross-section. There was a marginally greater ratio of penetration area to fillet-weld area and this may account for the absorption of more energy in the second fillet weld.

11.3 Plates with double-sided continuous fillet-welded single stiffeners: computationally efficient thermo-mechanical treatment

Section 9.4 described the distortion behaviour of the small-scale, relatively narrow, stiffened plates, discussed in the previous section. It was also shown (Table 9.1) that the TCS deformation algorithm tended to underestimate the angular deformation. Transient deformation records (Fig. 9.13) show that approximately the same angular deformation developed during each weld pass and that there was little further deformation in the post-completion period. The dimensions of these small test specimens, in terms of width and height, mean that they would probably behave mechanically as 'beams', i.e. plane sections remaining plane and with linear strain distribution. The small values of final longitudinal sagging curvature shown in Table 9.1 (of the opposite sign from the predictions) suggest that the effective axis of longitudinal contraction was probably fairly close to the neutral axis of the cross-section, due to stiffener cooling.

Figure 11.6 shows a more typical, large-scale, stiffened plate assembly of much greater width and length, where the outer regions of the plate would have been less effective in terms of developing longitudinal bending stiffness, due to large-displacement effects.[3] Figure 11.6 also shows the adaptation of

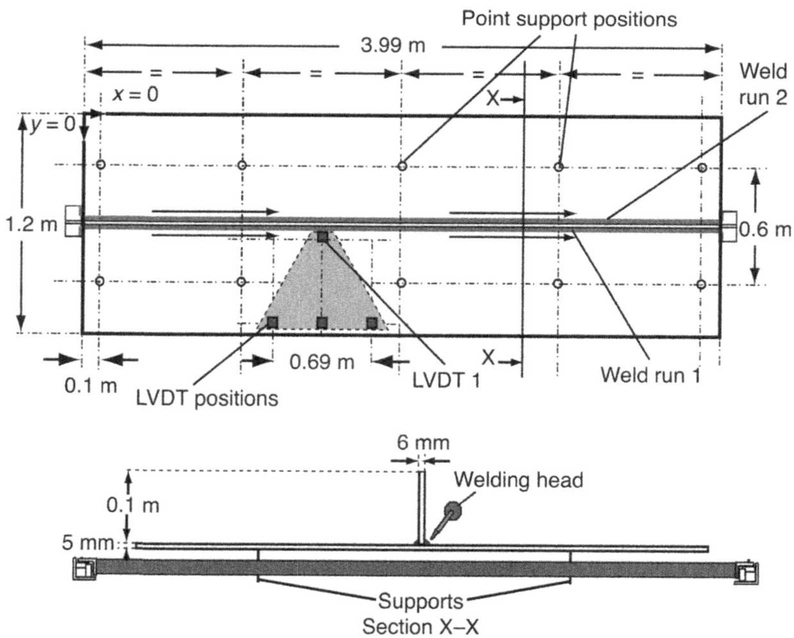

11.6 Large-scale stiffener welding and deformation measurement rig.

the moving support table for the fillet-welding trials. Reduced solution methods are particularly useful for large-scale fabrication models, given the much reduced computational budget. The computational application of algorithms (see section 8.2.3) to two double-sided fillet-welded large-scale cases (designated LS1 and LS2) is now described. These large-scale trials differed from the small-scale tests in various ways: smaller fillet welds were used, with an energy input of 1.14 kJ/mm relative to the input of 1.28–1.41 kJ/mm used in the small-scale study; the base-plate material was significantly thinner, at 5 mm; and the stiffener height was less, at 100 mm instead of 150 mm.

Figure 11.7 shows distinctly different transient behaviour from the small-scale test results in Fig. 9.13. Very little deformation occurs in the monitored transient-displacement region during the weld pass itself, but around 5 minutes after weld completion, a steady growth of angular and longitudinal deformation develops. Figure 11.8 shows a strong fold in the hogging direction in the transient measurement region and the apparently large angular deformation is one-sided, due to the asymmetric folding process, rather than being similar to the conventional symmetrical angular contraction seen in the butt-welding tests.

Deformation during the second pass is also interesting (see Fig. 11.9 and Fig. 11.10). Expansion of the stiffener as the weld passes has the effect of flattening the previous fold, but then the previous distortion mechanism develops after weld completion, leading to an increase in deformations in the same sense as for the first fillet pass. Note, however, that longitudinal bending during the second fillet pass is not as large as for the first pass,

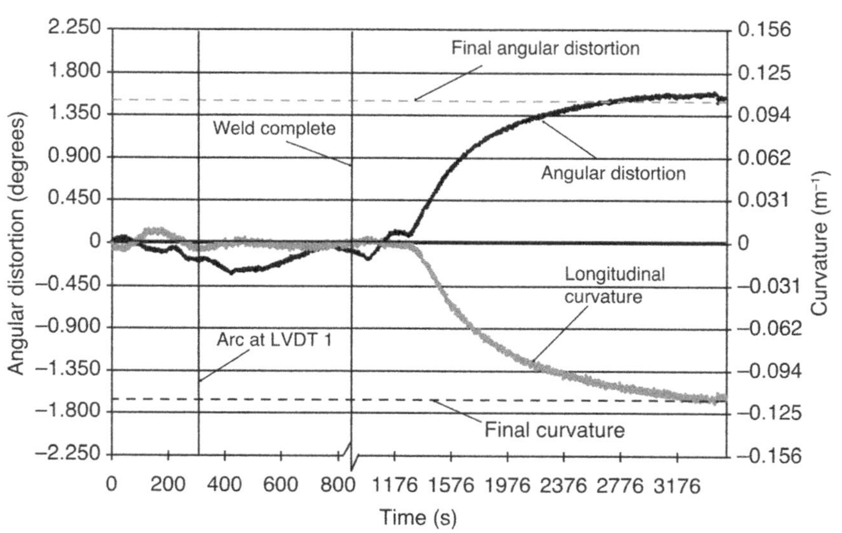

11.7 Transient distortion – large-scale test specimen LS1, first pass.

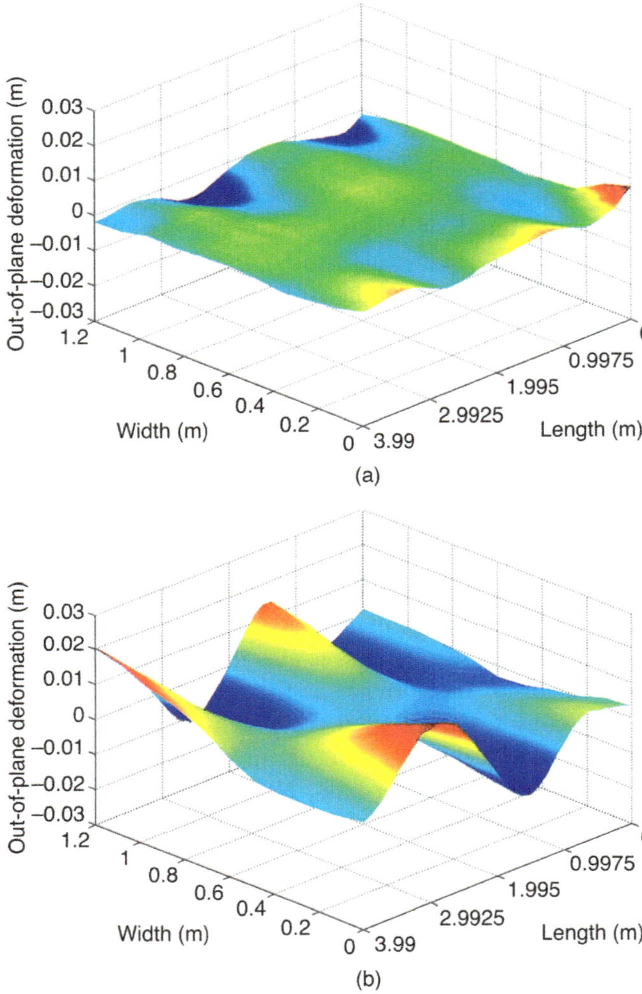

11.8 Out-of-plane deformation due to first pass on test specimen LS1: (a) initial out-of-plane deformation; (b) out-of-plane deformation due to first weld pass.

supporting the 'overlapping weld' algorithmic treatment described previously in Chapter 9.

11.3.1 Transverse contraction

The weld cross-sections were established through averages of direct measurements on three macrographs for each test case, as shown in Fig. 11.11. The dimensional key is given in Fig. 11.12 (expanded relative to Fig. 11.5 to show penetrated areas) and the results are shown in Table 11.2. Note

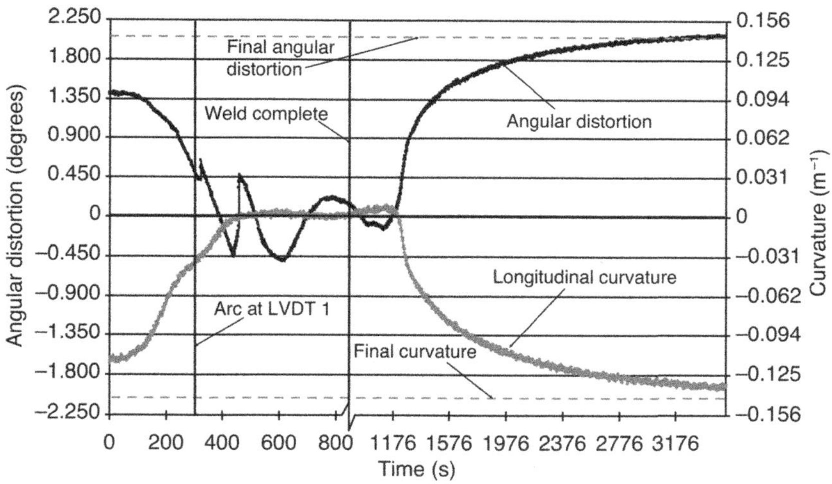

11.9 Transient distortion results – specimen LS1, second pass.

Table 11.2 Fusion zone dimensions and transverse thermal loads for test specimens LS1 and LS2

	LS1		LS2			
Weld pass	1	2	1	2	Average	
Vertical weld size L_1 and L_5 (mm)	7.60	7.17	8.70	7.10	7.64	Fillet
Horizontal weld size L_2 and L_6 (mm)	5.73	6.75	4.82	6.12	5.86	size 6.75
Penetration on plate L_3 and L_7 (mm)	0.34	0.52	0.35	0.48	0.42	
Penetration on stiffener L_4 and L_8 (mm)	1.01	0.92	1.00	0.97	0.98	
Fillet weld area A_{wd} (mm^2)	25.27	27.57	26.14	25.08	26.02	Total
Plate penetration area A_{wp} (mm^2)	1.30	2.32	1.14	1.97	1.68	fusion
Stiffener penetration area A_{ws} (mm^2)	5.16	4.38	5.79	4.58	4.98	area 32.68
Model A $T_T - T_{ambient}$ (°C)	−1069	−1063	−1071	−1068	−1068	
Model B $T_T - T_{ambient}$ (°C)	−664	−752	−528	−708	−663	
Model A+B $T_T - T_{ambient}$ (°C)	−1069	−752	−1071	−708	−900	
Model C $T_T - T_{ambient}$ (°C)	−140	−140	−140	−140	−140	

also that the average weld sizes are smaller than in the small-scale tests shown in Table 11.1.

Different models for the transverse deformation can be applied, reflecting different assumptions about the restraints on fusion zone contraction. The assumption in model A is that the deformation arises from free contraction of the triangular weld zone A_{wd} (similar to the pattern for full-penetration butt welds) plus bending of the stiffener and the plate in the un-penetrated areas

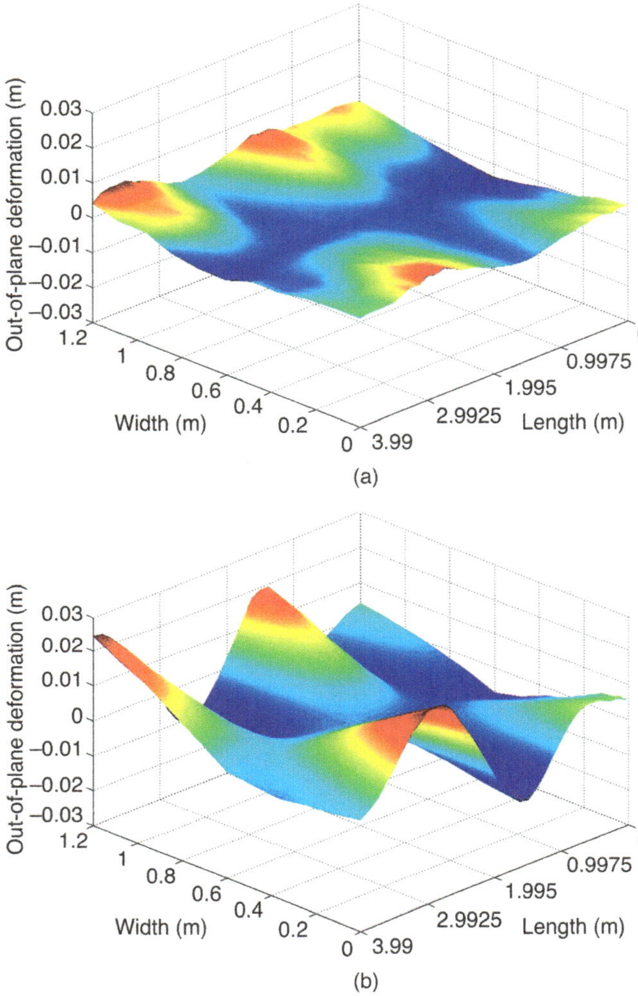

11.10 Out-of-plane deformation due to double-sided fillet welding – test specimen LS1: (a) out-of-plane deformation due to second weld pass; (b) total out-of-plane deformation.

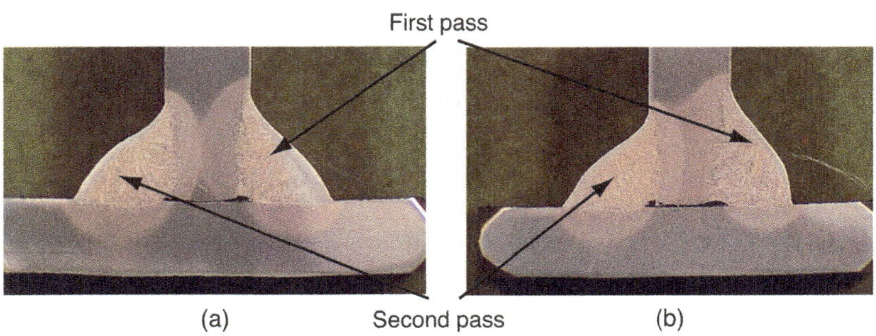

11.11 Macrographs of test specimen: (a) LS1 and (b) LS2.

below the fused zones A_{ws} and A_{wp} respectively. Model A is expressed as

$$T_T = T_{ambient} - T_s\left(1 + \frac{A_{ws}}{A_{wd}}(1 - k_s) + \frac{A_{wp}}{A_{wd}}(1 - k_p)\right) \qquad [11.1]$$

Equation 11.1 identifies the artificial thermal load applied to A_{wd} elements in the fusion zone to give the required transverse thermal strains. The k factors reflect the constraints on the penetrated zones and depend on the ratio of the penetration depths (e.g. L_3 and L_4 for the first fillet weld) to the local perpendicular thicknesses t. The values of k are given by the nomogram in Fig. 9.10. Model A should give the largest temperature loads and angular deformations as this should generate the least constraint (see the four bottom rows in Table 11.2).

Model B is based on the assumption shown in Fig. 11.13, whereby the

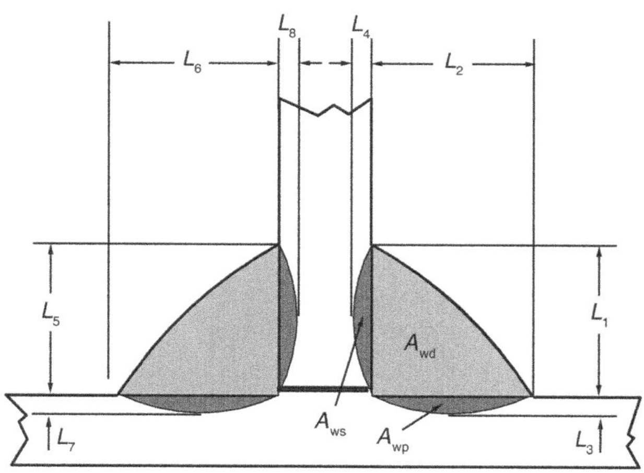

11.12 Dimensional key for Table 11.2.

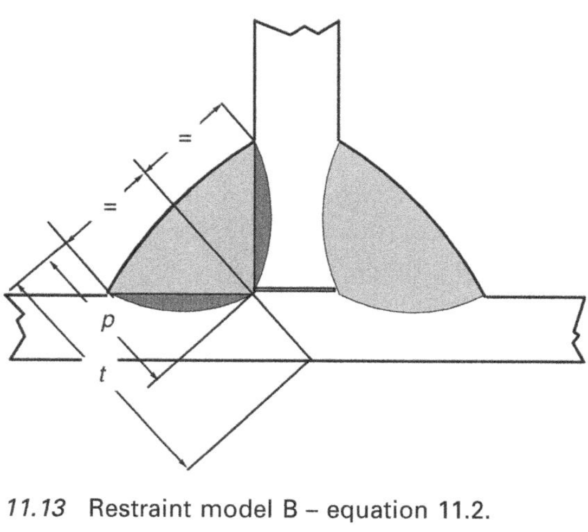

11.13 Restraint model B – equation 11.2.

fusion zone is treated as an inclined partial-penetration weld through equation 11.2

$$T_T = T_{ambient} - T_s\left(1 + \frac{A_{wp}}{A_{wd}} + \frac{A_{ws}}{A_{wd}}\right)(1 - k_f) \qquad [11.2]$$

where k_f is again derived from Fig. 9.10 in terms of the ratio p/t. Note that both models A and B could be applied to either or both welds. A third possibility is to apply model A to the first weld pass and model B to the second. The argument would be that restraint is probably small during the first weld, especially if the stiffener is only lightly tacked to the plate, but restraint is greater for the second weld, where completion of the first weld should have created a stiff structural connection between the stiffener and the base plate. This third assumption is described as model A + B.

Finally, a condition of maximum restraint is embodied in model C, where both welds are assumed to develop a transverse thermal contraction corresponding to yield strain, given by

$$T_T = T_{ambient} - (\varepsilon_y/\alpha) \qquad [11.3]$$

The artificial thermal loads resulting from these different models are shown also in Table 11.2. Despite fairly large variations in weld dimensions, the variation of temperature loadings derived from the TCS algorithm is not high. The effectiveness of these assumptions will be considered later, after discussion of longitudinal contraction modelling.

11.3.2 Longitudinal contraction

The mismatched thermal strain (MTS) algorithm is applied as for butt welds, by defining three regions – an inner zone at yield level, an outer elastic zone and a region between the two where the maximum temperature is approximately inversely proportional to the transverse distance. The critical difference for fillet welds is that the algorithm and the thermal loads are applied also to the stiffener. Moreover, the loading is not symmetric, as the temperature profiles are different for each side of the plate and for the stiffener. This is illustrated schematically in Fig. 11.14 and a flow chart for the process used to find the boundaries of each region in the algorithm and the associated thermal loads is given in Fig. 11.15.

As explained previously in section 9.4 and shown in Fig. 9.11, if there is intermediate cooling to ambient temperature between each weld, the contraction force field developed by the second weld overlaps the first and the increase in force due to the second weld is very much less than the force that would have been developed in isolation. The overlap principle also applies to the thermal loads used for the stiffener. Therefore, as the artificial

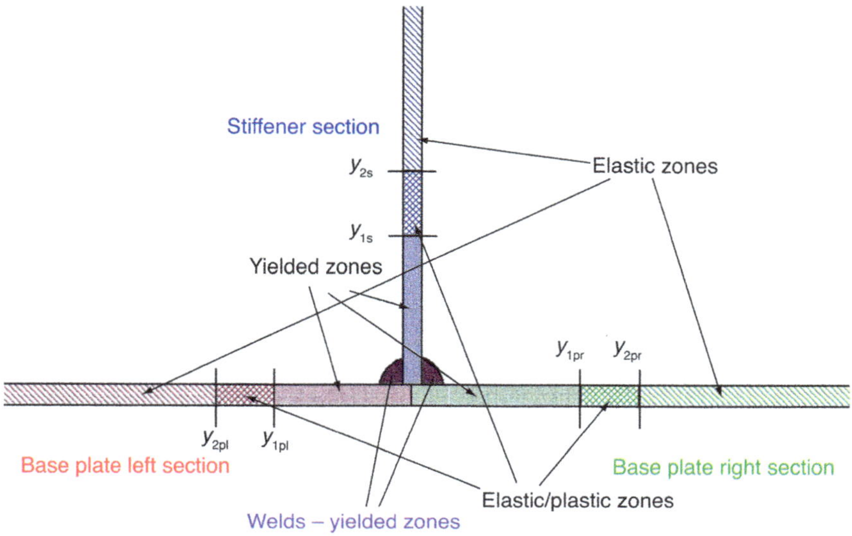

11.14 Thermal loads in fillet welding.

thermal loads arising from the MTS algorithm can be determined for both welds from the outset, it would be possible to apply both temperature loads in a single load step. That would neglect the non-linear effect arising from the fact that the first weld generates a distorted shape, which in turn acts as a starting point for the deformation process caused by the second weld, but the error in using a single step is probably small. A flow chart for the required computations is given in Fig. 11.16.

If the fabrication is *not* allowed to cool to ambient temperature before commencing the second pass, as may often be the case, a new thermal analysis has to be carried out to reflect that and different contraction forces will result. This will be treated in Chapter 12, as a fabrication-related issue.

11.3.3 Computational results and test comparisons

Firstly, it is useful to have a baseline result, based on an initially flat, stiffened plate, although this perfect shape is disturbed slightly in practice by small gravity loading deformations at the support positions. Figure 11.17 and Table 11.3 show the computational results for first and second weld passes, using model A (equation 11.1) for the transverse deformation. These results may be compared with the experimental test results for the two large-scale stiffened assemblies, given in Table 11.4. The distorted shapes of the test specimens at various stages were given in Fig. 11.8 and Fig. 11.10 for specimen LS1; the results for specimen LS2 are in Fig. 11.18.

Several features are evident in Fig. 11.17. The distortion pattern following

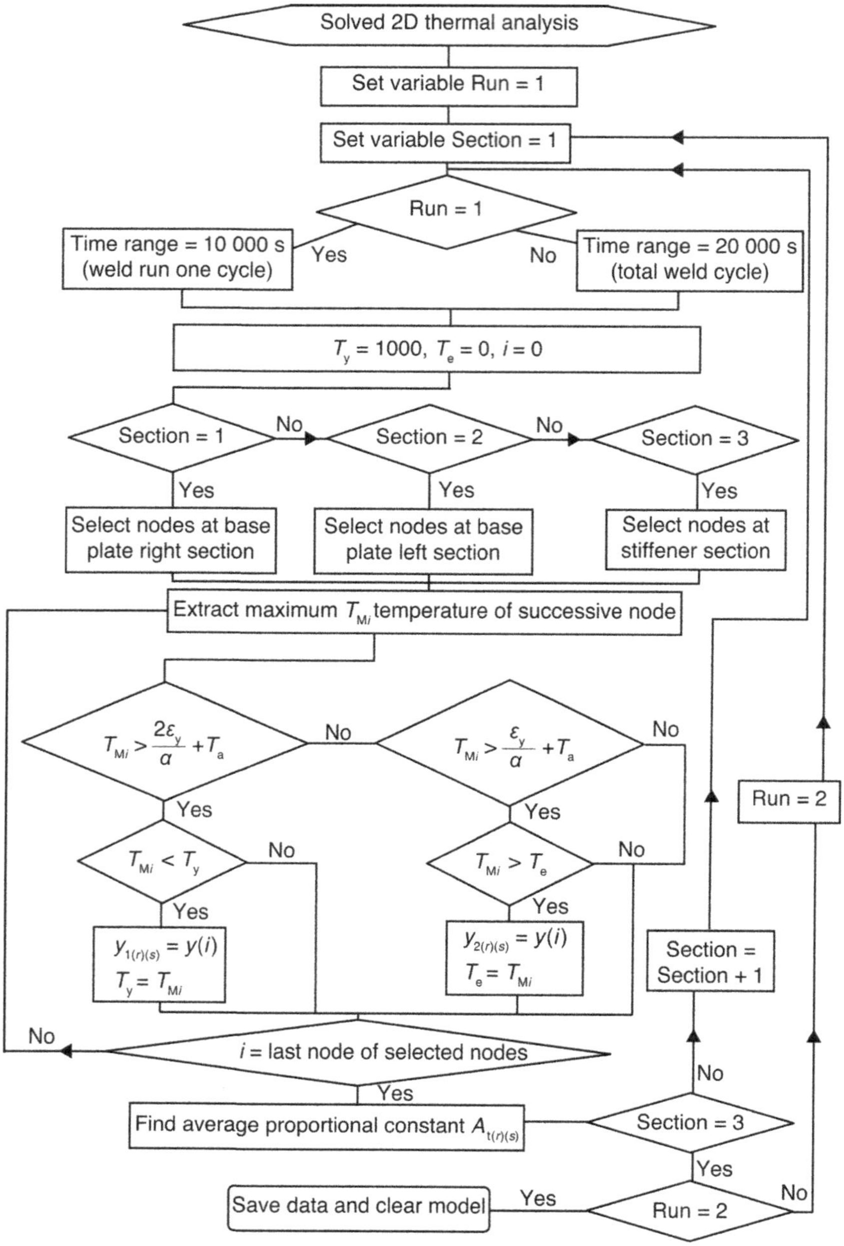

11.15 Algorithm to find the mismatched thermal strain in fillet welding.

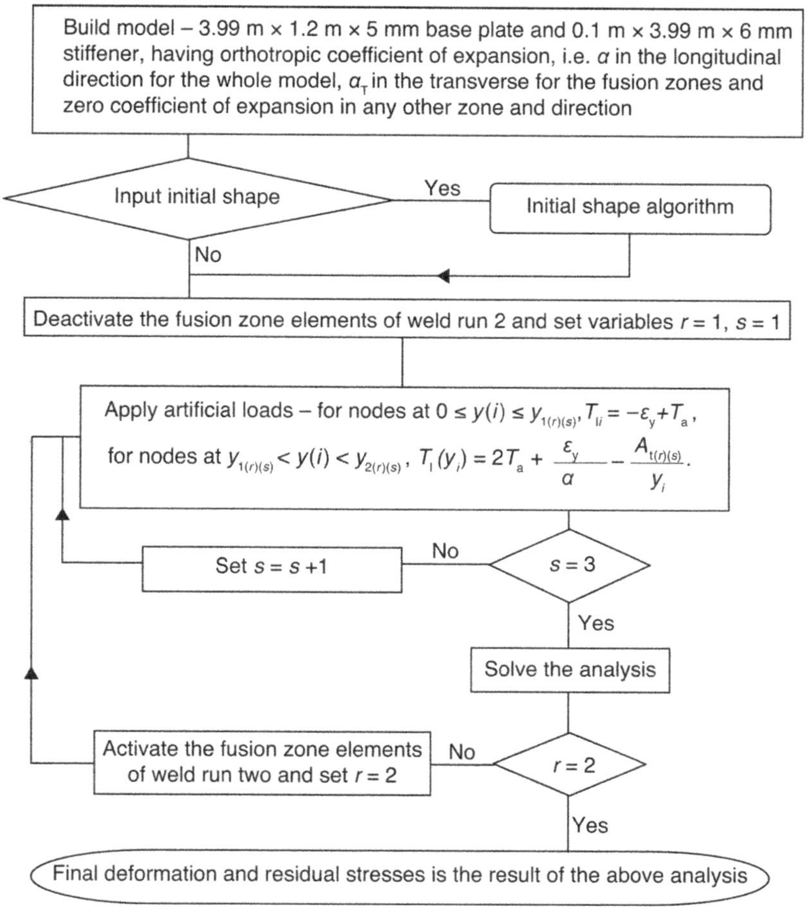

11.16 Structural finite-element algorithm for double-sided fillet welding.[3]

Table 11.3 CE model distortion of initially flat stiffened plate

Weld pass	RMS distortion (mm)	RMS angular distortion (degrees)
First	1.546	0.453
Second	4.584	1.114

Table 11.4 Experimental distortions of fillet-welded stiffened plates

Specimen	Weld pass	RMS distortion (mm)	RMS angular distortion (degrees)
LS1	First	6.63	0.916
	Second	8.12	1.314
LS2	First	5.00	0.90
	Second	6.57	1.30

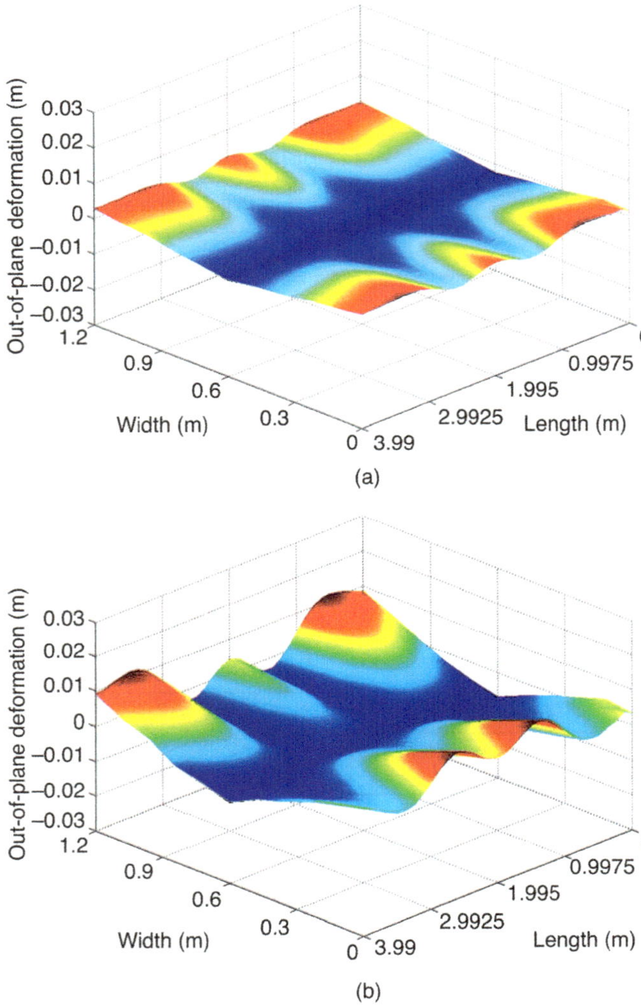

11.17 Predicted out-of-plane deformation for an initially flat plate: (a) first weld pass; (b) second weld pass.

the first pass is characterised by magnification of the initial deformations local to the supports, while the area close to the stiffener remains relatively flat in the longitudinal direction. The angular distortion is more than doubled in the second pass, as might be expected from the use of model A for both welds. Greater angular distortion has a consequent effect on the overall root mean square (RMS) out-of-plane distortion, although there is little sign of overall longitudinal bending, especially local to the central stiffener. Increased deformation in the second pass has generated

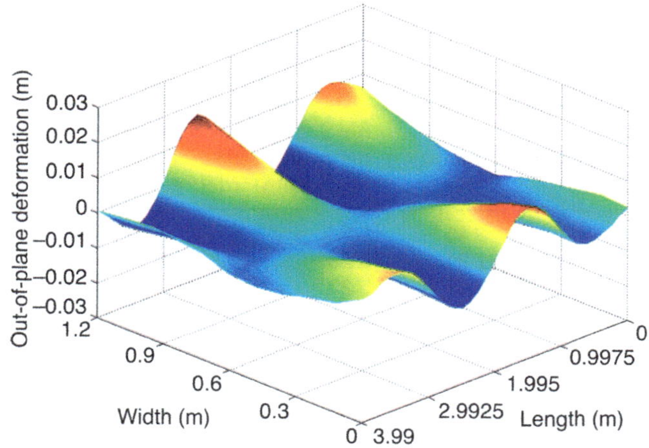

(a)

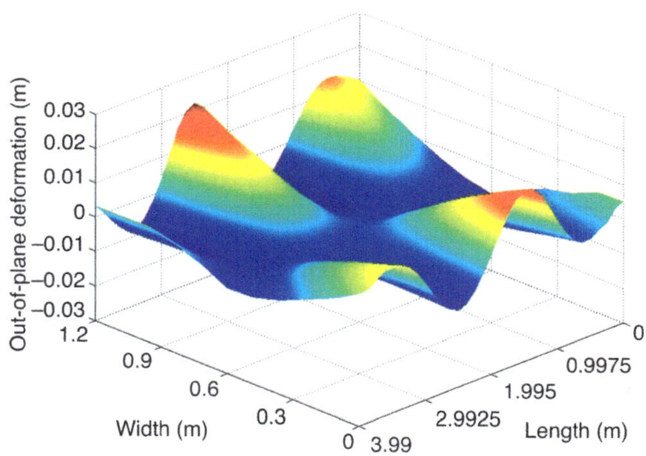

(b)

11.18 Out-of-plane deformation due to double-sided fillet welding – test specimen LS2. (a) First weld pass. (b) Second weld pass.

asymmetric buckling at mid-length, which also increases the average distortion measurements.

Figures 11.8, 11.10 and 11.18, showing the measured distortion patterns of the two test specimens, indicate qualitative similarity to the baseline result, but the average out-of-plane and angular measurements were substantially larger, as is evident by comparing the results in Tables 11.3 and 11.4. It was concluded that the initial shapes are critical to the outcomes, as in the case of the butt-weld trials. Moreover, the doubling of angular deformation from the first pass to the second, seen in the stiffened,

Table 11.5 Computed deformations using initial deformed shapes and alternative transverse deformation models. Comparison with experimental results based on RMS point errors

	Transverse deformation model	RMS distortion (mm)	RMS distortion error (mm)	RMS angular distortion (degree)	RMS angular distortion error (degree)
LS1	A	3.15	5.22	0.645	0.663
First pass	B	2.91	5.25	0.520	0.659
	A+B	3.15	**5.22**	0.645	**0.663**
	C	2.90	5.20	0.443	0.716
LS1	A	5.08	5.59	1.283	0.727
Second pass	B	4.28	5.65	0.949	0.679
	A+B	4.64	**5.59**	1.106	**0.677**
	C	3.59	6.24	0.562	0.976
LS2	A	3.82	2.78	0.624	0.476
First pass	B	3.72	3.02	0.523	0.563
	A+B	3.82	**2.78**	0.624	**0.476**
	C	3.79	3.25	0.506	0.652
LS2	A	4.20	3.55	0.826	0.613
Second pass	B	4.84	3.44	0.854	0.606
	A+B	5.06	**3.18**	1.015	**0.511**
	C	4.74	4.32	0.642	0.957

ideally flat plate simulation, was not found in the tests (see Fig. 11.7 and Fig. 11.9). From this, it was concluded that model A is not correct for the second pass, where the restraint on transverse contraction is greater due to the stiffening effect of the first-pass weld fillet. Hence, further simulations were constructed to take account of the initial out-of-plane shapes and to explore the alternative effects of using the different transverse deformation models. Using a single RMS value for this purpose, as in Tables 11.3 and 11.4, does not discriminate well, as many different shapes can give rise to the same global RMS out-of-plane and angular deformations. Hence the comparisons were based on a different averaging calculation, whereby the experimental result at each point on the test specimen was subtracted from the corresponding computational result. The RMS values of the resulting *errors* were then calculated over the complete plate surface. These results are shown in Table 11.5.

The differences between the outcome errors using the first three transverse deformation models are not large, but model A + B tends to produce the smallest errors (shown in bold in the table). Direct comparison of experimental shapes for test specimen LS2 and the corresponding distortions, recalculated using model A + B, are shown in Fig. 11.19. The computational results capture the actual shape well, including the

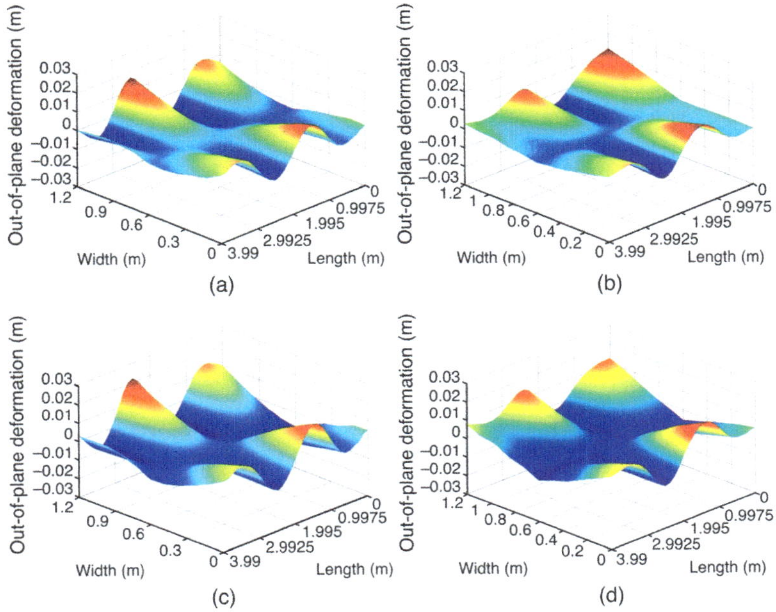

11.19 Result for specimen LS2. (a) First weld pass, experimental result. (b) First weld pass, predicted result. (c) Second weld pass, experimental result. (d) Second weld pass, predicted result.

asymmetric and buckling effects, although the range of distortion displacements is still underestimated by the computation.

11.4 Multiply-stiffened plates: case study on welding sequence

Most stiffened-plate fabrications incorporate more than one stiffener and, given the structural function of the stiffeners, their spacing is likely to be in the range where there will be interaction between attachment welds. The main effects will be related to incremental distortion and residual stress fields. Design configuration questions concerning the thickness of plate to be used and the size, thickness and spacing of stiffeners are therefore not easy to resolve without considering the role of the attachment welds. It is not reasonable to ignore the welding distortion implications, as the design options that will lead to least weight, such as using thin plate and closely spaced stiffeners, are the very conditions that will lead to greater distortion and potential buckling. There are also fabrication questions to be considered, such as the size of welds to be used, the stiffener welding sequence and assembly clamping or tacking procedures. Given the potential range of options and the likelihood that large-scale structures will be

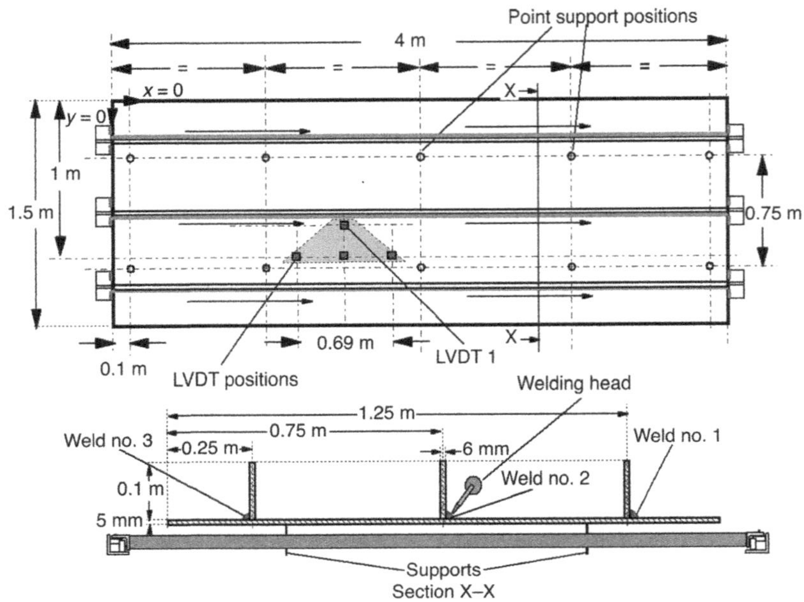

11.20 Multiply-stiffened test plate assembly – welding rig.

involved, computational simulation and reduced solutions have much to offer in terms of guidance information for design and fabrication.

The following study considers a configuration with three stiffeners, fabricated according to three different completion sequences.[4] Single fillet welds are used here, although double-sided fillet attachments are often preferred to create a better structural connection and reduce sensitivity to fatigue failure and corrosion. However, as seen in the previous section, a double-sided fillet attachment can be treated largely as a single weld if proper account is taken of the increase in contraction force and assuming complete cooling to ambient temperature between consecutive weld passes. The computational study was supported by full-scale tests on 4 m × 1.5 m assemblies, formed from 5 mm thick plate and 100 mm × 6 mm thick stiffeners, attached by 6 mm single fillets (see Fig. 11.20). The three attachment welding sequences used were as follows.

- Sequence A: centre stiffener first – weld order 2, 1, 3 using numbering on Fig. 11.20.
- Sequence B: sequential – weld order 1, 2, 3.
- Sequence C: outer stiffeners first, then centre – weld order 1, 3, 2.

All the stiffeners were first tack-welded to the plate at 307 mm intervals (stiffeners were attached using sequence A). Clamping was necessary in this operation, to bring the stiffeners and plates into contact, as the as-received

Table 11.6 Fusion zone dimensions and thermal loads for multiple-stiffener tests

Weld pass and order	Sequence A			Sequence B			Sequence C			Average
	2	1	3	1	2	3	1	3	2	
Vertical size L_1 (mm)	7.04	6.83	5.95	6.78	6.46	5.85	6.10	7.19	6.69	6.5
Horizontal size L_2 (mm)	6.14	6.58	7.32	5.51	6.37	7.40	6.19	5.90	5.89	6.4
Plate penetration L_3 (mm)	0.57	0.57	0.91	0.52	0.50	0.89	0.64	0.66	0.53	0.64
Stiffener penetration L_3 (mm)	1.07	0.82	0.48	1.16	0.84	0.68	0.93	1.07	1.27	0.92
Fillet area A_{wd} (mm^2)	25.0	25.1	23.6	22.5	22.9	23.1	23.1	25.2	22.2	23.6
Plate pen. A_{wp} (mm^2)	2.33	2.49	4.43	1.92	2.13	4.37	2.62	2.59	2.06	2.8
Stiff. pen. area A_{ws} (mm^2)	5.02	3.72	1.89	5.23	3.61	2.64	3.79	5.11	5.67	4.1
Model A $T_T - T_{ambient}$ (°C)	−1077	−1062	−1062	−1085	−1062	−1077	−1070	−1079	−1095	—
Model B $T_T - T_{ambient}$ (°C)	−719	−739	−787	−670	−725	−794	−720	−697	−718	—
Model B+C $T_T - T_{ambient}$ (°C)	−140	−739	−787	−670	−140	−794	−720	−697	−140	—

plates showed more out-of-plane deformation than plates used in previous studies.

The nominal weld process conditions were 217 A, 25.9 V and 5 mm/s travel speed, giving a gross heat input of 1.12 kJ/mm. There were small deviations from these conditions, but these variations were reflected in the fusion zone dimensions and thermal inputs used in the corresponding computational analyses, as were the initial out-of-plane profiles of the tacked assemblies. The weld dimensions, determined as before via the averages of three macrographs for each weld, are shown in Table 11.6, together with the corresponding artificial thermal loads for the different constraint models on the single fillet weld. The transverse contraction models (models A, B, etc.) are applied in a slightly different manner than for the single-stiffener, double-sided fillet example previously and this will be discussed in section 11.4.2.

11.4.1 Experimental study

The transient deformations associated with sequence A fabrication are shown in Fig. 11.21. These traces show very small transient and final distortion levels compared with the pattern shown in Fig. 11.7 for a single, central stiffener. This can readily be attributed to the much greater out-of-plane bending stiffness of the tacked three-stiffener assembly, particularly so for the weld on the centre stiffener. It should be noted that this effect has inhibited angular contraction as well as longitudinal bending, due to the large-displacement interaction between longitudinal and transverse deformations. Complete out-of-plane deformation profiles confirm the improved resistance to distortion exhibited by the multiple-stiffener configuration and these will be shown later in the context of computational/experimental comparisons.

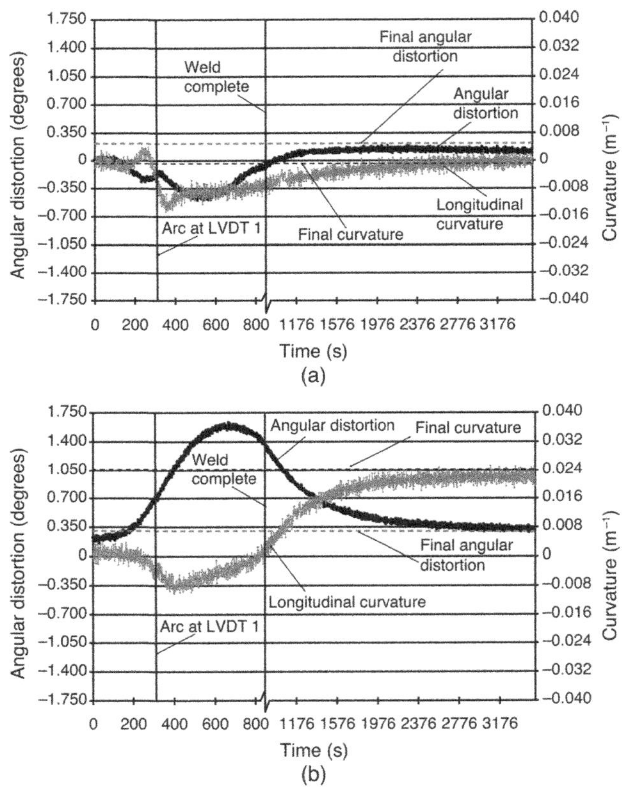

11.21 Transient deformations due to sequence A: (a) centre weld 2; (b) right-hand outer weld 1; (c) left-hand outer weld 3.

(Continued)

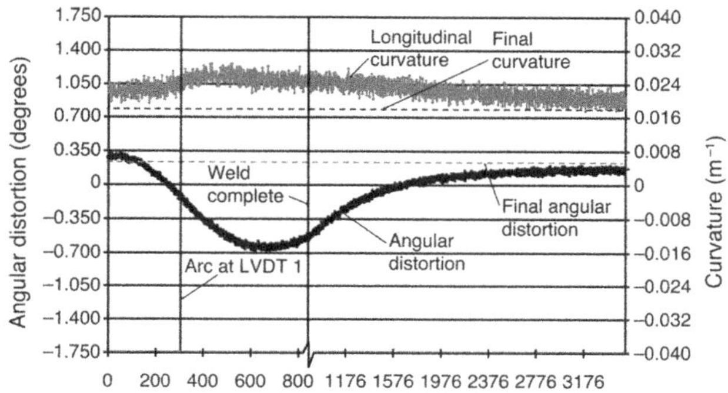

(c)

11.21 Continued.

The effects of welding according to sequence B (sequential right-to-left) are shown in Fig. 11.22. The main effect here is that the left- and right-hand edges of the plate show substantial distortion at each stage (right-hand particularly) whereas the central region remains relatively flat. The free outer edges make these regions particularly vulnerable to local buckling and the periodic deformation along these edges is most likely stimulated by the support reactions.

Behaviour in sequence C (outer welds first) was particularly complex, as one of the tack welds on the central stiffener fractured during welding of the outer stiffeners. The subsequent thermal input in the region of this failed tack directed itself mainly towards the base plate as a result. Fortunately, a thermocouple array had been placed in this region and it was therefore possible to allow for the accidental thermal and structural discontinuities in the computations.

11.4.2 Computational study and comparison with tests

The computational modelling requirements for fabrications with several sequentially executed longitudinal welds in parallel were laid out in section 10.6 in the case of butt welds. The first requirement is to take account of the longitudinal residual stress fields generated by previous welds, along the line of the weld currently being considered. This is dealt with in the algorithmic approach through equation 10.1, repeated here for convenience as equation 11.4.

$$\frac{F_r}{F} = \frac{\ln[2\varepsilon_y/(\varepsilon_y + \varepsilon_i)]}{\ln 2} \qquad [11.4]$$

As previously, F is the contraction force in the absence of an initial stress, F_r is the force altered by the residual stress presence and ε_i is the initial strain

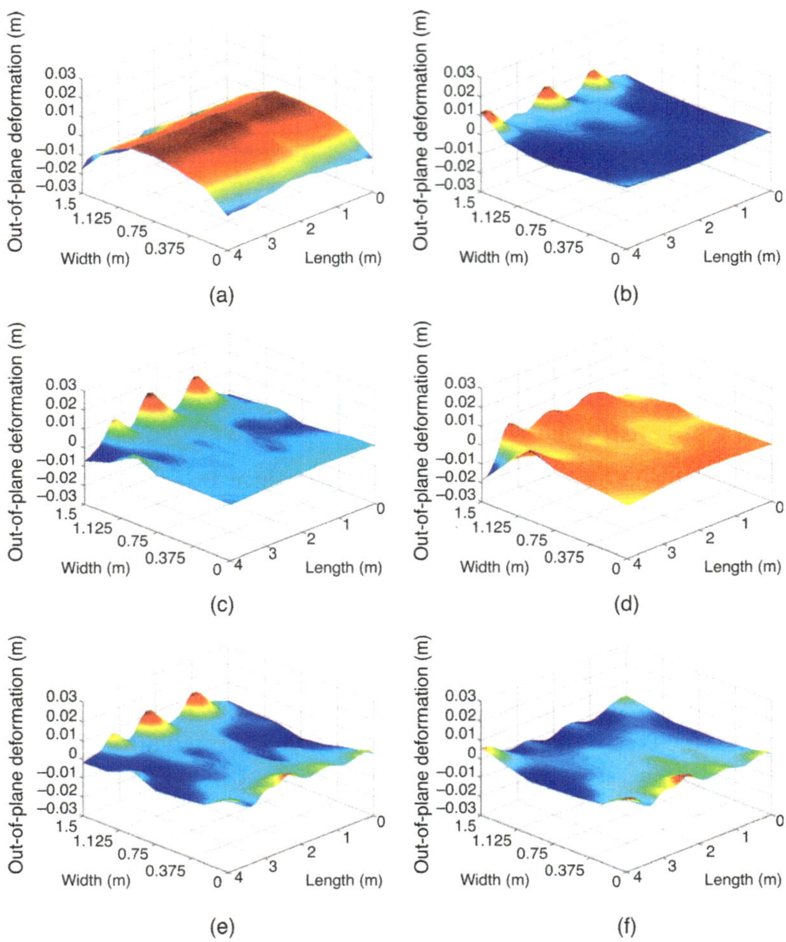

11.22 Experimental out-of-plane deformations due to sequence B (sequential order): (a) initial out-of-plane deformation; (b) deformation after right-hand weld; (c) total deformation after centre weld; (d) deformation due to centre weld; (e) total deformation after left-hand weld; (f) deformation due to left-hand weld.

along the line of the new weld. The primary contraction forces generated by each of the three welds, when applied in a stress-free region, are therefore extracted from the respective thermal analyses. Each stiffener–plate joint is divided into three sectors, as described previously in section 11.3.2 and Fig. 11.14 and Fig. 11.15, to obtain the bounded regions $y_{1(r)(s)}$ and $y_{2(r)(s)}$, together with $A_{t(r)(s)}$ (r is the weld number). Note that the local coordinates are altered such that the origin is moved to the intersection of the stiffener–plate joint under consideration. The mismatched thermal loads are then found in terms of nodal positions and nodal numbers. This process is

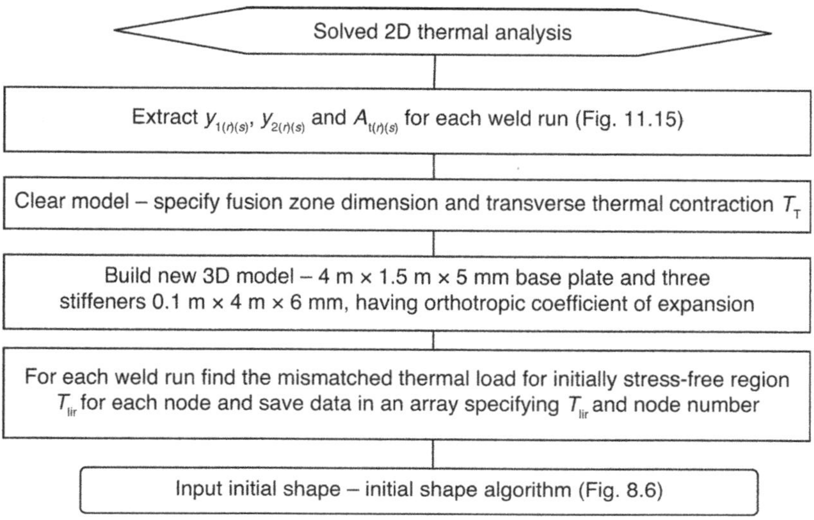

11.23 Algorithm to find mismatched thermal loads and input for initial shape.

illustrated in the flow chart shown in Fig. 11.23. The thermal loads, the fusion zone dimensions with the respective transverse thermal loads and the initial shapes are then entered into the computational process given in Fig. 11.24.

Different models for transverse contraction were investigated, as for the single stiffener case in section 11.3.3. Model A is based, as before, on equation 11.1, on the assumption that the angular distortion is unrestrained and arises entirely from the fusion zone elements. Model B is based on equation 11.2, on the assumption that there is some restraint, as in an inclined partial penetration weld. Model B + C recognises a difference in constraint between the central stiffener weld and the outer stiffener welds: model B is applied to the outer stiffeners, whereas the maximum restraint condition, model C, is applied to the central weld in terms of equation 11.3.

Stronger agreement with the experimental results was found in the case of model B + C, applied as discussed. However, although the distortion patterns were well predicted, the deformation magnitudes consistently underestimated the experimental results. Dilatometer tests on the plate material showed a higher average coefficient of expansion than for previous plate materials used (16.8×10^{-6} per °C) with an elevated contraction characteristic in the 500–200°C range of cooling. Figure 11.25 shows a direct comparison of the experimental and computational simulations for sequence A, using model B + C and a higher expansivity coefficient of 20×10^{-6}. The match is reasonable at each stage but the computations still underestimate the peak levels of deformations on the longitudinal edges.

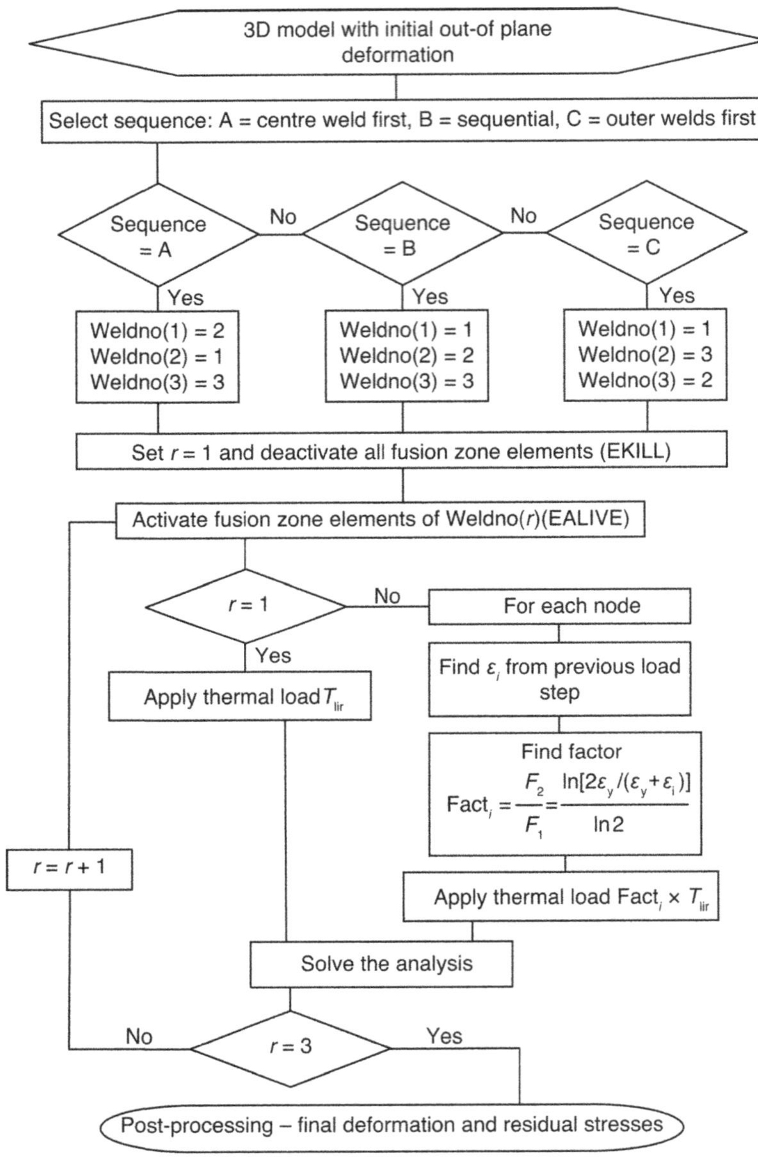

11.24 Flow chart for overall algorithmic simulation of multiple-stiffener fabrication.[4]

Table 11.7 provides a numerical comparison for each attachment stage and sequence. Note that the initial shape for each attachment sequence was different, reflecting the respective experimental cases.

The predictions in each case overestimate the distortions for the first weld attachment but progressively underestimate distortions for the remaining

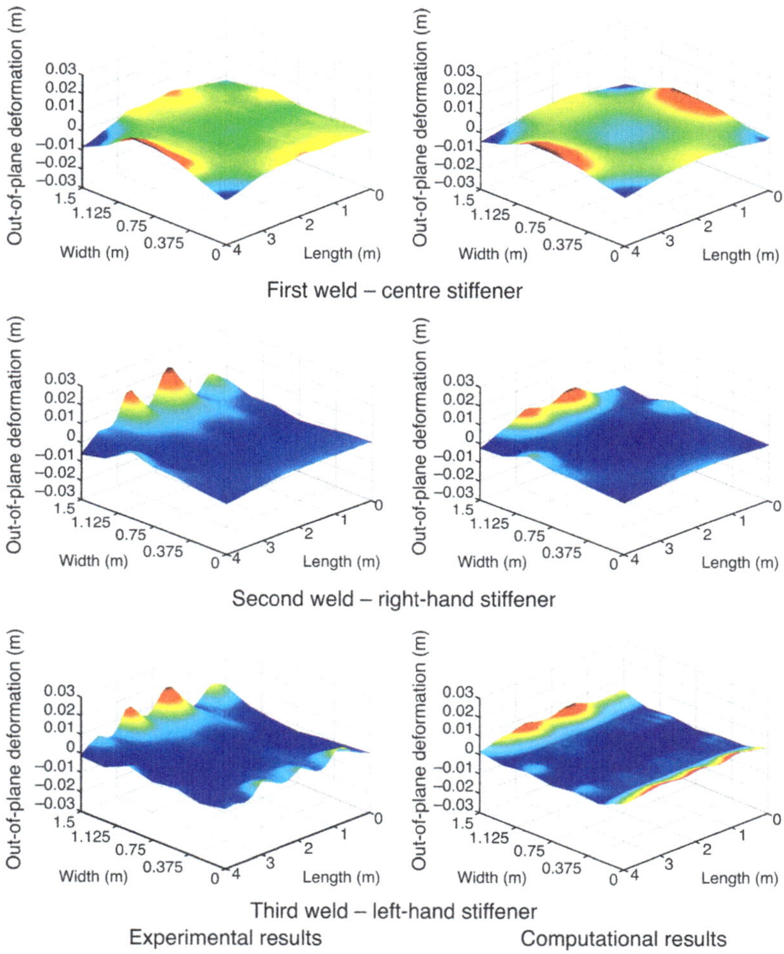

11.25 Comparison of experimental and computational results, sequence A – centre stiffener first then outer stiffeners.

two welds. In the case of sequence A (Fig. 11.25) and sequence B (Fig. 11.22) it is evident that the sharply deformed areas local to the supports on the right-hand side contribute substantially to the RMS deformations. This suggests that these supports are carrying a greater proportion of the weight, due to the initial out-of-flatness of the tacked assembly. The algorithmic approach will not necessarily reflect such effects fully, as transient deformations and consequent dynamic load transfers are not included in the analysis. However, given the strong suggestion from the poor initial flatness that the as-received plates may not have been residual stress-free, the quality of the computational prediction is surprisingly good over all comparisons.

Table 11.7 Comparison of experimental and computational distortions

			Sequence A		Sequence B		Sequence C	
First weld	RMS distortion (mm)	Experiment	1.923	Centre	2.541	RHS	2.540	RHS
		Computation	2.923		2.941		2.957	
	Angular distortion (degree)	Experiment	0.419		0.701		0.684	
		Computation	0.980		0.986		0.987	
Second weld	RMS distortion (mm)	Experiment	4.116	RHS	3.143	Centre	2.459	LHS
		Computation	2.577		2.571		2.569	
	Angular distortion (degree)	Experiment	1.017		0.839		0.769	
		Computation	0.873		0.866		0.870	
Third weld	RMS distortion (mm)	Experiment	4.139	LHS	3.278	LHS	3.292	Centre
		Computation	2.418		2.523		2.685	
	Angular distortion (degree)	Experiment	1.028		0.820		0.819	
		Computation	0.778		0.799		0.815	

This study does not provide a conclusive answer to a search for the 'best' attachment sequence, other than that there are not large differences in outcome between the different sequences. Sequence A would be judged to be the poorest on experimental grounds, but marginally the best from the computational results. The non-linear differences attributable to attachment sequence are largely swamped by differences in initial out-of-flatness, which are not modelled well by the algorithmic method. Some clues to the progressive build-up of distortion in the different sequences can be found by studying the longitudinal residual stress and strain patterns in the computational case. (Note that the algorithmic method does not provide an accurate portrayal of residual stress but the calculated values at each stage form the basis of the allowance for the influence of residual stress fields on subsequent welds, through equation 11.4.) Figures 11.26 and 11.27 show the algorithmic-based residual stress fields for sequences A and C, reflecting the extremes of distortion behaviour.

The first point to note is that the residual stress levels generated in the region where the next weld will be placed are generally small. In the case of the first pass in sequence A (Fig. 11.26(a)) the stress induced in the region occupied by the right-hand stiffener lies at the higher end of the range −38 MPa to +2 MPa and therefore will have little effect on the contraction force. Likewise, after completion of the second pass (Fig. 11.26(b)), the compressive stress in the region of the third attachment is approximately

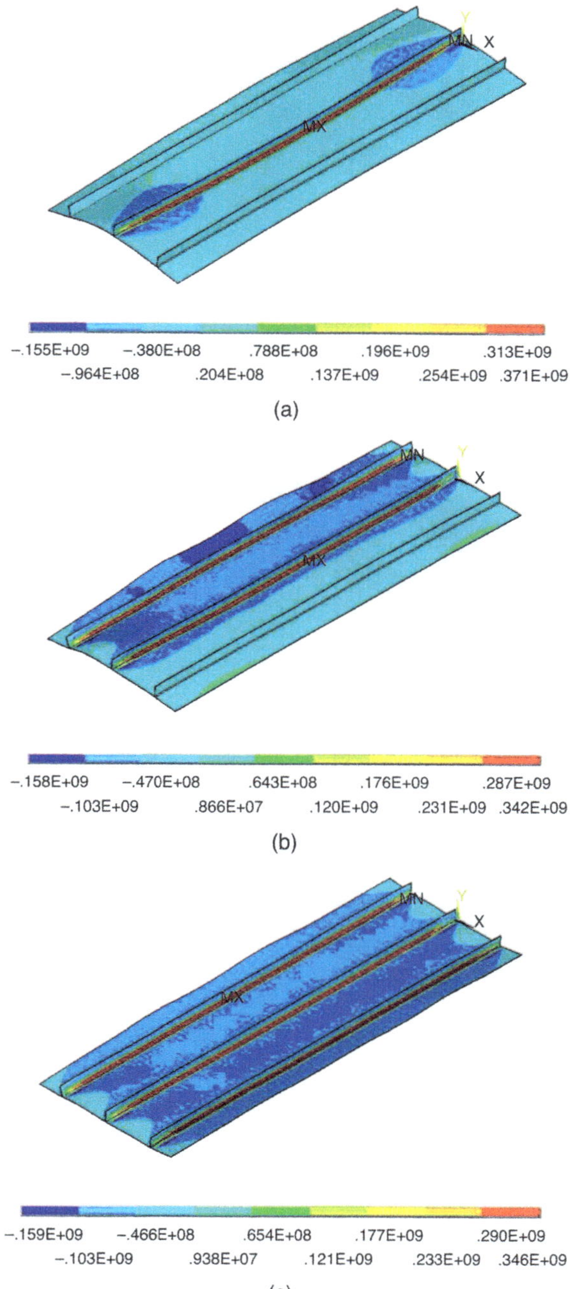

11.26 Sequence A – build-up of longitudinal stress: (a) centre weld, first pass; (b) right-hand weld, second pass; (c) left-hand weld, third pass.

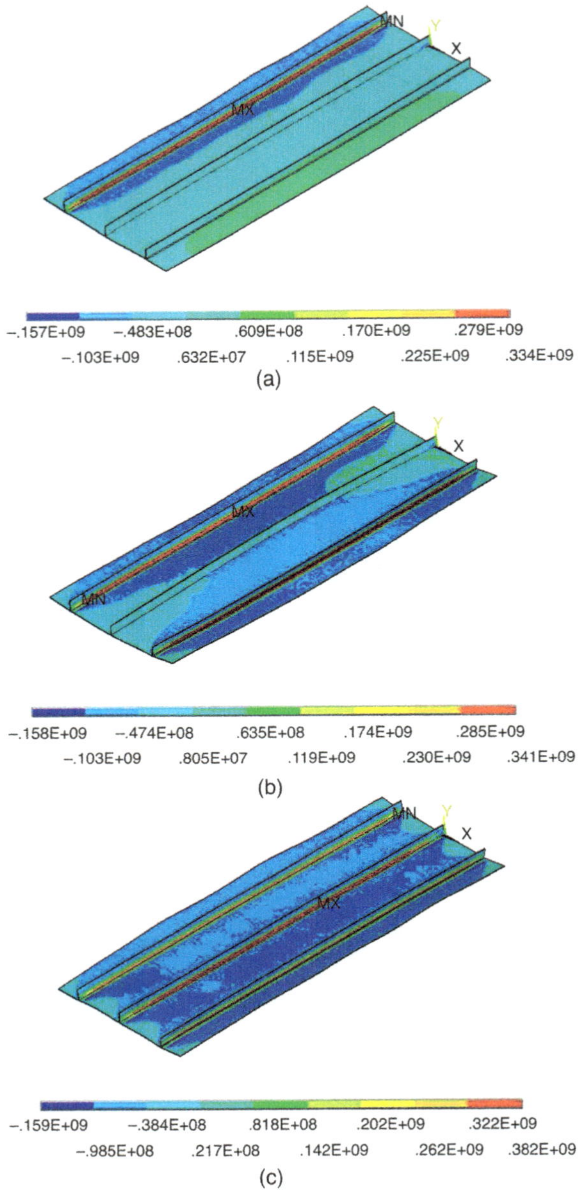

11.27 Sequence C – build-up of longitudinal stress: (a) right-hand weld, first pass; (b) left-hand weld, second pass; (c) centre weld, third pass.

−20 MPa, which – through equation 11.4 – would give only a 7% increase in the contraction force. Note, however, that the combined effect of the first and second passes has produced higher local compressive stresses along the outer right-hand-side edge, leading to local non-linear deformation. This is removed to a large extent by the third pass, due to in-plane bending across the width of the plate. Hence the conclusion might be that compressive residual stress from previous passes has had little effect in terms of increasing the contraction forces. (These graphics also show deformations and, in particular, transverse hogging is noted in Fig. 11.26(a)); this is probably due to the initial out-of-flat shape, rather than arising from transverse deformation during the first pass.)

In the case of sequence C (Fig. 11.27), in-plane bending, during the first pass on the right-hand side, generates a small tensile residual stress over most of the length at the site of the second pass on the left-hand side (Fig. 11.27(a)). Hence, there should at least be no *increase* in contraction force for the second weld. However, the combined compressive residual stress fields from the two outer welds (Fig. 11.27(b)) generate a higher compressive residual stress along the central region, 47.4 MPa at minimum, which is enough to increase the contraction force by at least 18% in that region. These differences are fairly marginal, but are probably enough to tip the balance towards the conclusion shown in Table 11.7 for the computational approach, namely that sequence A produces slightly less distortion.

The overall conclusion that can be drawn from these studies on multiple stiffeners is that the effects of sequence are highly non-linear and not at all straightforward to predict without computational evidence. As always, initial out-of-flatness and build-up of distortion during successive welds will be the major factors, unless the welds are rather closer together than in this example. In that case, residual stress would have more effect. Note also that the distortion could have been much worse, but for the fact that the structure was first fully assembled and tacked to provide as much out-of-plane bending stiffness as possible.

11.5 Conclusion

Thermal flows during fillet welding of stiffener attachments are more complex than for butt welds between plane plates, and experimental information on thermal aspects was laid out in section 11.2. Thermocouple and thermographic data were used to suggest appropriate heat flow models for each pass in a double-sided fillet-welded stiffener. Reduced solution strategies for single-stiffener welding in large-scale fabrications were given in section 11.3 and were contrasted with the behaviour of equivalent small-scale joints, discussed previously in section 7.8. The factors controlling angular distortion are more difficult to treat in the TCS algorithm than for butt welds and different modelling assumptions were tested against experimental results. It is concluded that different transverse constraint assumptions are required for first- and second-pass welds in a double-sided fillet welded stiffener joint.

Section 11.4 considered plates with multiple parallel stiffeners and the modelling assumptions required. If the stiffeners are closely spaced, welding distortion effects will be non-linear, as the residual stress fields will interact and the distortion induced by a given weld will depend on the distortions generated by preceding welds. A case study dealing with a three-stiffener panel was used to demonstrate an appropriate treatment that takes account of sequence effects. The results were set against experimental data.

11.6 References

1. Camilleri, D., Mollicone, P. and Gray, T.G.F., 2006, Alternative simulation techniques for distortion of thin plate due to fillet-welded stiffeners, *Modell. Sim. Mat. Sci. Eng.*, 14, pp 1307–1327.
2. Camilleri, D., Comlekci, T. and Gray, T.G. F., 2004, Use of thermography to calibrate fusion welding procedures in virtual fabrication applications, *Proc. Inframation Conference, Las Vegas, USA*, 5, pp 121–131.
3. Camilleri, D., Comlekci, T. and Gray, T.G.F., 2006, Thermal distortion of stiffened plate due to fillet welds – computational and experimental investigation, *J.Therm. Stresses*, 29, pp 111–137.
4. Camilleri, D., Comlekci, T. and Gray, T.G.F., 2005, Design support tool for prediction of welding distortion in multiply stiffened plate structures: experimental and computational investigation, *J. Ship Prod.*, 21, 4, pp 219–234.

12
Exploiting welding distortion models: examples of design and manufacturing strategies to optimise fabrication

DOI: 10.1533/9780857099327.265

Abstract: Information from all preceding chapters in the book is applied to support design and manufacturing strategies, showing how simulation can be applied to explore alternative design configurations and manufacturing sequences. In terms of design, configuration and manufacturing sequence questions in the fabrication of multi-stiffened plates are examined. Studies also include the use of multiple welding heads, where computation is applied to show that an optimum spacing can be established, avoiding the risk of buckling. Simulation is also applied to investigate three techniques that have been proposed to mitigate distortion.

Key words: design optimisation, stiffener spacing, weld sequence, multiple stiffeners, buckling, multiple heat inputs, transient thermal tensioning, low-transformation-temperature filler, cryogenic cooling.

12.1 Introduction

Chapters 4 to 11 have dealt primarily with the computational processes that can be applied to simulate typical welding operations used in fabricating thin-plate structures, the particular aim being to predict distortion outcomes and residual stress fields. The benefit in using these methods is not so much that they will provide exact predictions, as certain sensitive input variables such as initial flatness, are usually too poorly known to expect a precise outcome. But, as the underlying mechanics are complex, simulations can often provide an understanding of the reasons for the different distortion patterns and magnitudes that arise from specific assembly and welding procedures. The relationship of these distortions to different design configurations and fabrication processes can therefore be understood and appropriate design revisions can then be introduced.

The particular merit of computational prediction methods is that they can be applied in advance during the design or fabrication planning sequence and can be used to explore alternative treatments that could not easily be investigated by trial and error through practical testing. In this concluding chapter, it is assumed that minimisation of distortion is always a desirable outcome. The methods described previously will therefore be applied to show how the actions required to reduce distortion can be found, as far as is possible in a given case.

Before turning to specific examples related to the chapter title, it is useful to index certain content in previous chapters that bears on the requirement to minimise distortion. Firstly, at a dimensional level, equations 4.5–4.8 in Chapter 4 and equation 8.5 in Chapter 8 provide very clear and simple messages concerning specification of weld size and welding process. For a given material with fixed basic thermo-mechanical properties, the contraction force and the resulting deformations and curvatures will increase directly with the heat energy input per unit length of weld. This parameter also relates closely to the weld cross-sectional size. Hence, the most benefit comes from minimising the size of the weld and by choosing a welding process that minimises the required energy input. Distortion due to multipass welds remains within this principle, at least to a first approximation, because the contraction stress pattern of each pass overlaps with neighbouring passes and the increase in overall contraction force is limited by the fact that the material yields (see Fig. 9.11 and associated text).

The most dominant material property influence comes from the coefficient of expansion or, more correctly, the contraction behaviour during cooling. Yield strength does not seem to have a strong influence, although information will be given later that suggests certain other properties may be important that are also affected by metallurgy, as is yield strength.

Within the overall requirement to reduce heat input, angular distortion depends primarily on the variation of transverse weld width in the through-thickness direction. Configurations that achieve a narrow-angle fusion zone will therefore be beneficial. Minimisation of angular distortion is particularly important for fabrications with low longitudinal bending stiffness such as flat plates, because of the strong interaction between longitudinal and transverse deformation, as demonstrated in Fig. 5.10.

One of the problematic aspects of welding distortion is the large variability in outcome, despite seemingly identical designs and process conditions. Section 10.2 uses computational methods to expose the effects of initial flatness deviations in the source material and thus emphasises the critical importance of starting with flat plate stock and of making sure that support arrangements maintain this ideal condition (contrast Fig. 10.3 and Fig. 10.5).

Some method will be required to maintain the relative position of elements in an assembly while welding and section 10.3 exposes the sometimes unexpected outcomes of using tack welds to do this. The studies on support arrangements and clamping in section 10.4 highlight the contradictions involved in establishing a fabrication procedure that will control the relative movement of assembly elements, but will not constrain the inevitable thermal movements (see Fig. 10.22).

Multiple welds are often required and this raises many questions for design and fabrication strategy in terms of spacing and welding sequence. The important message is that the effects are non-linear and not simply additive. The basic interactive effects, whereby the residual stress field generated by one weld will influence the deformation of a following weld, are dealt with in section 10.6. This gives pointers to design and fabrication principles that will minimise the effects of residual stress fields. Section 11.4 provides a specific case study on the effects of carrying out multiple welds according to different sequences and establishes some principles to minimise distortion in such an instance.

The distinction between 'design' and 'fabrication' is fairly arbitrary in most contexts and the two sets of issues should ideally be considered together, in an integrated manner. The design process itself tends to be led by consideration of geometrical factors and structural analysis, but it is also a stage where the likely effects of alternative manufacturing processes and routes might be freely explored through simulation. The simulation process, as seen in the preceding chapters, is largely based on structural mechanics analysis techniques and is therefore appropriate for consideration at the design stage. The examples in considered sections 12.2–12.4 are primarily design-led, in that issues of structural form or configuration provide the starting point, but the effects of manufacturing processes and procedures are not ignored, as they can have a profound effect on the outcome. Sections 12.5–12.7 treat cases that are primarily fabrication-led.

12.2 Optimising multi-stiffener configuration

12.2.1 Structural model

Three target design criteria were specified for this study, namely a steel structure with longitudinal second moment of area not less than 325 cm^4 and mass per longitudinal metre not more than 48 kg/m. The base plate of the structure is one metre square and 4 mm thick. Nine configurations were investigated, covering two different stiffener sizes and five different stiffener

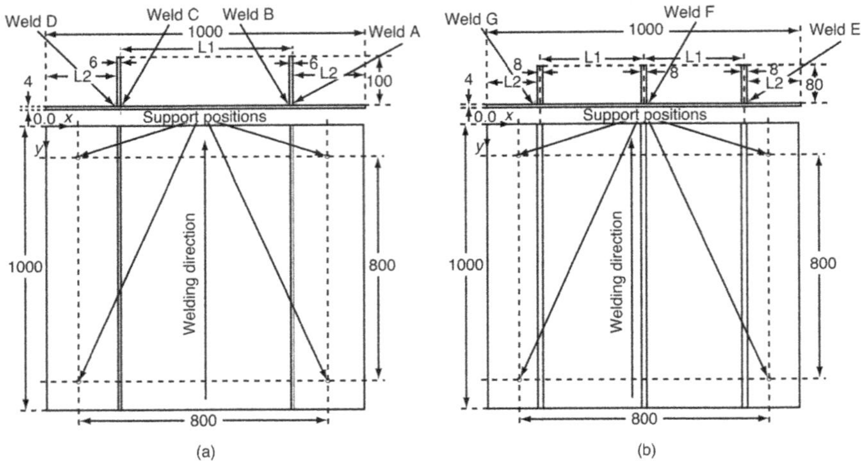

12.1 Multiple-stiffener fabrications: (a) two-stiffener configuration; (b) three-stiffener configuration. Dimensions in mm.

Table 12.1 Stiffened plate dimensions and fabrication procedures

Test plate	No. of stiffeners	Stiffener size (mm)	Weld placement	Welding sequence	L_1 (mm)	L_2 (mm)
1	2	100 × 6	Outer	A, D	516	242
2	2	100 × 6	Inner	B, C	516	242
3	2	100 × 6	Inner/outer	B, D	516	242
4	2	100 × 6	Inner/outer	B, D	344	328
5	2	100 × 6	Inner/outer	B, D	688	156
6	3	80 × 8	Outer	E, G, F	344	156
7	3	80 × 8	Outer	F, E, G	516	156
8	3	80 × 8	Outer	E, F, G	516	156
9	3	80 × 8	Outer	E, G, F	258	242

spacings as detailed in Fig. 12.1 and Table 12.1. These dimensions were typical of designs used by a collaborating fabricator. Single-sided fillet welds were applied, with variations in weld placement and sequence as shown. Fillet placements designated 'inner' are on inward-facing stiffener faces and vice versa for 'outer'; L_1 is the inner width spacing of stiffeners and L_2 is the spacing to the outer edge. Test plates were supported on four fixed points near the outer edges and subject to natural free surface cooling on all sides. All stiffeners were tacked to the plate before embarking on the full fillet welding operation.

Welding energy input, using a GMAW process, was nominally 0.92 kJ/mm, giving a 45° weld of 5.5 mm nominal leg-length. Thermal transients in the plate and stiffeners were measured by thermocouple arrays, placed at cross-sections one-third and two-thirds along the plate length. Averaged

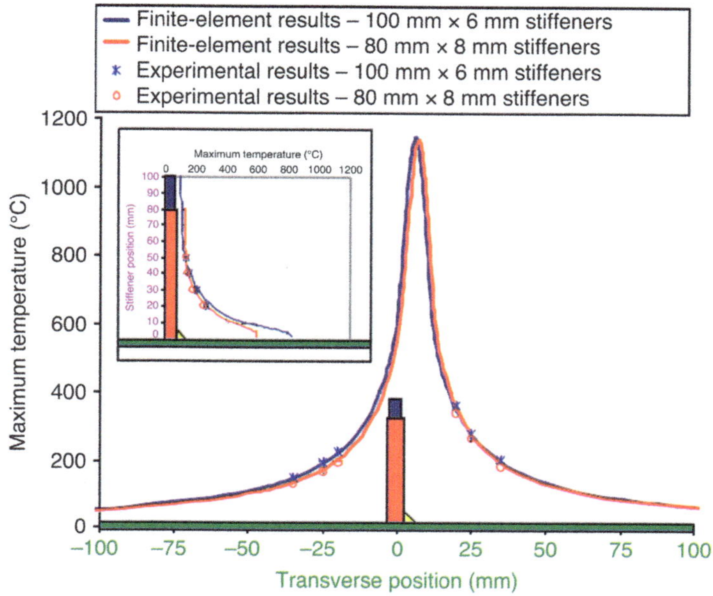

12.2 Comparison of finite-element and experimental results – 90% efficiency.

maximum temperature results are compared in Fig. 12.2 with finite-element determinations from a two-dimensional (2D) model, matched by assuming 90% weld efficiency (plate temperatures are shown in the main graph and stiffener temperatures in the inset figure). Differences in the temperature profiles of the two stiffener sizes are evident in the results.

A hybrid (simultaneous) analysis was used for the thermo-mechanical part of the computation, each weld being treated in sequence, using a single step. In this study, the primary objective was to explore the influence of issues such as sensitivity to support arrangements and the consequent effects due to gravity loading together with the influence of initial flatness and its effect on the variability of outcomes. These computational variations are shown in the headings under 'Finite-element results' in Tables 12.2 and 12.3. The 'no gravity – perfectly flat' condition represents an ideal support where any variations in outcome are purely due to configuration factors, such as stiffener size and spacing, together with weld placement and sequence. The 'gravity – perfectly flat' condition adds the influence of initial sagging and gravity displacement during welding and the 'no gravity – initial shape' case explores the influence of initial out-of-flatness, relative to the ideally supported case. Therefore, both the 'experimental' and the 'no gravity – initial shape' columns embody a random and less visible factor in the form of initial out-of-plane profile. The spacing between stiffeners (L_1) and distance to the outer edges (L_2) are broadly categorised as 'narrow', 'medium' and 'wide'.

Table 12.2 RMS out-of-plane distortion comparisons

Test plate	Experimental results (mm)	Finite-element results (mm)			Stiffener spacing	
		No gravity – perfectly flat	Gravity – perfectly flat	No gravity – initial shape	L_1	L_2
Two-stiffener						
1	4.24	1.94	3.87	3.60	Medium	Medium
2	4.05	1.17	3.76	2.90	Medium	Medium
3	3.87	1.64	3.81	3.34	Medium	Medium
4	3.89	3.25	5.72	3.40	Narrow	Wide
5	1.56	0.80	1.66	0.59	Wide	Narrow
Three-stiffener						
6	1.97	0.66	1.99	0.68	Medium	Narrow
7	1.69	0.74	2.10	0.77	Medium	Narrow
8	1.46	0.65	1.99	0.73	Medium	Narrow
9	5.24	1.93	4.82	2.01	Narrow	Medium

Table 12.3 RMS angular distortion comparisons

Test plate	Experimental results (degrees)	Finite-element results (degrees)			Stiffener spacing	
		No gravity – perfectly flat	Gravity – perfectly flat	No gravity – initial shape	L_1	L_2
Two-stiffener						
1	1.37	0.57	0.99	0.95	Medium	Medium
2	1.14	0.44	0.94	0.72	Medium	Medium
3	1.00	0.52	0.96	0.82	Medium	Medium
4	1.06	0.74	1.21	0.64	Narrow	Wide
5	0.74	0.29	0.82	0.26	Wide	Narrow
Three-stiffener						
6	0.86	0.39	0.82	0.38	Medium	Narrow
7	0.76	0.40	0.85	0.40	Medium	Narrow
8	0.62	0.38	0.82	0.39	Medium	Narrow
9	1.58	0.71	1.34	0.67	Narrow	Medium

12.2.2 Effect of initial shape

Comparisons on the basis of root mean square (RMS) results are not always adequate, as quite different modes of deformation can generate identical RMS values. Mode shapes found experimentally in test plates 1–5 and 9 were not well-predicted by the computations based on initially flat plates subject to gravity forces, although the RMS predictions were reasonable (compare Fig. 12.3(a) and 12.3(b)). Modelling of the asymmetric buckling

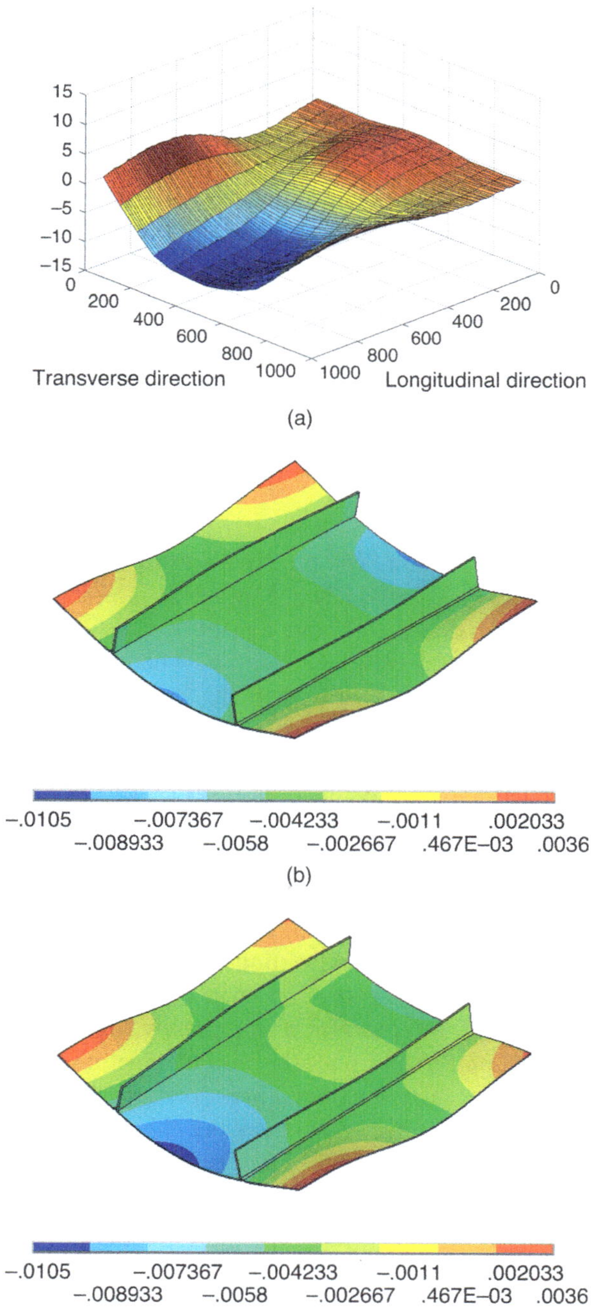

12.3 Out-of-plane deformation (m) of test plate 1: (a) experimental result; (b) finite-element result – flat with gravity; (c) finite-element result – initial shape, no gravity.

mode was better if the initial deformation was used as the starting point, as in Fig. 12.3(c), even though gravity was omitted.

12.2.3 Effects of support and gravity

Comparison of the first two columns of finite-element results in Tables 12.2 and 12.3 highlights the critical role of fabrication support during welding. In the two-stiffener cases, distortion is approximately doubled if the fabrication is supported only at the edges. The out-of-plane RMS value is approximately trebled in the three-stiffener cases. Similarly, the experimental cases (where gravity is present) show larger deformations than the 'no gravity – initial shape' results.

12.2.4 Stiffener spacing and weld sequence

The configurations of test plates 5 and 8 show the least distortion in theory and practice, and by a fair margin. The principal feature of these test plates is that the stiffener welds are as widely separated as is practicable. Weld sequence is a further variable factor and constitutes the only difference between the three-stiffener test plates 6, 7 and 8, where the 'outer–inner' sequence of plate 6 gives poorer results than the 'sequential' sequence of plate 8. This apparent effect of sequence confirms the results of the large-scale multi-stiffener trials reported in section 11.4. (Note that, in the earlier case, the simplest of the computationally efficient (CE) strategies was used, rather than the hybrid approach used here.)

These findings, taken together, suggest that the non-linear effects of longitudinal residual stress may be responsible and this possibility is examined through Fig. 12.4 and Fig. 12.5. In Fig. 12.4(c), the first weld B leaves no appreciable residual stress at location D and the subsequent weld D is not therefore affected. In test plate 4 shown in Fig. 12.4(b) the stiffeners are much closer and weld B generates a compressive residual stress at location D, which generates greater distortion on completing weld D. This was made worse by the concentration of compressive residual stress between the stiffeners, which created a longitudinal buckling effect. The distortion of plate 3, shown in Fig. 12.4(a), lies between the other two cases. Similar effects can be seen in the comparison of test plates 6 and 9, shown in Fig. 12.5, where spacing constitutes the only difference.

12.2.5 Effects of weld placement

The effects of placing welds on inward- or outward-facing sides of stiffeners can be seen consistently in all computational results for test plates 1–3, where this is the only variable, aside from initial shape. In test plate 2 (the

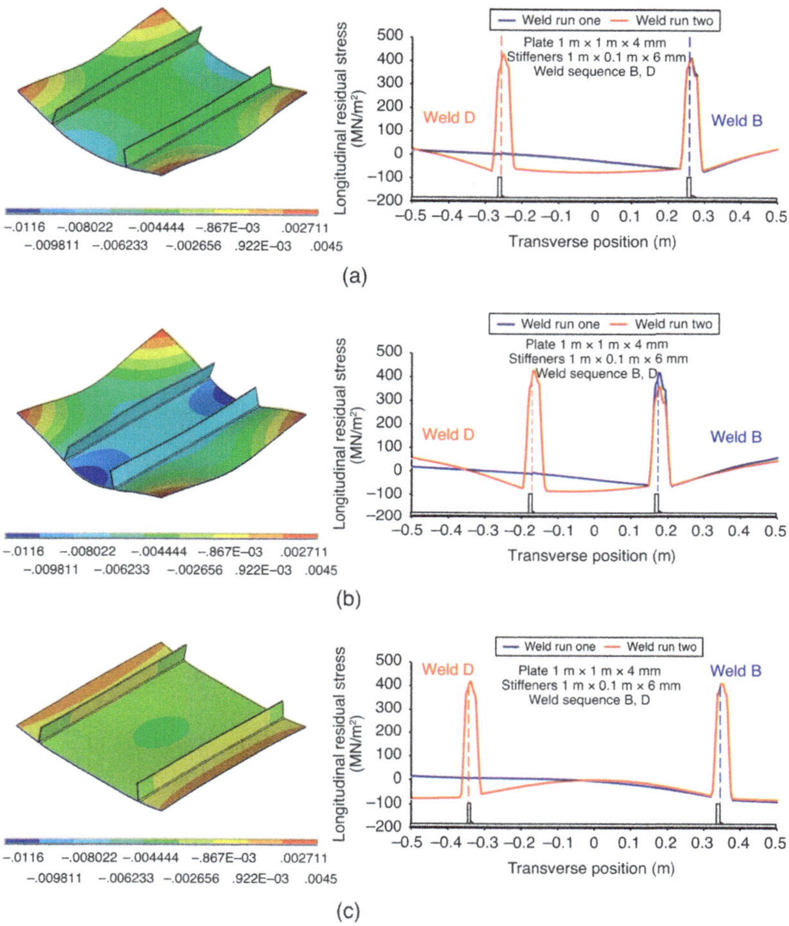

12.4 Effect of spacing and residual stress on out-of-plane deformation for two-stiffener case: (a) test plate 3; (b) test plate 4; (c) test plate 5.

best result) the focus of the contracting forces is along the inward-facing sides of the stiffeners and both stiffeners therefore act to resist the contracting force. The restraining structure is like a plate, reinforced at both edges. In the case of the outward-facing welds in plates 1 and 3, on the other hand, the restraining structure comprises a free edge, which develops less restraint. Hence, as a general rule, distortion can be reduced for single-fillet attachments by placing the welds on inward-facing sides of the stiffeners.

As regards the structural targets, the longitudinal second moment of area and specific mass are respectively 350 cm^4 and 41 kg/m in the two-stiffener arrangement and 332 cm^4 and 47 kg/m in the three-stiffener case. They both therefore meet the design targets of 325 cm^4 and 48 kg/m. The stiffener spacing, weld placement and sequence adopted in test plates 5 and 8 were

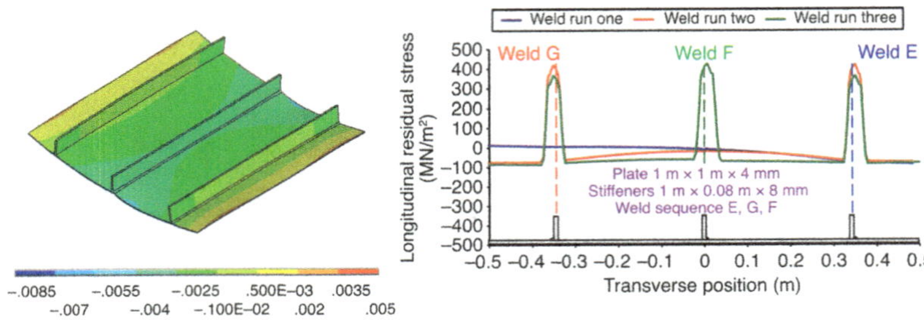

(a)

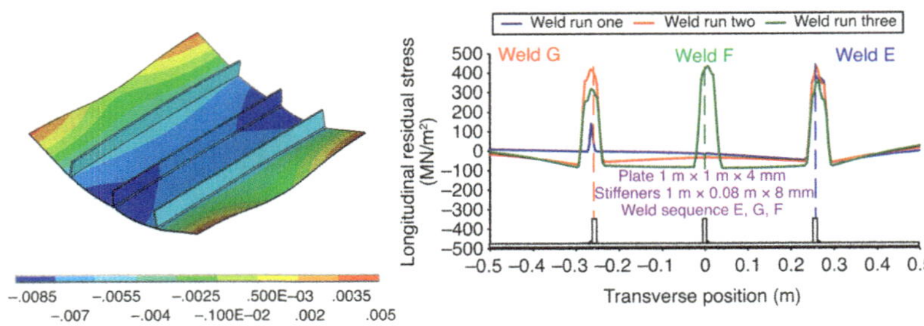

(b)

12.5 **Effect of spacing and residual stress on out-of-plane deformation for three-stiffener case: (a) test plate 6; (b) test plate 9.**

clearly identified as the best of the combinations studied and the corresponding arrangements in test plates 4 and 9 gave the worst results. Some further consideration of the structural aspects is needed however, as the longitudinal bending stiffness will not correspond exactly to the second moment of area due to the fact that the simple theory of bending does not apply to a wide, thin plate, where the 'flange' element is not fully effective. Hence, although the two-stiffener design appears to be more structurally efficient and would presumably cost less to manufacture, the much wider spacing of the stiffeners may result in a loss of effective width.

Aside from structural aspects, the results tables show that the effect of gravity loading during welding is a major distortion driver. Supporting the assembly at more points should therefore be beneficial. However, as the experimental transient deformation traces show, deformations during the passage of the welding source will lead to loss of contact with some supports and further variability in outcome. The ideal 'no gravity' state will not then be achieved unless constant force supports can be applied to maintain a uniform reaction force over the fabrication at all times.

12.3 Optimising the design in terms of weld position

12.3.1 Introduction

Many of the decisions made in the engineering design process are essentially arbitrary and do not always take into account important factors of which the designer may be unaware. Welding distortion is often one such factor. In the case studies shown in Fig. 12.6 and Fig. 12.7, the butt welds might be placed at any point across the width of the respective fabrications, from the structural point of view.

In the case of the built-up plate in Fig. 12.6, the dimension x might be chosen in practice to minimise wastage in the cutting operation. The study in Fig. 12.7, on the other hand, considers a multi-stiffener fabrication, where the location of a butt weld has to be determined in relation to the bulb stiffeners, which are attached using double-sided fillet welds. The design

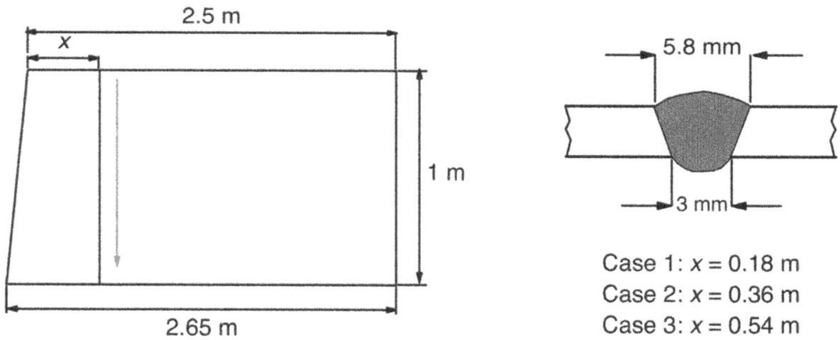

12.6 Butt weld position study.

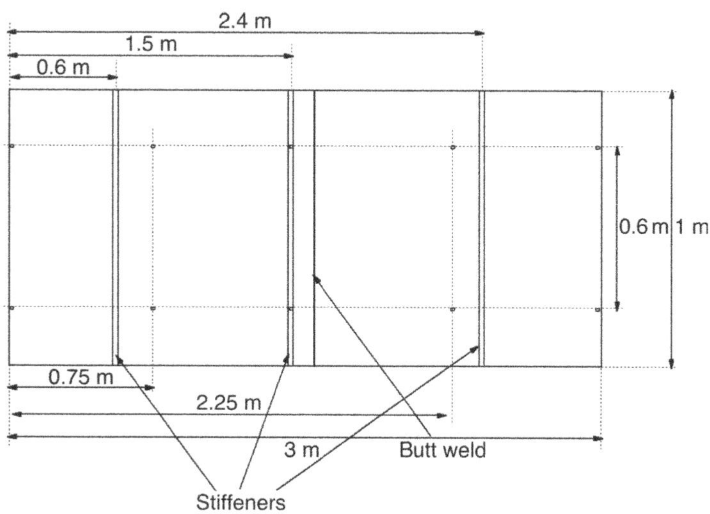

12.7 Multiple-stiffener fabrication study.

constraint applied to minimum offset distance had been set according to a Lloyd's Register Standard, which limits this dimension to 25 mm, although the reasoning behind this restriction was not known. Both studies considered 4 mm thickness CMn steel plates and welding process conditions were assumed according to existing practices for such fabrications.

The relatively large scale and complexity of these cases suggested the adoption of a 'stepwise hybrid' computational strategy (see sections 5.8 and 8.3). This approach reduces the computational budget relative to a full thermo-elastic-plastic model, but retains some recognition of transient effects. These will arise through the thermal response of the tapered plate piece in the first case study (Fig. 12.6) and due to progressively changing support effects in the second (Fig. 12.7).

12.3.2 Built-up plate study

The thermal analysis for this case assumed a welding energy of 0.835 kJ/mm and 80% efficiency, giving fully penetrated weld dimensions as shown in Fig. 12.6. A preliminary 3D thermal analysis was carried out to explore the effect of the tapering geometry and the nearness to the left-hand edge of the three proposed weld locations (cases 1–3). These different widths gave effectively identical results. This implies that the zones exceeding the yield limit of the material ($\alpha T \geq \varepsilon_y$) would also be identical and any differences in distortion shown by the analysis should be due to mechanical effects rather than differing thermal transients.

Previous studies had shown that support configuration and initial lack of flatness can have a significant influence on distortion outcomes. Three different support schemes, reflecting possible production cases, were therefore investigated, as shown in Fig. 12.8 (case 3 value for x (Fig. 12.6)). These ranged from relatively uniform support in case A (given an initially flat or flexible plate) to rather poor support in case C, where all supports are likely to be fully loaded but with more sagging between.

All cases were therefore subject to gravity forces and supports were simulated by link elements to simulate loss of contact at any point due to out-of-plane deformation. Cases B and C were assumed to be supplied as flat and stress-free plates, whereas in case A, two forms of initial out-of-flatness were investigated, assumed to be due to tacking distortion (see section 10.3). The initial shape was generated by applying a typical saddle-shaped form defined by alternative quadratic expressions as follows

$$\text{Initial negative out-of-plane deformation} = 0.0025u^2 - 0.02v^2$$

$$\text{Initial positive out-of-plane deformation} = 0.02v^2 - 0.0025u^2$$

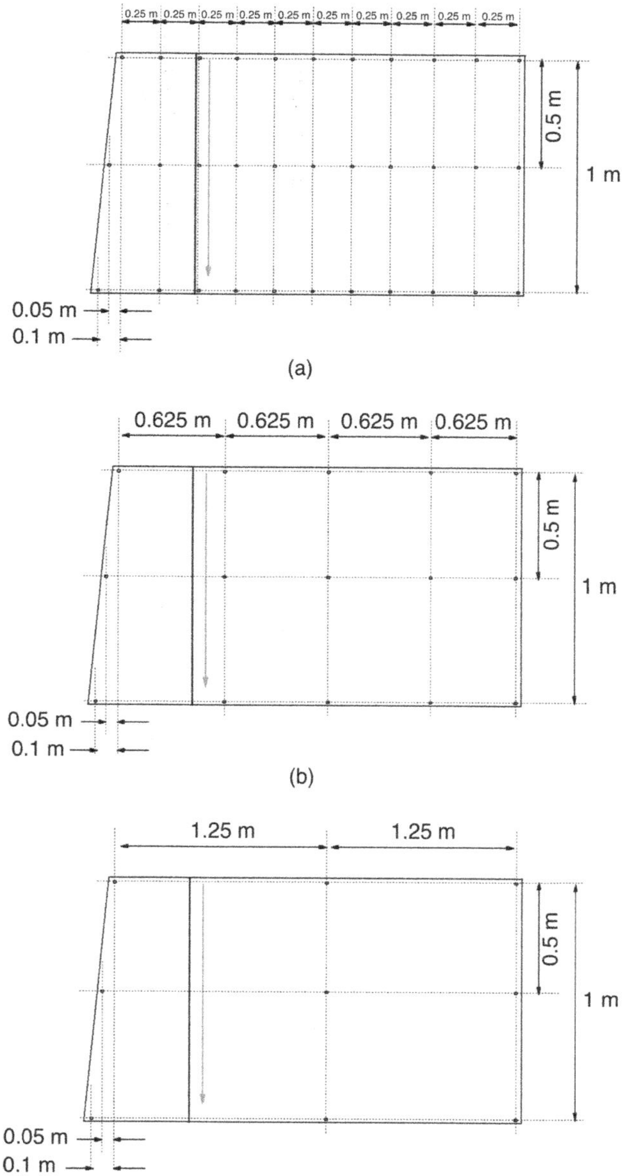

12.8 Alternative support arrangements: (a) case A; (b) case B; (c) case C.

where u is the transverse nodal position relative to the weld line and v is the longitudinal nodal position relative to the mid-length of the weld. As indicated in section 10.3, negative deformation has been found to accompany tacks placed on the same (upper) side of the plate as the

Table 12.4 RMS out-of-plane distortion for butt-weld examples for three different cases of weld-line offset

Initial shape	Support configuration	RMS out-of-plane distortion (mm)		
		Case 1, 0.18 m	Case 2, 0.36 m	Case 3, 0.54 m
Negative	A	13.4	10.3	8.4
Positive	A	6.7	3.1	2.5
Flat	B	7.9	4.9	3.5
Flat	C	18.9	11.5	16.0

subsequent butt weld and positive initial deformation for the reverse case. However, in all three support cases, the tapered plate was assumed to be weld-tacked to the main plate at 50 mm intervals along the weld length. 'Activation/deactivation' elements were used, as previously described, to simulate fusion of the tacks as the weld progresses.

In this application of the stepwise hybrid method, angular distortion is analysed through a 3D elasto-plastic model, where high-temperature strength and expansion/contraction properties are included through measured material properties. However, it is assumed that only elements within the fusion zone are active in terms of contraction. Full elasto-plastic behaviour is attributed both to the fusion zone and to the surrounding parent material in the case of longitudinal actions. In this case, a single step was used to reduce the computational budget.

The resulting out-of-plane deformations for the different weld-line offsets, initial out-of-plane assumptions and support configurations are shown in Table 12.4 and a specific result for support case A and alternative initial shape assumptions is given in Fig. 12.9.

The first clear outcome seen in Table 12.4 is that placing the weld as far away as possible from the outer edge generates the least distortion, in all but one case. The reason is that the smaller longitudinal bending stiffness of the narrow tapered plate in case 1 permits greater angular movement and this has interacted with longitudinal deformations to give an overall poorer result. Support of the narrowest plate is also poor. Figure 12.9 and the tabulated results also show that the use of reverse-side tacking has had the effect of reducing both local and overall distortion. However, the gravity effect in support case C and the resulting large initial sag between widely spaced supports is the more dominant factor in that case.

This is a specific example with its own special features but it nevertheless demonstrates that the most useful role of simulation is to provide an understanding of the conflicting effects and thereby generate an optimum strategy for weld positioning, support configuration and tacking assembly procedure.

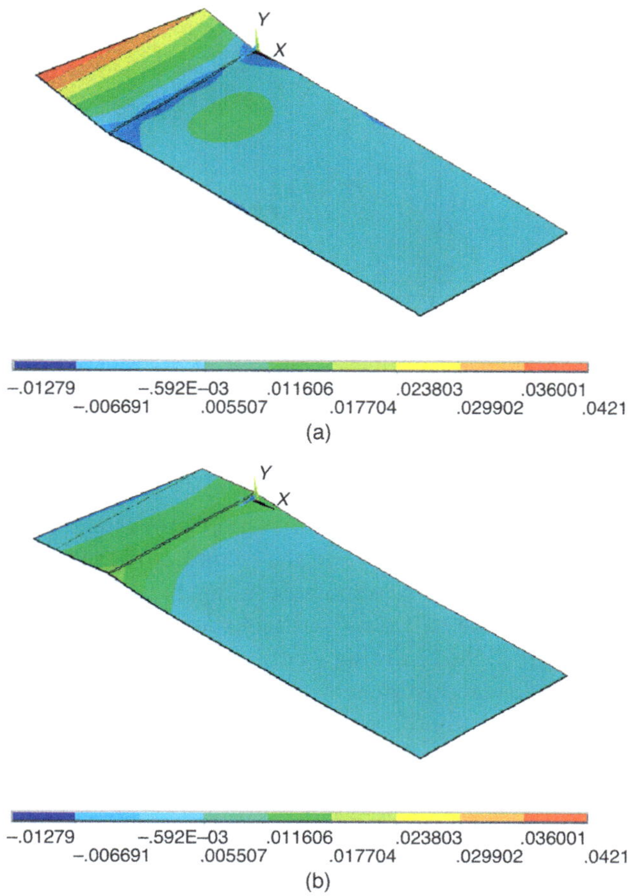

12.9 Out-of-plane deformation, 0.36 m weld-line offset, support configuration A: (a) initial negative shape; (b) initial positive shape.

12.3.3 Butt weld position in multiple-stiffener plate

This case, shown in Fig. 12.7, with weld details in Fig. 12.10, is potentially very complex, as it involves several welds, which could be sequenced in a variety of orders. Usual practice would be to carry out the butt weld first, to create a built-up plate and then assemble and tack the stiffeners. The butt-welding operation will distort the plate to some extent and external forces will therefore be required to draw the stiffeners and the plate together progressively, whereupon weld tacks can be used to keep the assembled structure in place for the stiffener fillet welds. This step is difficult to quantify for the purposes of simulation.

In an alternative approach, which may be feasible with some fabrication facilities, the stiffeners would firstly be tacked to the separate plates, which

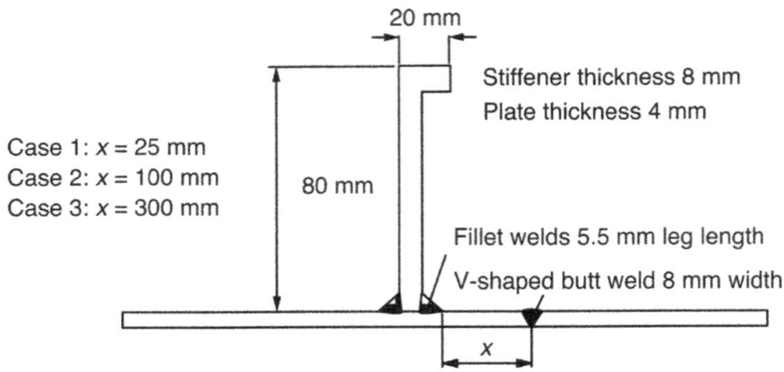

12.10 Cross-sectional detail showing relative position of butt weld and stiffener.

would then be butt-welded together and finally the main fillet welds on the stiffeners completed.

The first approach, where the plate is first built up using a butt weld, is designated 'stiffeners not pre-tacked' and the second, 'pre-tacked stiffeners'. Both of these approaches can be simulated using the same finite-element model. In the first case, the stiffeners and tacks are included in the model from the outset, but 'killed' during the butt-welding stage. Elements representing fillet welds and stiffeners are also 'killed' appropriately to simulate fillet weld completion on each side of the stiffener and sequential attachment of the three stiffeners (left, centre and right). A 'thermal break' between the stiffener and the plate was included, as discussed in sections 7.8 and 11.2. The butt-welding operation in this case was simulated using a heat input of 0.91 kJ/mm and 92% efficiency and stiffener welds were simulated using 1.05 kJ/mm 80% efficiency on both sides. As before, a hybrid stepwise analysis was applied.

The computational results for the alternative welding sequences and the three different weld offset cases are given in Tables 12.5 and 12.6. Three different measures of the out-of-plane deformation outcomes are given, as the significance of a given measure depends somewhat on which tolerances are critical for the application. In the 'RMS1' columns, deformations are referenced to the support positions as zero. For 'RMS2', deformations are referenced to the lowest point on the plate as it sits on the supports and 'range' gives the absolute difference between the maximum and minimum deformations, wherever they occur. The result with smallest range may not, however, coincide with a minimum RMS value if the maximum out-of-plane deformation peaks in a small region.

For both fabrication strategies (butt weld applied to plain plates or to plates with pre-tacked stiffeners) the smallest out-of-plane deformation range is associated with the smallest permitted distance between the butt weld and the centre stiffener (case 1). However, there are at least three

Table 12.5 Stiffeners not pre-tacked (built-up plate first) – out-of-plane deformation (mm) for three different weld offset cases

Fabrication stage	Case 1, 25 mm			Case 2, 100 mm			Case 3, 300 mm		
	RMS1	RMS2	Range	RMS1	RMS2	Range	RMS1	RMS2	Range
Butt weld	4.07	10.13	19.4	3.44	12.59	21.8	5.25	22.07	26.9
Left stiffener	4.90	12.35	22.4	4,48	12.85	23.3	5.93	22.16	28.2
Centre stiffener	4.29	11.15	20.3	4.17	16.98	24.4	5.92	22.36	28.7
Right stiffener	4.64	11.22	19.8	4.40	15.66	23.8	5.76	22.02	28.1

Table 12.6 Pre-tacked stiffeners (before butt welding) – out-of-plane deformation (mm) for three different weld offset cases

Fabrication stage	Case 1, 25 mm			Case 2, 100 mm			Case 3, 300 mm		
	RMS1	RMS2	Range	RMS1	RMS2	Range	RMS1	RMS2	Range
Butt weld	2.82	3.92	13.2	4.19	15.47	24.6	2.43	15.43	20.7
Left stiffener	3.50	9.51	17.7	5.27	15.82	26.4	3.16	15.78	27.7
Centre stiffener	3.72	8.67	20.7	4.92	17.86	27.1	3.59	17.82	30.8
Right stiffener	4.47	10.71	22.7	5.37	18.76	27.2	4.21	18.72	31.7

factors that influence the outcomes in a complex manner. These are: the effects of the stiffeners in terms of resisting deformations (especially for the butt-welding operation in the pre-tacked case); the effects of longitudinal residual stress fields on deformations caused by following welds; and the position of the butt weld relative to the supports. Figures 12.11 and 12.12 provide a clearer overview of deformation development in the alternative fabrication strategies.

When the plate is built up without the presence of tacked stiffeners (top row of Fig. 12.11) the principal influence arises from the support spacing and gravity effects. In case 1, the butt weld is near to the central support pair (see Fig. 12.7) and the deformations are very much smaller than in case 3, where there is less support adjacent to the butt weld. This is also reflected in the 'range' values in the top row of Table 12.5.

The addition of pre-tacked stiffeners prior to butt welding (top row of Fig. 12.12 and Table 12.6) produces some improvement, due to the additional constraint, but angular deformation local to the butt weld in case 3 is in the *opposite* direction to that shown in the stiffeners not pre-tacked fabrication route. This form of butt weld deformation is unusual, but has been noted previously in Fig. 10.22(d) where the effect of clamped boundaries was examined. What happened in that case was that the natural tendency of the plate to 'hog' upwards as the heat source passed was suppressed and the constrained expansion relieved by downward movement

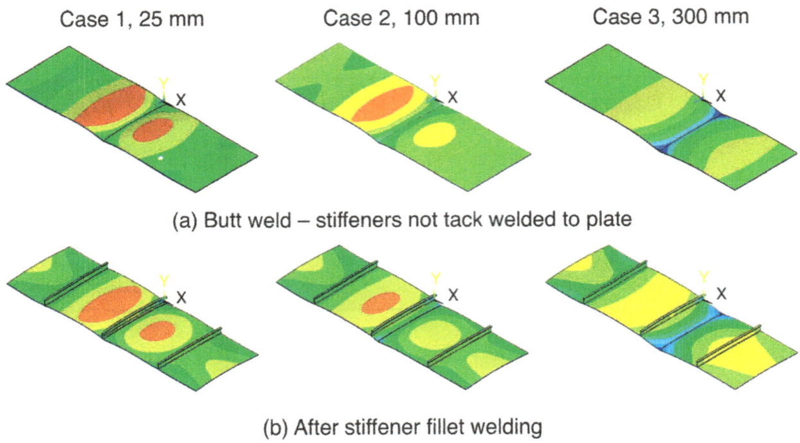

12.11 Development of out-of-plane deformation for the stiffeners not pre-tacked cases.

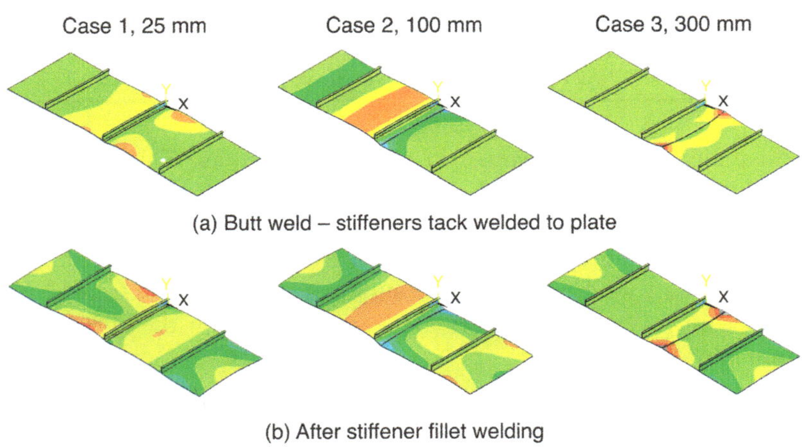

12.12 Development of out-of-plane deformation for the pre-tacked stiffeners cases.

and negative angular deformation. In the present case, it would appear that the structural restraint of the pre-tacked stiffeners has had a similar effect.

The remaining graphics in Fig. 12.11 and Fig. 12.12 show that the deformation patterns local to the butt weld after the first stage remain during welding of the three stiffeners. In general, welding the stiffeners tends to increase the average deformation levels and there is not, in the end, a great deal to choose between the two fabrication routes.

The final deformations do not increase monotonically with butt weld offset distance, especially for the 'pre-tacked stiffeners' strategy, but examination of the residual stress patterns gives some insight into this

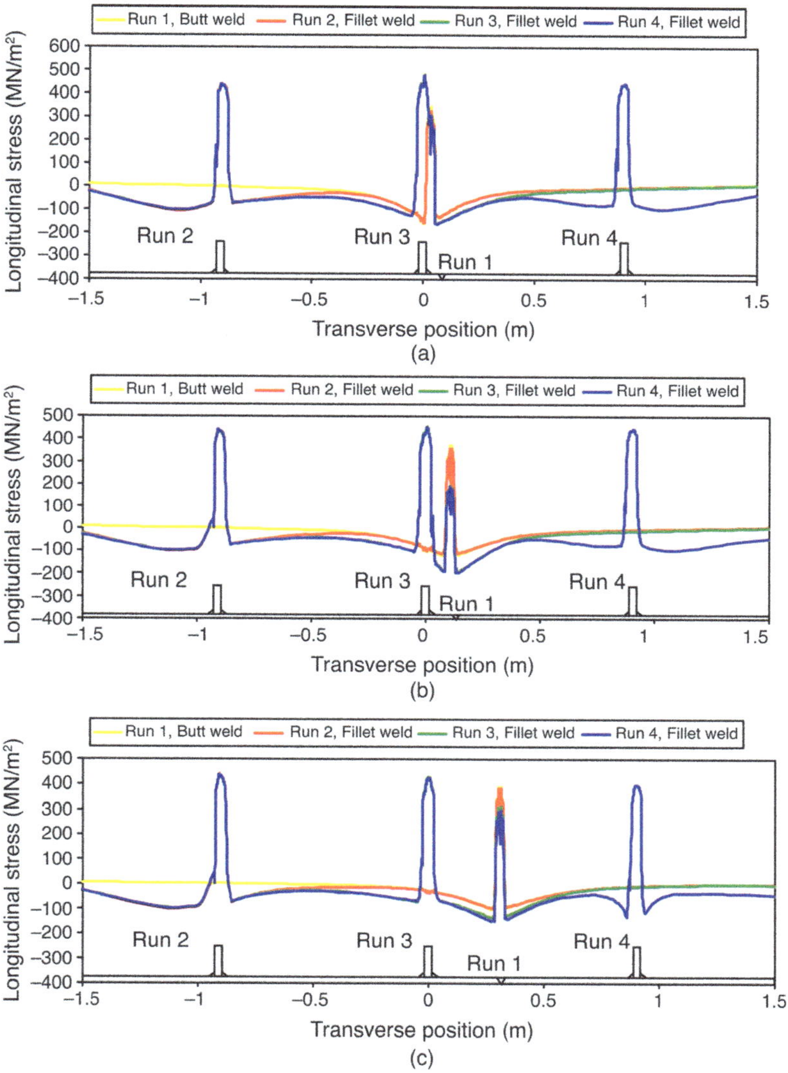

12.13 Pre-tacked stiffener assembly – evolution of residual stresses: (a) case 1, butt weld offset 25 mm; (b) case 2, butt weld offset 100 mm; (c) case 3, butt weld offset 300 mm.

outcome. Figure 12.13 shows the indicated residual stress fields for the three offset butt weld cases. Firstly, it is evident that the 25 mm offset places the central stiffener welds within the tensile residual stress zone of the earlier-completed butt weld, which will have the effect of *reducing* the contraction force and deformations relative to the other two offset cases. In case 2, the central stiffener welds lie within the *compressive* residual stress field of the butt weld, leading to an *increased* contraction force and deformations. The effect is in fact worse than case 3, where the butt weld and the centre stiffener are further apart and the residual stress at the stiffener is much smaller. Similar effects of residual stress field interaction are apparent in the 'stiffeners not pre-tacked' cases, but do not outweigh the other effects of gravity and constraint, as the residual stress field at the centre stiffener is not so different for cases 2 and 3.

These arguments show that even a rudimentary understanding of residual stress interaction is essential to providing a consistent explanation of the deformations that accompany different fabrication procedures. Also, interestingly, the 25 mm Lloyd's Register Standard for the minimum offset is shown in this instance to be a desirable *maximum*.

12.4 Limiting heat input to avoid buckling

12.4.1 Introduction

The final design-related case study considers the interaction of buckling behaviour with multi-weld manufacturing procedures where the fabrication is not allowed to cool to ambient temperature before applying consecutive welds or where there is simultaneous use of multiple welding sources (which has the same effect). Such operations give rise to greater heat input, with consequently higher longitudinal contraction forces and higher compression stresses in most of the plate width. As discussed in section 4.8 and noted in many of the previous examples, this may well lead to a non-linear, out-of-plane response, which could be stable or unstable in character. The two 'LS1' and 'LS2' double-sided fillet-welded stiffener test specimen cases in section 11.3 are used to consider a situation where a twin-arc welding machine is used, with welding heads run on either side of the stiffener. The heads may be run simultaneously – directly opposite on the stiffener or the second head may be staggered at some distance behind the first – or, in the limit, the second fillet completion may be delayed until the fabrication has cooled to ambient temperature, which is the case treated in sections 11.3 and 7.8.

12.4.2 Buckling analysis

The classical analytical buckling formula for rectangular plates subject to equal uniform compression on opposite ends (equation 12.1)[1] provides a benchmark to index the critical buckling compressive stress σ_{crit}. K_β, the buckling coefficient, depends on the length-to-width ratio $\beta(= a/b)$ and the boundary conditions, and d_p is the plate thickness.

$$\sigma_{\text{crit}} = K_\beta \frac{E}{1-\upsilon^2}\left(\frac{d_p}{b}\right)^2 \qquad [12.1]$$

Young[1] does not provide a buckling coefficient value for the particular boundary conditions in this case, but Timoshenko[2] treats the case of a long plate subject to a uniform axial compressive stress and reinforced by a central rib. The buckling coefficient depends additionally on the ratio of the bending stiffness constants of the stiffener and the plate γ and the ratio of the cross-sections δ where

$$\gamma = \frac{EI_s}{(1/12)[E/(1-\upsilon^2)]bd_p^3} \qquad [12.2a]$$

$$\delta = A_s/bd_p \qquad [12.2b]$$

A_s and I_s being the cross-sectional area and the second moment of area of the stiffener.

The buckling coefficient is then given by

$$K_\beta = \frac{\pi^2}{12}\frac{(1+\beta^2)^2 + 2\gamma}{\beta^2(1+2\delta)} \qquad [12.3]$$

Applying the dimensions of the single stiffener assembly shown in Fig. 11.6 to equations 12.2 and 12.3 gives a buckling coefficient K_β of 14.0. Following through equation 12.1 then gives a critical buckling stress of 55 MN/m^2. This stress level corresponds to a contraction force of 330 kN applied by the central weld area to both 0.6 m wide plates.

An improved analysis, appropriate to the boundary conditions, can be made through a two-stage finite-element analysis. The contraction loading is found for the maximum temperature profile generated by the particular sequence of thermal input being considered, through the mismatched thermal strain (MTS) algorithm – for example where the interpass temperature is reduced to ambient, as in the previous treatment, or where the temperature profile corresponds to dual heat source inputs that are staggered at a given distance. This loading is applied firstly to a small-deformation, zero gravity, initially flat analysis, to evaluate a stiffness matrix $[K]$ and a stress stiffness matrix $[S]$. Secondly, the Block Lancoz, eigenvalue/eigenvector extraction method offered by the ANSYS software is applied, following equation 12.4, to identify critical buckling loads and modes associated with the structural configuration and boundary conditions

$$[K]\{\phi\} = \lambda_1[S]\{\phi\} \qquad [12.4]$$

12.4.3 Deformation analysis

The computational approach described for this geometry in section 11.3 was applied to find the deformations associated with dual heat inputs staggered at various distances, ranging from 4 m spacing down to zero distance, where the two welding heads are assumed to be at the same position on opposite sides of the stiffener. The initial shape profiles of specimens LS1 and LS2 described previously were used as the starting points and gravity forces were included. Substantial solution convergence difficulties were encountered for stagger distances less than 4 m, due to plate instability, such that the analyses fluctuated from one deformation mode to another. Convergence was much improved when the artificial thermal load derived from the MTS algorithm was incremented over three load steps, rather than the single-step treatment, which had been adequate for analysis of consecutively laid welds. The variation in contraction force associated with each stagger distance is shown in Fig. 12.14, together with the force levels for each buckling deformation mode identified through the analysis embodied in equation 12.4. This figure also includes the previous result for sequential welding, placed notionally at infinite stagger distance.

RMS out-of-plane deformations versus stagger distance are shown in Fig. 12.15, together with indications of the deformation profiles at each solution point. It is immediately apparent that a large increase in deformation occurs for stagger spacings less than about 2.5 m. This corresponds to just over 300 kN contraction force, which approximates to the analytical buckling load. The severe distortion levels in this contraction force regime are exemplified by profiles 1 and 2 which show torsional buckling. The length-to-width ratio and its effect on buckling is more clearly seen in Fig. 12.16,

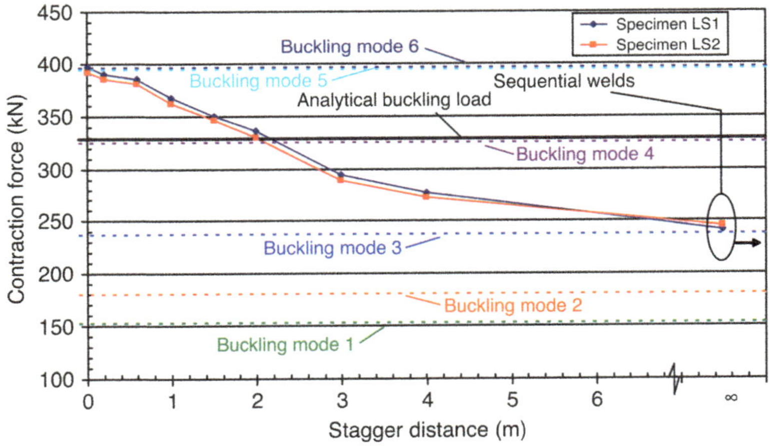

12.14 MTS contraction force versus stagger distance between welding sources.

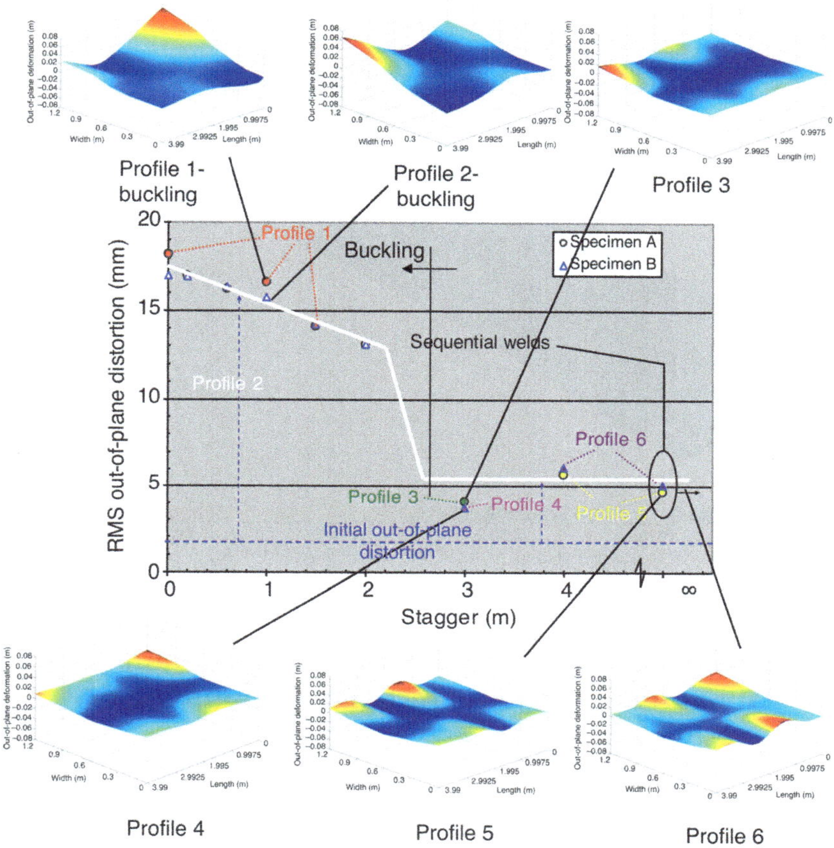

12.15 Computational prediction of buckling for staggered twin-arc welds.

which is taken directly from the finite-element analysis output and shows the plate at true scale.

Profiles 5 and 6, at the opposite extreme, reflect the initial out-of-plane shapes of the specimens, including the effects of gravity, whereas at slightly higher contraction force levels, buckling modes swamp the initial out-of-plane shapes, leading fortuitously to slightly flatter profiles 3 and 4.

This application of the algorithmic CE method shows how it can be used to provide insight into the effects of a simple variation in manufacturing procedure that might seem insignificant, but could have severe consequences for welding distortion.

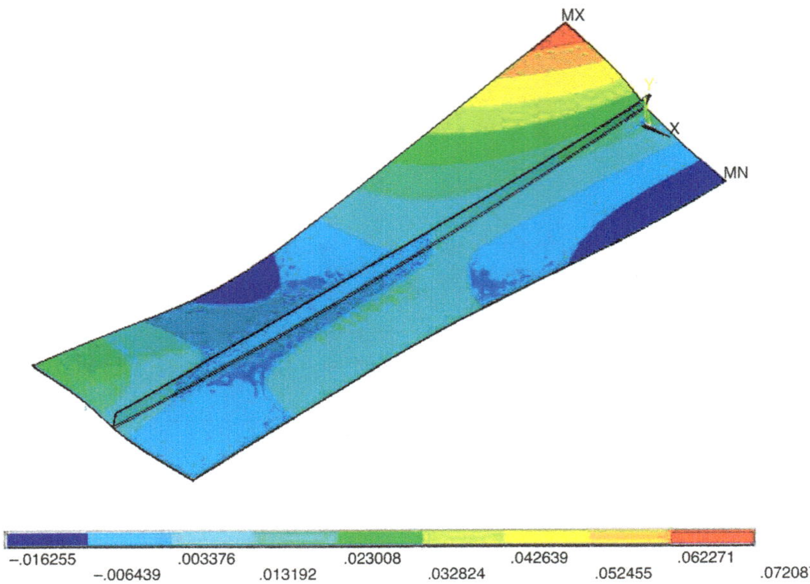

12.16 Profile 1. Finite-element analysis deformation, zero stagger.

12.5 Simulation of transient thermal tensioning: fabrication-related distortion reduction study

12.5.1 Introduction

Several examples featuring multiple, parallel, butt and fillet welds have been given in Chapters 10, 11 and earlier in this chapter. These show that, if the welds are located at transverse distances such that their thermal and residual stress fields interact, then the effects on distortion can in some cases be beneficial, depending on a variety of factors. This principle has prompted several welding process applications where heaters are applied along parallel paths to a weld, with the aim of mitigating distortion and buckling. Further information on these methods can be found in the literature.[3–5] The transient thermal tensioning (TTT) method, mentioned previously in Chapter 3, was suggested initially by Michaleris and Sun[6] and patented by Dull *et al.*[7] Figure 12.17 shows the basic principle of the method, where the lines of heating may be oriented transverse to the welding direction as in (a) or parallel, as in the longitudinal option (b). Within these two configurations, several parameters can be set – the heated length, the energy input, the transverse separation of the heater pair and the longitudinal position of the heaters relative to the point of welding heat input. There is no obvious qualitative principle to guide the choice of parameters. Nagy *et al.*[8] carried out some practical experiments using both of these configurations and the aim of the simulation that follows (based on

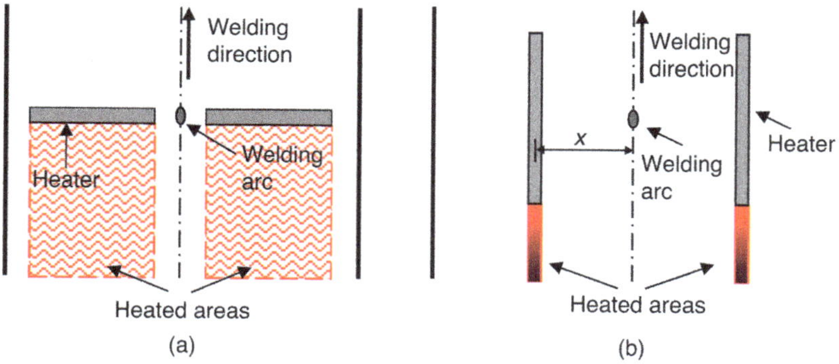

12.17 Alternative heater orientation in TTT: (a) transverse; (b) longitudinal (courtesy M. Scicluna).

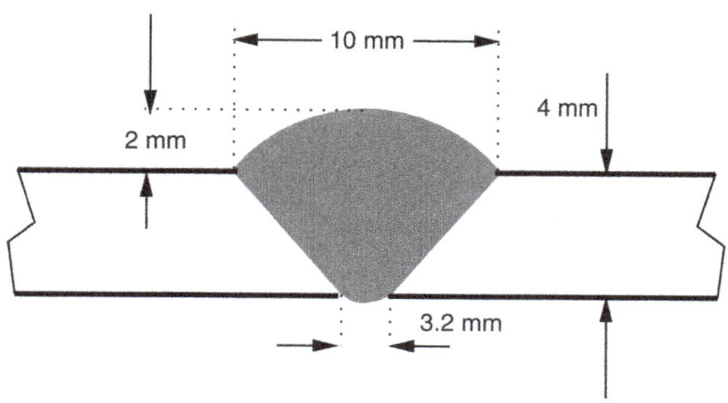

12.18 Fused zone dimensions.

the work of Scicluna and Camilleri[9]) is to provide a parallel and extended computational study of their results.

The experiments reported by Nagy *et al.*[8] were based on 700 mm × 250 mm × 4 mm thick DH36 CMn steel plates that were butt-welded using a pulsed, metal active gas welding process at an electrical energy level of 0.84 kJ/mm. Using a welding efficiency of 80%, a fully penetrated weld zone was predicted, as in Fig. 12.18. This is very similar to the test results of Fafiolu[10] from where data for the study of Nagy *et al.*[8] were taken.

The computational trials investigated two lengths of test specimen, as shown in Fig. 12.19 and Fig. 12.20. The transverse heater configuration was simulated, at a single spacing distance and at two levels of heater input energy, namely 2.0 and 2.5 MW/m^2, as shown in Fig. 12.19. Longitudinal heater configurations were investigated for both specimen A and specimen B, as shown in Fig. 12.20, covering three different longitudinal heater spacings and several heater energy levels from 0.7 to 3.5 MW/m^2. The longer specimen (B) corresponded to the referenced experimental test case and the

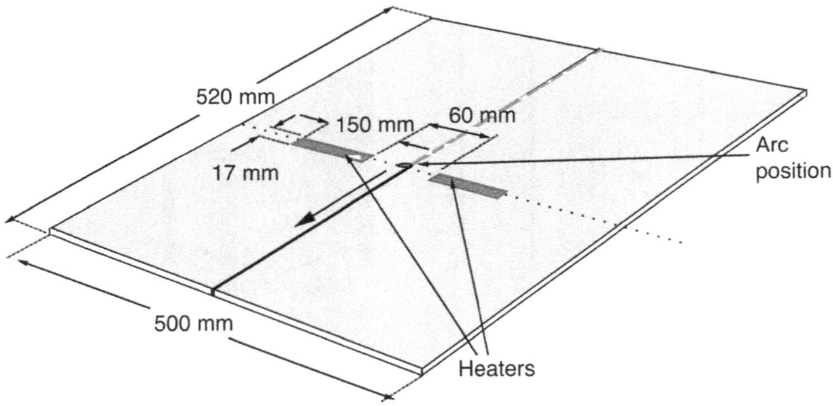

12.19 Transverse heater configuration – test specimen A.

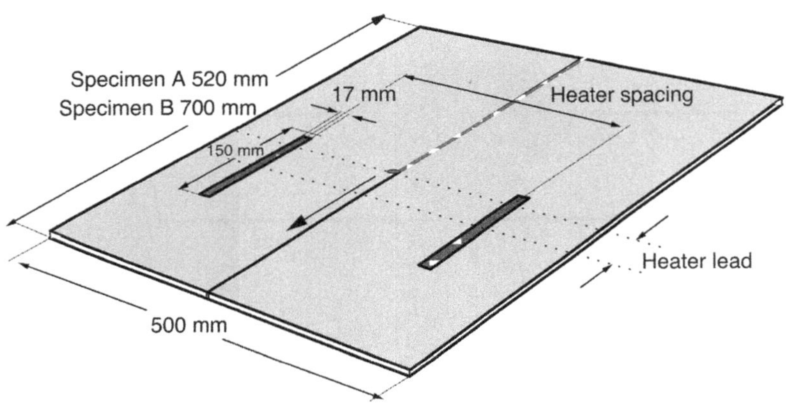

12.20 Longitudinal heater configuration – test specimens A and B.

heater spacing of 420 mm used in that test was adopted here with heater inputs between 0.4 and 1.5 MW/m^2. In all but two trials, the centreline of the heater pairs was aligned with the welding heat source; in two trials using the longer specimen, the heater position was set to lead the weld source position by 75 mm, corresponding to the cited trials.[8]

12.5.2 Thermal model

The heat input to the weld zone was modelled through volumetric heat inputs and the input to the heater zone was treated as a heat flux through the surface, covering the heater area. No heater energy losses were applied. Two-dimensional and three-dimensional thermal models were established. Figures 12.21 and 12.22 show temperature profiles across half the plate

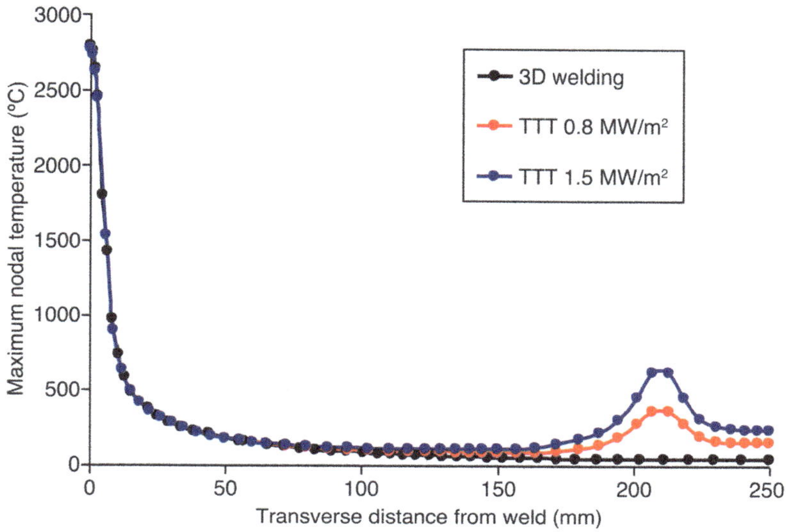

12.21 Maximum temperatures reached at welds with longitudinal TTT (courtesy M. Scicluna).

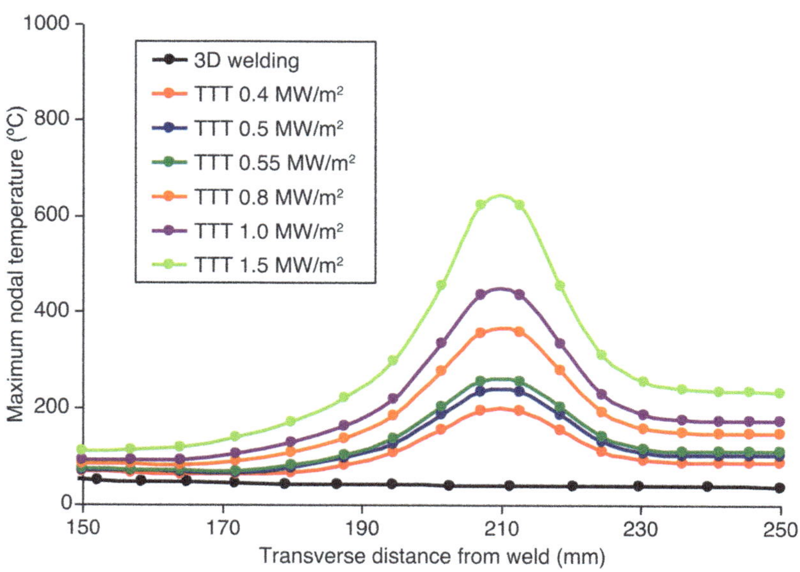

12.22 Maximum temperatures reached local to heater region with longitudinal TTT (courtesy M. Scicluna).

width, for various longitudinal heater energy inputs, taken from a 3D analysis. The overall heater spacing, as shown in Fig. 12.20 was 420 mm.

Direct comparison with the experimental data[8] is difficult as the heater input was determined by calibrating the burner arrangement to generate maximum plate temperatures in the range 270–350°C, in the absence of simultaneous weld energy input. From Fig. 12.21, this would correspond to an energy input in the range 0.6–0.8 W/m^2. In practice, temperatures much higher than 600°C for steel would probably be discouraged, which would limit this technique to around 1.5 MW/m^2. Nevertheless, it is useful to determine the effects over a larger range, as will be shown later.

12.5.3 Thermo-mechanical model

These cases could well be treated via a CE model, where the maximum temperature envelope in Fig. 12.21 and Fig. 12.22 would be applied to construct artificial cooling loads in the weld zone and the two heater zones. However, in this instance, a transient thermo-elastic-plastic analysis was applied, as this should give a more accurate description of residual stress patterns, which may be important in explaining distortion phenomena. The thermal strains extracted from the thermal analysis were applied as body-load forces in the structural models, ramped over 140 load steps in the case of the 700 mm long model B. Two-dimensional and three-dimensional models were constructed, the 2D models being supported against gravity forces at each transverse end and the 3D models at all four corners. However, all results quoted here will be for 3D models. The 3D models included three equally spaced tack welds and the 'birth-and-death' procedure was used to simulate weld filling. The 2D models were solved in approximately five minutes, whereas the 3D models required a computing time of around 24 hours.

12.5.4 Distortion and residual stress results

Figure 12.23 shows out-of-plane displacement plots for specimen A and 420 mm longitudinal heater spacing. Weld completion without TTT generates angular V-shaped distortion (greater at the finishing end) and a 'hogging' longitudinal bow. The lowest level of heater application (0.9 W/m^2) flattens the plate considerably but reverses the angular distortion to an inverted V and the longitudinal distortion to a 'sagging' mode (Fig. 12.23 (b)). Increasing the heater input to 1.5 W/m^2 (Fig. 12.23(c)) flattens the plate to a slightly greater extent in the same pattern, but further increase of heat input produces poorer results. Although the bowing distortion decreases, the angular distortion increases.

Quantitative comparisons are given in Table 12.7. This table also lists the internal longitudinal force developed by the residual stress system and the significance of this will be discussed later. The effect of varying the heater spacing can be seen in the results given in Table 12.8 (results of 420 mm spacing are repeated for ease of comparison). This shows that the narrow

Table 12.7 Out-of-plane displacement – specimen A; 420 mm heater spacing; zero heater lead

Heater input (MW/m²)	Out-of-plane displacement (mm)				Temp. reached at heater location (°C)	Internal longitudinal force (MN)
	Minimum	Maximum	Range	RMS		
0	−11.60	2.24	13.83	4.12	30	2.27
0.70	−1.78	6.34	8.12	1.89	329	−2.24
0.90	−1.33	5.47	6.80	1.56	410	−1.74
1.50	−0.87	5.76	6.63	1.66	647	−0.92
2.00	−0.44	8.09	8.83	2.66	812	−0.03
2.50	0.00	8.94	8.94	3.41	1140	1.20
3.00	0.00	9.73	9.73	5.07	1253	3.74

Table 12.8 Out-of-plane displacement – effect of heater spacing

Heater input (MW/m²)	Heater spacing (mm)	Out-of-plane displacement (mm)		
		Minimum	Maximum	Range
0.80	186	−4.12	15.78	19.90
1.00		−16.42	3.92	20.34
0.70	420	−1.78	6.34	8.12
0.90		−1.33	5.47	6.80
1.50		−0.87	5.76	6.63
2.00		−0.44	8.09	8.83
0.90	470	−1.16	5.79	6.95
1.20		−0.90	4.29	5.19

spacing of 186 mm actually *increases* the distortion from the situation where no heating is used. The wider spacing, which brings the centrelines of the heaters to within 15 mm of the plate edges, gives marginally poorer results.

An understanding of the reasons for the apparent effects of heater spacing can be gained from the residual stress patterns associated with each heater configuration. Figure 12.24 shows the residual stress distributions across the plate at mid-length and mid-thickness, for the 420 mm spacing. The maximum stress at the heater location reaches yield for heat inputs of 0.9 MW/m^2 or greater and the associated tensile stress zone enlarges in width, balanced by an increase in compressive stress in the intermediate plate zone. The integrated internal force reduces with increasing heater input, as listed in Table 12.7. This reduces the longitudinal bending effect and, with it, the main source of out-of-plane distortion.

It appears that the main effect of the longitudinal heaters is to generate an outside pair of in-plane, counteracting longitudinal moments, which reduce the longitudinal bending effect of the weld and flatten the plate, except at the ends, where the longitudinal forces cannot be sustained. This effect is not

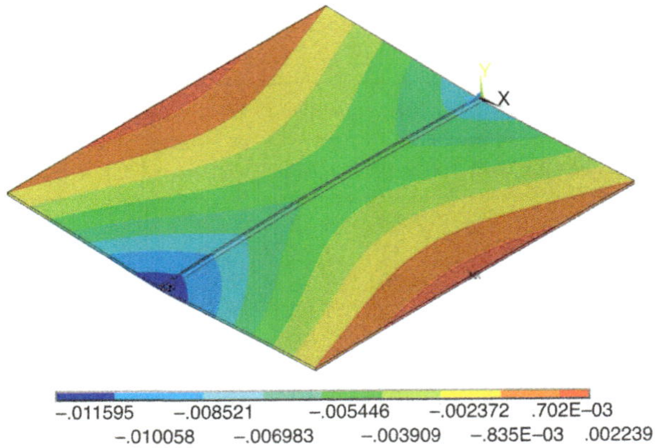

(a)

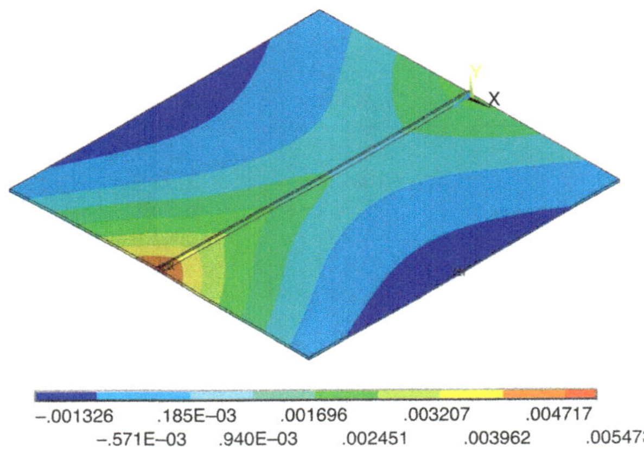

(c)

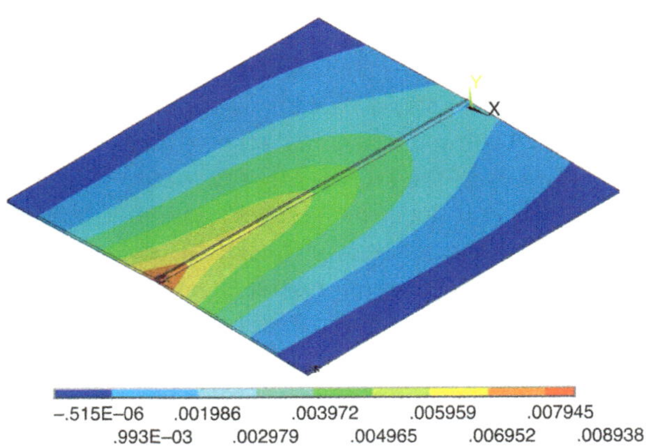

(e)

12.23 Out-of-plane displacement profiles for weld completion without TTT (a) and with increasing heater input energy: (b) 0.9 W/m^2; (c) 1.5 W/m^2; (d) 2.0 W/m^2; (e) 2.75 W/m^2; (f) 3.0 W/m^2 (courtesy M. Scicluna).

(*Continued*)

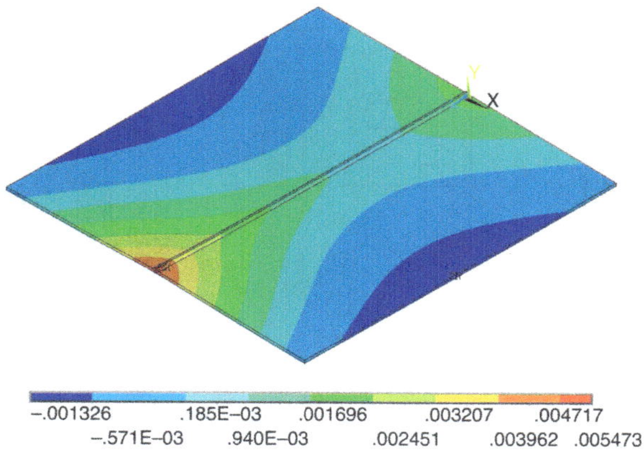

(b)

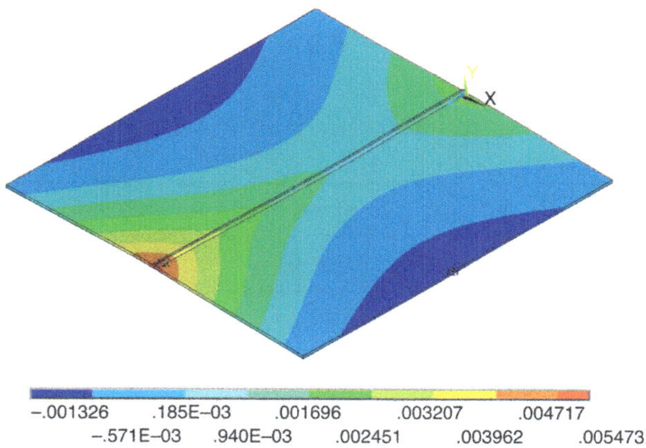

(d)

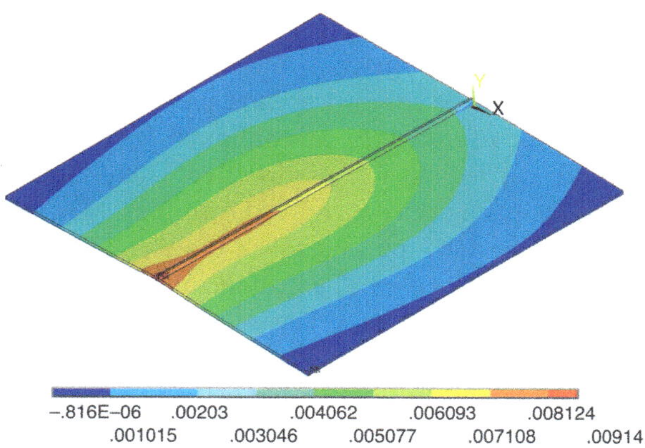

(f)

12.23 Continued.

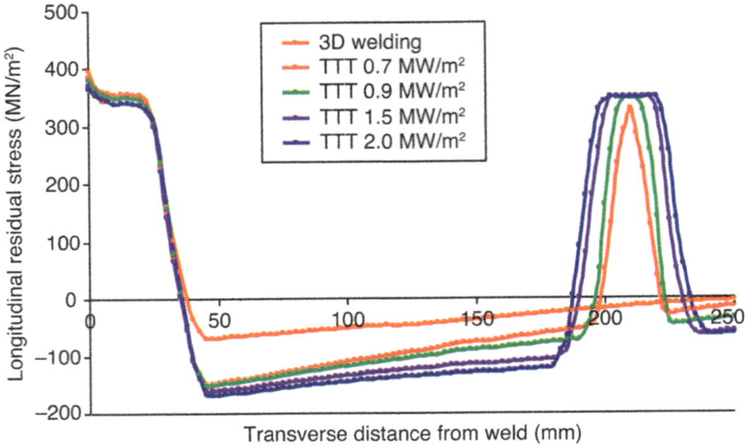

12.24 Longitudinal residual stress distributions for specimen A (courtesy M. Scicluna).

Table 12.9 Out-of-plane displacement – specimen B; 420 mm heater spacing; zero heater lead

Heater input (MW/m^2)	Out-of-plane displacement (mm)			
	Minimum	Maximum	Range	RMS
0	−15.14	5.94	21.08	4.12
0.40	−8.65	9.94	18.59	3.40
0.50	−7.99	9.44	17.43	3.17
0.80	−5.74	8.89	14.62	2.49
1.00	−4.86	9.10	13.96	2.37
1.50	−3.30	9.94	13.25	2.60

obtained in the case of narrow spacing and less contraction force is developed in the case of the wide spacing, so that the intermediate 420 mm spacing appears to be close to optimum, as confirmed by Table 12.8.

A limited analysis was carried out to find any effect related to heater lead spacing (see Fig. 12.20), using specimen A dimensions, 420 mm heater spacing and two levels of heater input, namely 1.0 and 1.5 MW/m^2. There was some flattening of the weld-only profile with these parameters but not to the same extent as the zero-lead results. Likewise, a limited investigation of transverse heater orientation (Fig. 12.19) over four values of heater input showed a negative influence as the out-of-plane deformations increased.

Analysis of the 700 mm plate configurations (specimen B) required much greater computational resources and was therefore limited to 420 mm longitudinal heater spacing. The results are given in Table 12.9 and Fig. 12.25. The out-of-plane displacements are greater than those for specimen A,

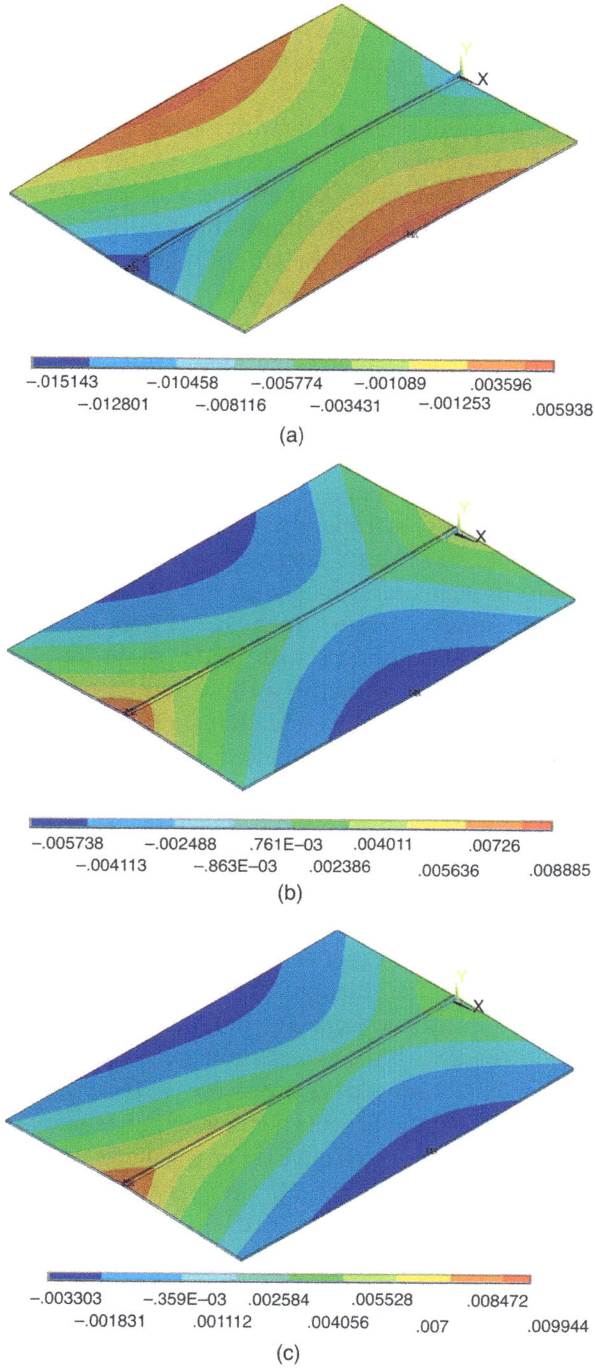

12.25 Out-of-plane displacement for specimen B (700 mm long): (a) no heater input; (b) heater input 0.8 MW/m^2; (c) heater input 1.5 MW/m^2 (courtesy M. Scicluna).

as one might expect in a longer plate, and the flattening effect of longitudinal thermal tensioning is less, but similar mechanisms as outlined previously seem to be present.

Comparison of the computational results with the experimental results[8] is uncertain, due in part to ambiguities in the data. Also, the specimens were spot-clamped to a supporting surface and not simply supported as assumed in the computations. With these qualifications, the distortion displacement range of the 700 mm long specimen B, without heaters, was given as 17.5 mm in one trial and 28.5 mm in a second trial, compared with 21.08 mm in the computations. However, the more important issue concerns the effectiveness or otherwise of the thermal tensioning principle. Firstly, the experiments and computations agree that transverse heating *increases* out-of-plane deformation. Similarly, there is agreement that a narrow spacing of heaters in the longitudinal configuration (250 mm in the experimental case and 186 mm in the computation) had a negative effect.

There is disagreement between experiment and theory with respect to certain other parameters. The tests appear to show deterioration of distortion correction if temperatures higher than 310°C are applied (equivalent to heater inputs greater than 0.7 MW/m^2), whereas the computations show increasing response with heater input and local plate temperature. In any event, the computations show that an optimum has to be sought in terms of both heater input and configuration.

However, both approaches show that longitudinal heater orientation can be used effectively to reduce out-of-plane distortion, a key factor being the transverse spacing of heaters relative to the plate width. An explanation of this factor has been given, related to the beneficial alteration of residual stress systems, as revealed by the computational approach. Computed results suggest a reduction in deformation of around 30% within the range of heat inputs investigated in the experiments, which themselves appear to show a greater reduction. The computational results show that the maximum out-of-plane deviations occur at both ends of the welded plate and this finding is consistent with the explanation given for the importance of the spacing parameter. Nagy *et al.*[8] provide a somewhat different explanation of the reasons for the beneficial effect and conclude that '... it is unlikely that this method will be effective on larger plates where it will be difficult to induce such lateral bending'. If the explanation given here is correct, there seems no reason to believe that the method cannot be applied to larger plates, other than due to any practical difficulties in scaling up the heater input. However, it is clear that the outcome depends critically on optimisation of system parameters and that computational simulation can provide a necessary degree of insight into the effects of parameter variation.

12.6 Simulated use of low-transformation-temperature filler material to reduce distortion

12.6.1 Introduction

It has been noted several times in previous chapters that the thermal contraction coefficient of a material being welded is a major driving parameter for residual stress development and distortion. Earlier, Fig. 4.5 showed that this coefficient is often not linear across the temperature range suffered by welds, but can show step changes in the relationship between thermal strain and temperature. These steps are evidence of structural phase changes where, for example in steels, austenite may transform to martensite and/or bainite over a particular temperature range, leading to a volume *expansion* and temporary increase in length. This is followed by further contraction at a different rate from that occurring at higher temperatures, as the metallurgical structure will then be different.

The temperature at which the transformation takes place is particularly important in relation to the strength of the material in the corresponding temperature range. In the case of the example material given in Fig. 4.5, the relevant temperature range is 690–530°C and Fig. 4.6 shows that the yield strength of the material in that temperature band is probably in the 150–250 MN/m^2 range at most, i.e. much less than the ambient temperature yield strength of 375 MN/m^2. Hence any longitudinal expansion during the phase change is likely to be successfully resisted by the surrounding cooler material, which has much higher yield strength. There is then a further period of cooling to ambient temperature, during which the material follows an extensive linear contraction profile. This will be sufficient to generate tensile residual stresses in the central fusion zone of the order of ambient temperature yield strength. In such a case, therefore, the phase change expansion should have little effect.

However, some materials experience this phase change at a much lower temperature than others, and there is therefore the possibility of developing a greater expansion response followed by a much shorter further range of cooling and contraction. Such materials are sometimes described as 'low-transformation-temperature-start' alloys and can be conveniently introduced to a fusion zone via the filler wire. The general effect is that strong expansion during the phase change will develop compressive residual stress in the weld zone, or at least substantially reduce the tensile residual stress developed during the other periods of linear contraction. Reducing the tensile component of residual stress will reduce distortion. This approach to residual stress reduction has been extensively researched by, for example, Henwood *et al.*[11] Bhadeisha,[12] Mikami *et al.*[13] and Murakawa *et al.*[14] in the context of ferritic steels and Shirzadi *et al.*[15] for stainless steels. Materials with this special attribute must of course also satisfy other performance criteria such as having adequate strength and fracture toughness.

The potential of this method to mitigate distortion is clear on qualitative grounds, but the complexity of interaction between different parameters in a practical fabrication context makes it difficult to be certain of the quantitative outcome without extensive live trials. The following study therefore uses computational simulation to investigate the likely outcomes of using three different welding filler wires, which exhibit different phase change, thermo-mechanical characteristics.

12.6.2 Model

This trial is based on the simulated butt welding of two 500 mm × 250 mm × 4 mm steel plates as shown in Fig. 12.26. The plates are assembled using three 25 mm long tacks, of the same dimensions as the main weld. The tacks are 'removed' during simulated welding by deactivating appropriate elements when the main welding heat source arrives. The assembly is assumed to be flat initially, but is then supported at all four corners and subject to gravity. (Note: tacking operation was not simulated.)

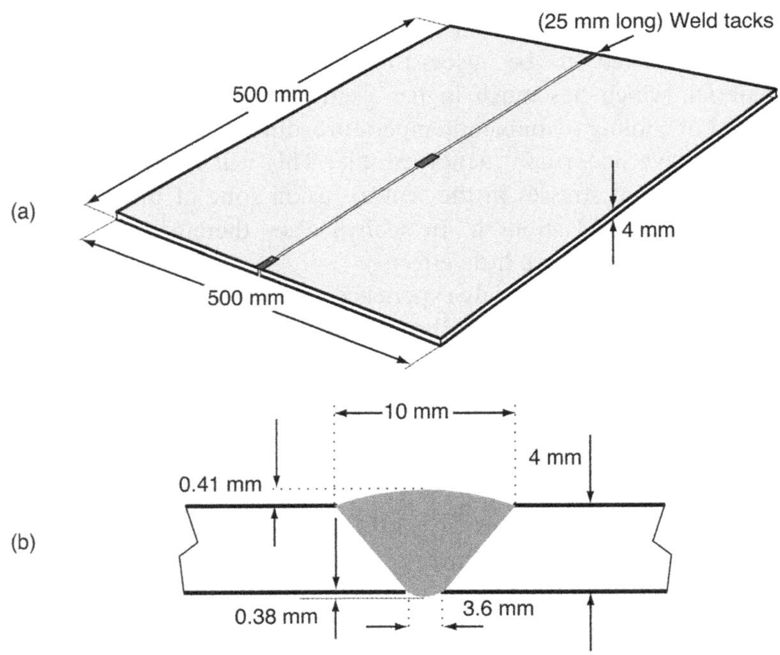

12.26 Dimensions of butt-welding model for low-transformation-start trial: (a) tacked plate dimensions; (b) weld dimensions.

Table 12.10 Chemical composition (%) of welding wires and parent material

Element	Filler wire: Metal-cored (MC)	Filler wire: Flux-cored (FC)	Filler wire: Low-transf. (LT)	Parent material DH36
C	0.06	0.05	0.04	0.17
Si	0.50	0.41	0.32	0.35
Mn	1.50	1.36	0.36	1.22
P	<0.015	0.01	0.008	0.019
S	<0.015	0.008	0.004	0.01
Cu	—	0.26	0.15	—
Ni	—	—	11.9	0.04
Cr	—	—	—	0.05
Mo	—	—	—	0.002
Al	—	—	—	0.004

Table 12.11 Welding process conditions

Filler wire	Current (A)	Voltage (V)	Travel speed (mm/s)	Energy input (kJ/mm)
Metal-cored (MC)	155	20.5	2.21	1.44
Flux-cored (FC)	235	19.0	2.27	1.97
Low-transf. (LT)	160	23.5	2.25	1.67

An uncoupled, transient thermo-elastic-plastic approach is used to simulate welding, following the methods in Chapters 9 and 10.

12.6.3 Materials and welding process conditions

Weld trials were conducted to determine the expansion characteristics of weld metals produced by three different filler materials applied in the butt welding of 4 mm thick Lloyd's grade DH36, CMn plates. The chemical composition of the wires and the parent material is given in Table 12.10. The filler wires included two commercial wires – 'metal-cored' (MC) according to EN758:T46 4 MM1 and 'flux-cored' (FC) according to EN758:T42 2 PM1 H5. The 'low-transf.' (LT) wire was a nickel-based solid wire. The high nickel content promotes the formation of martensite and reduces the martensite formation start temperature.

A metal inert gas welding process, using a shielding gas mixture comprising 80% Ar, 20% CO_2 was applied for all three welding wires. The plates were square-edged with a 4 mm gap. It was not feasible to apply the same welding process conditions for each filler wire and, as shown in Table 12.11, the energy inputs were not identical.

In order to distinguish the influence of the low-transformation-start more closely, it was assumed in the model that the same fusion zone cross-section, as in Fig. 12.26, would apply to each of the welds. An identical welding efficiency of 85% was applied for the same reason. In order to minimise the differences even further, a second set of simulations was carried out for all three materials at a standard level of 0.63 kJ/mm net heat input, corresponding to welding process conditions of 190 A, 24.5 V, 5.5 mm/s and 75% efficiency. Both sets of results will be shown.

The important phase-transformation data for the three fusion zone weld metals, determined by dilatometry tests, are given in Fig. 12.27. The corresponding transformation start and finish temperatures, together with the ambient temperature yield strengths are noted in Table 12.12. 'Integrated expansivity' in Table 12.2 was determined by integrating the contraction strain over each region from 1000°C to ambient temperature. This result suggests that there is little overall difference in expansivity between these materials.

The results for the two cored wires suggest that there is no martensite formation in these fusion zones. The transformation start temperature for the low-transformation-temperature material (LT) is indeed much lower

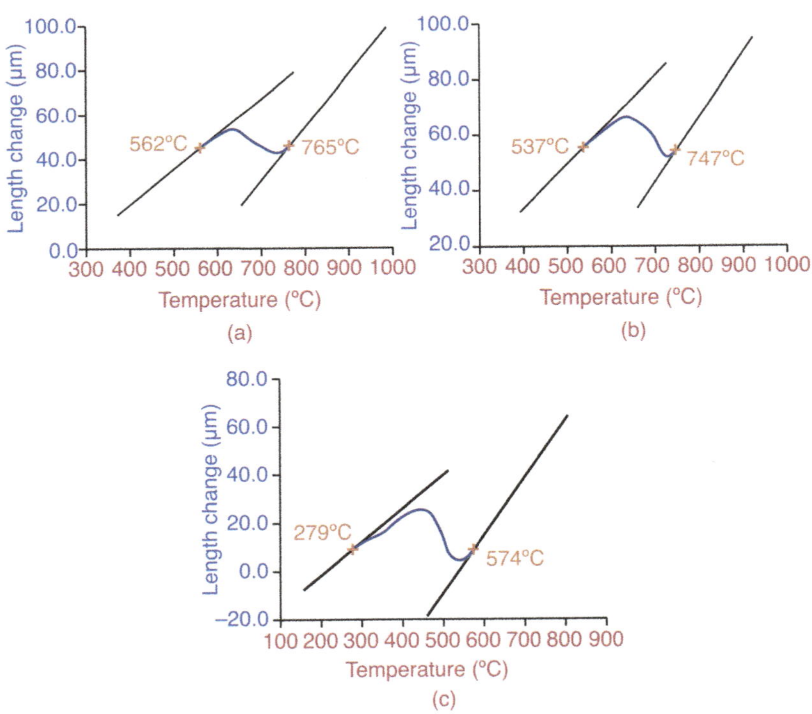

12.27 Dilatometry test results on all-weld-metal samples: (a) metal-cored and DH36; (b) flux-cored and DH36; (c) low-transformation wire and DH36.

Table 12.12 Material data for fusion zones generated by welding wires

	Transformation start temperature (°C)	Transformation finish temperature (°C)	Integrated expansivity ($°C^{-1} \times 10^{-6}$)	Ambient temperature yield strength (MN/m^2)
Metal-cored and DH36	765	562	14.25	418
Flux-cored and DH36	747	537	13.87	471
Low-transf. and DH36	574	279	13.87	500
DH36 parent material	700	515	16.00	354

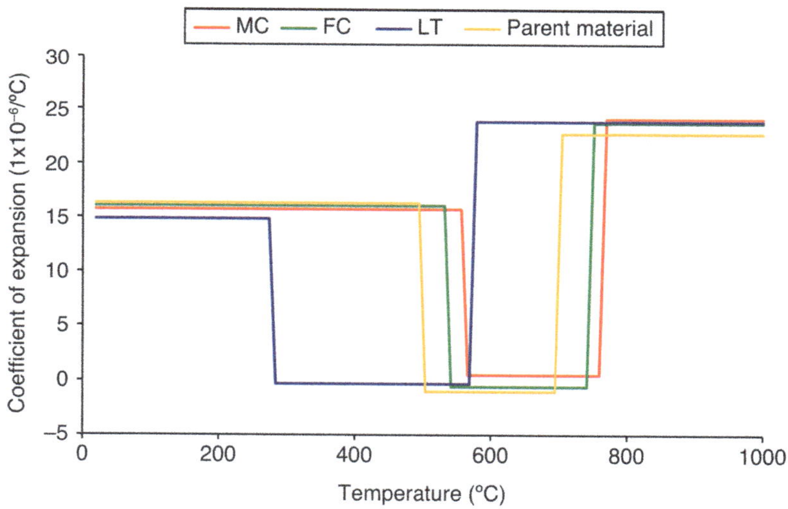

12.28 Simplified temperature-dependent expansivity for trial materials.

than for the other two but the subsequent contraction rate after the transformation is typical of a bainite/pearlite/ferrite mixture, suggesting that the proportion of martensitic material is small.

Figure 12.28 shows how the dilatometry results were interpreted in terms of expansivity, for the purposes of the computational models. The phase change region is modelled simply by setting an expansivity corresponding to the slope of the line connecting the transformation start and finish temperatures, i.e. ignoring the non-linear path between the two points. In practice, the fusion zone expansivities will be affected by parent material dilution. Successful prediction of residual stress and distortion also requires other temperature-dependent material properties such as yield strength, Young's modulus and plastic modulus, which were unfortunately not available for these materials. Therefore, the temperature-dependent data previously established for DH36 material, as noted in Chapter 4, were used, with alteration of the ambient temperature yield strength to reflect the values in Table 12.12.

12.6.4 Results

Figure 12.29 shows the predicted out-of-plane deformation profile for a plate welded using low-transformation-start welding wire. Noting that there were no signs of asymmetry or local buckling deformation, it was decided to characterise the deformation behaviour in terms of simple, overall measures of displacement at discrete points, to give global determinations of angular distortion and longitudinal bowing. The index points are shown in Fig. 12.30. The plates are supported at points D and F, which therefore have zero displacement and displacements were predicted for the remaining points A, B, C and E. The relative displacements of the points were then processed to determine the angular deformation at the starting end (AF), the middle (BE) and the finishing end (CD). Longitudinal bowing was calculated similarly along the weld line (ABC) and along the outer edge (DEF). Table 12.13 lists the results.

The differences in predicted distortion between the three filler materials are relatively small. In the cases where identical welding parameters and heat inputs were applied, longitudinal bowing is reduced when low-transformation-start filler (LT) is used, by a small but identifiable amount. Angular distortion is increased, however. The results for distortion range

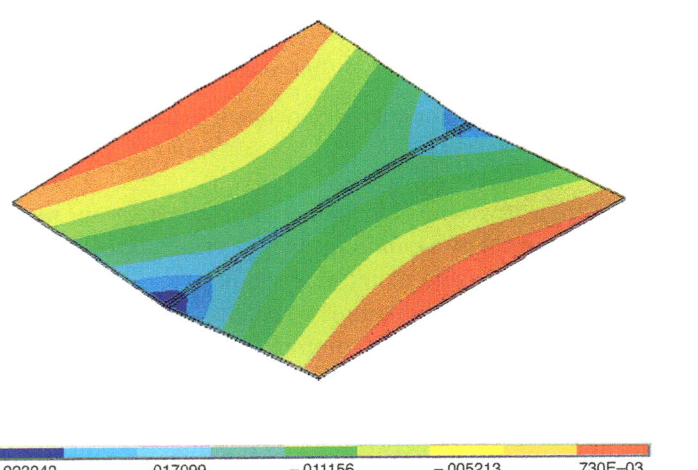

12.29 Predicted out-of-plane deformation profile for low-transformation-start weld.

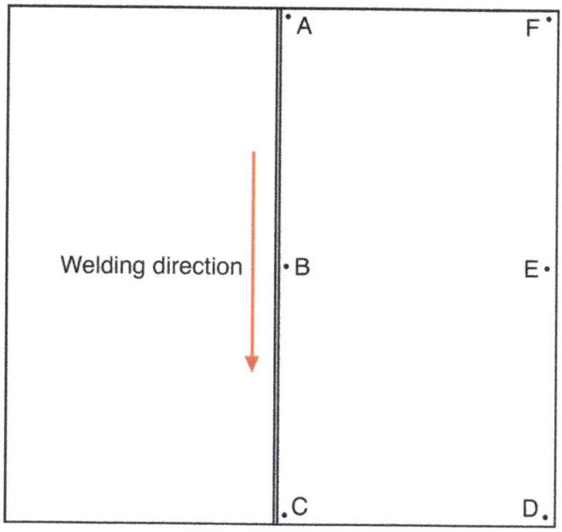

12.30 Deformation index points.

Table 12.13 Predicted out-of-plane deformation for three filler wires and two welding process sets

	Actual welding parameters			Identical welding parameters		
	MC	FC	LT	MC	FC	LT
Distortion range (mm)	26.03	27.07	26.73	22.61	22.50	22.78
Angular distortion (degrees)						
Start edge (AF)	8.39	9.08	8.91	7.15	7.08	7.27
Middle (BE)	7.23	7.81	7.88	6.47	6.42	6.68
Finish edge (CD)	10.05	10.50	10.53	8.99	8.94	9.14
Average	8.56	9.13	9.10	7.54	7.48	7.70
Longitudinal bow (m^{-1})						
Outer edge (DEF)	−0.129	−0.131	−0.118	−0.094	−0.095	−0.090
Centre weld (ABC)	−0.269	−0.270	−0.248	−0.206	−0.206	−0.197

reflect the angular distortions at the finish end. A similar pattern is seen in the solutions where actual welding parameters were applied, when due allowance is made for the differences in heat input. Given that the overall expansivity is almost the same for all three materials (see Table 12.12) the differences in angular distortion would seem to be related to the fact that the phase change expansion for the low-transformation-start material is occurring in a cooling regime where it is developing greater yield strength.

It is again instructive to examine the residual stress distributions for 'actual' welding inputs, as in Fig. 12.31. The low-transformation-start material (LT) shows a clear stress reduction in a narrow central band, which reduces the longitudinal contraction force relative to the other cases and is

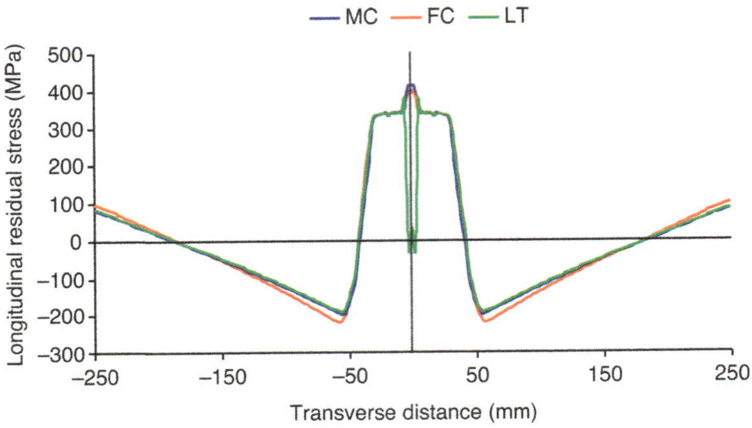

12.31 Predicted longitudinal residual stress for different filler materials.

responsible for the reduction in longitudinal curvature. It is likely that the simplified expansivities shown in Fig. 12.28 do not do full justice to the effect, as the interaction of volume expansion and developing strength occurs at a lower temperature level than the preceding contraction.

The apparent increase in angular deformation in the low-transformation-temperature material case is more problematic to explain, as the overall or 'integrated' contraction between the cut-off temperature and ambient temperature is not significantly different from the other materials (see Table 12.12). The higher yield strength of this material may be a factor. This partly contradicts the principle established previously that transverse contraction of a fully penetrated weld is unconstrained but, in this case, there is some angular constraint through the tack welds.

12.6.5 Conclusion

One conclusion of this study is therefore that the beneficial use of a welding process involving low-transformation-temperature filler is identifiable in the results, although the benefit is, in this instance, marginal. A critical factor is the 'strength' of the phase-change volume expansion. If it takes place at a temperature level where the fusion zone is at a low enough temperature to develop significant compressive yield strength, then it will have a significant effect in terms of reducing the contraction force and therefore the longitudinal bending distortion. There seems to be little effect on angular distortion; indeed, in this example, the special filler material has generated a slight increase in angular distortion.

This study also highlights the need for more sensitive data on materials if meaningful results are to be found. Phase change behaviour is known to depend on cooling rate and dynamic stress, whereas expansivity information is most readily obtained from simple dilatometry tests where the specimen is unconstrained and the cooling rates are relatively slow. Likewise, the interaction between phase-change expansion and yield strength is of critical importance for the effects of low-temperature-transformation to be realised and, in this instance, accurate modelling was compromised by insufficient information on the temperature dependence of yield strength in the transition region and during further cooling to ambient temperature.

12.7 Simulated use of weld-trailing cryogenic cooling process to reduce distortion

12.7.1 Introduction

In the cryogenic cooling strategy, the welding head is closely followed by a source of cooling material (see Li[16] for a fuller description). In the case considered here, a nozzle that delivers solid CO_2 to the plate surface is used.[17] This extracts heat from the plate, mainly by sublimation of the solid CO_2 to gas, and thus reduces the heat dissipated into the structure, which – as noted previously – is the main driver for distortion. Not every material can tolerate the more rapid cooling profile that results, particularly those that are hardenable, but it is otherwise a feasible technique for many materials. The method introduces several additional variables to the fabrication process, including the trailing distance set behind the welding heat source, the rate of heat extraction, which depends on the solid CO_2 delivery rate together with the efficiency of the nozzle arrangement, and the profile of the CO_2 impingement on the workpiece. This again is a situation where computational simulation can provide an indication of outcomes and, even more importantly, lead to an understanding of the physical effects that underlie these outcomes.

The study described here was carried out in support of a project to develop the CO_2 delivery and extraction part of the system[18] and therefore the heat extraction rate, the profile of the CO_2 impingement and the distance of the cooling patch behind the welding heat source were specified as given parameters. The aims of the study were to investigate different simulation analyses and determine the likely effects on the distortion of fabrications comprising butt welds and single fillet welds on a stiffener.

12.7.2 Model

The study considered butt welds in 2 mm thick steel plates and single-sided fillet welding of 5 mm thick stiffeners to 5 mm thick, steel base plates (see Fig. 12.32 and Fig. 12.33). The fabrications were assumed to be supported at each corner and subject to gravity.

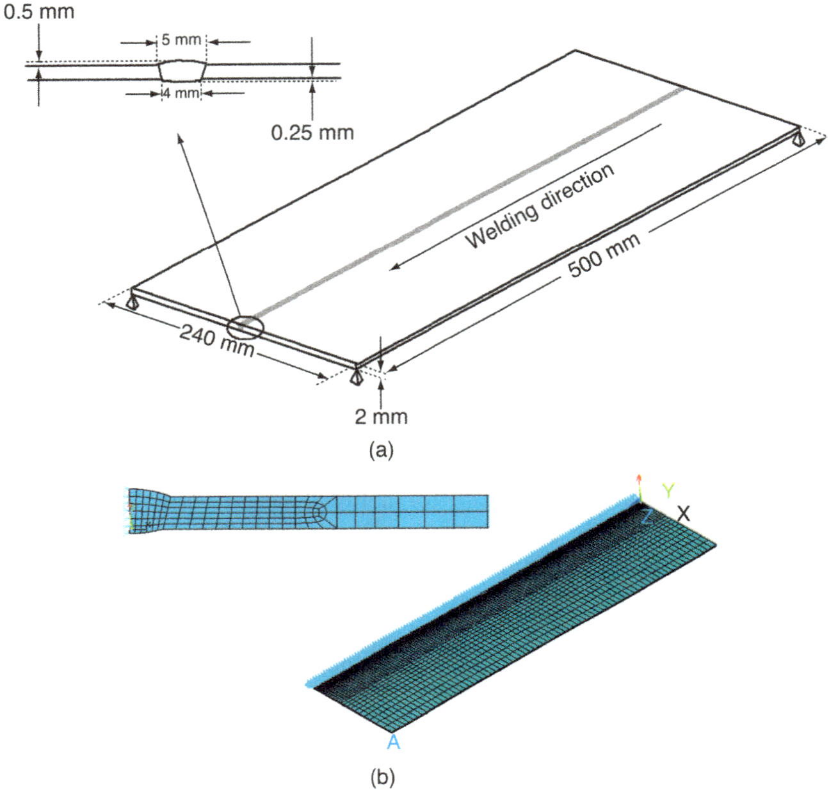

12.32 Butt-weld model: (a) dimensions; (b) finite-element formulation.

12.7.3 Thermal model

In the butt-welding case, process conditions with an input energy of 0.2 kJ/mm at 2 mm/s travel speed were assumed, together with an efficiency of 80%. In the 4 mm fillet-weld case, 0.37 kJ/mm input energy at a travel speed of 10.8 mm/s was assumed, again with 80% efficiency. Trials of the cooling system suggested that 6×10^{-3} kg/s delivery of solid CO_2 could be delivered to extract 1.2 kJ/s heat energy through the surface over a 90 mm diameter circular patch and a normal radial distribution of heat transmission was applied as a heat flux. The resulting welding heat input and cooling extraction profiles are shown for the butt- and fillet-welded models in Fig. 12.34.

Figure 12.35 shows that cryogenic cooling at the assumed conditions in the butt-weld model should have a substantial effect on thermal gradients and maximum temperatures across the width of the test plate. The maximum temperatures at transverse positions greater than the 4.75 mm location are significantly reduced, which means that the yielded regions will

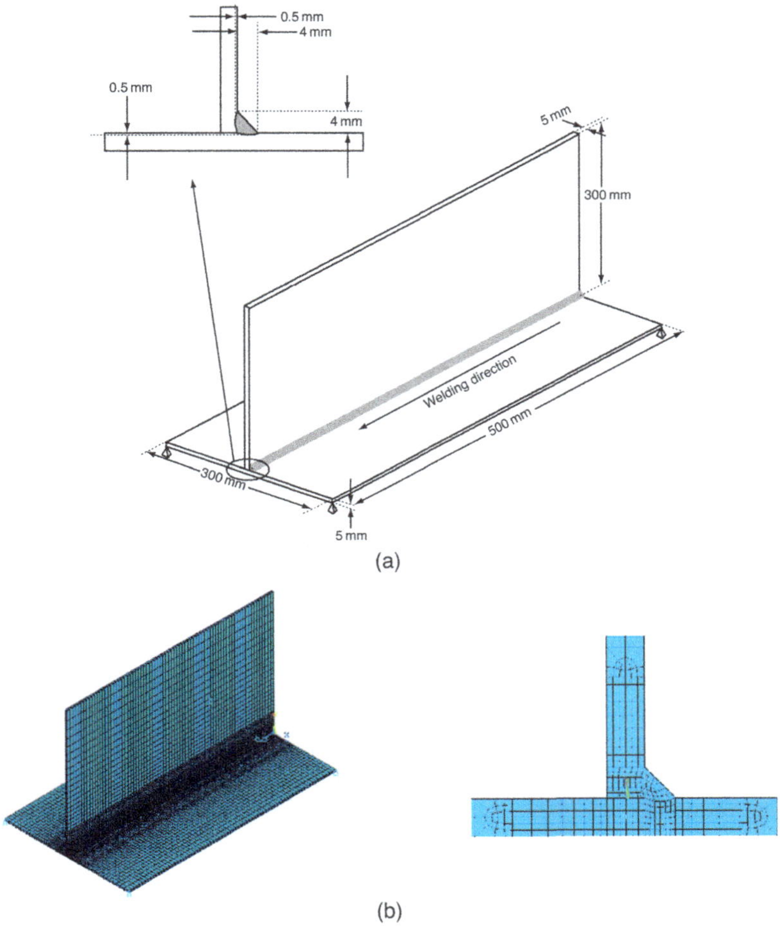

12.33 Base plate with fillet-welded attachment model: (a) dimensions; (b) finite-element formulation.

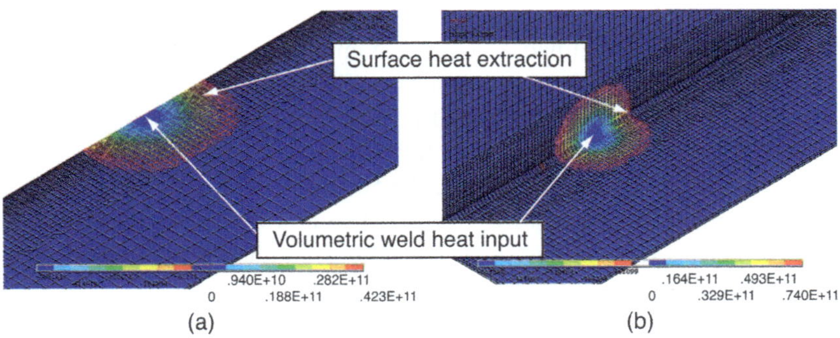

12.34 Welding heat input and cooling profiles: (a) butt weld; (b) fillet weld.

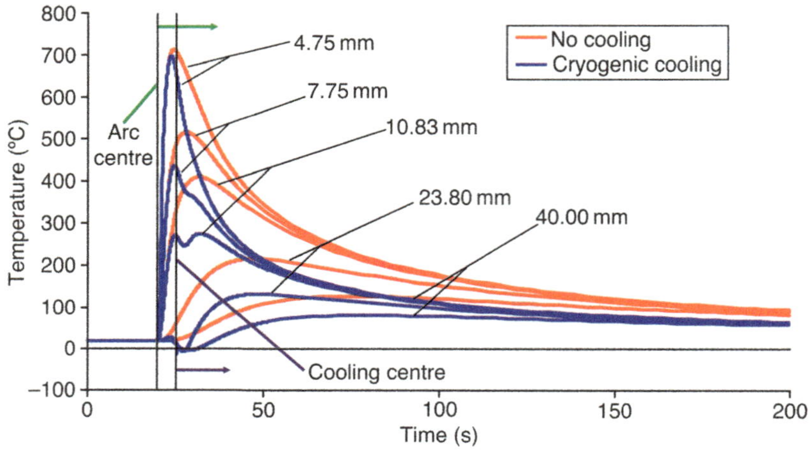

12.35 Thermal transients in butt-weld case.

be smaller and the contraction force less. The overall effect on thermal contours is effectively as if a welding process with much lower heat input had been used, as shown in Fig. 12.36.

Similar reductions in the width of thermal contours are seen in the fillet-welding case, as shown in Fig. 12.37. Figure 12.38 shows the thermal transients on the weld side of the stiffener and on the opposite side. The maximum temperatures are again much reduced when cooling is applied and it should be expected that the yielded zones and contraction forces will be reduced.

12.7.4 Thermo-mechanical analysis

All three computational strategies depicted in Fig. 5.3 were trialled, namely the CE model based on elasto-plastic algorithms, a hybrid model (both described in Chapter 8) and a computationally intensive model (CIM), as described in Chapter 9. The first two models are based on the maximum temperature profile determined across the complete welding and cryogenic cooling span, and this represents a substantial simplification of the true transient characteristic. Nevertheless, it is interesting to compare the results with those generated by the transient temperature description in the CIM model. In the case of the hybrid model, the 500 mm long plates were divided into five sections in the stepwise treatment. The non-linear expansivity and strength properties were represented in the hybrid and CIM models by the values shown in Fig. 12.39.

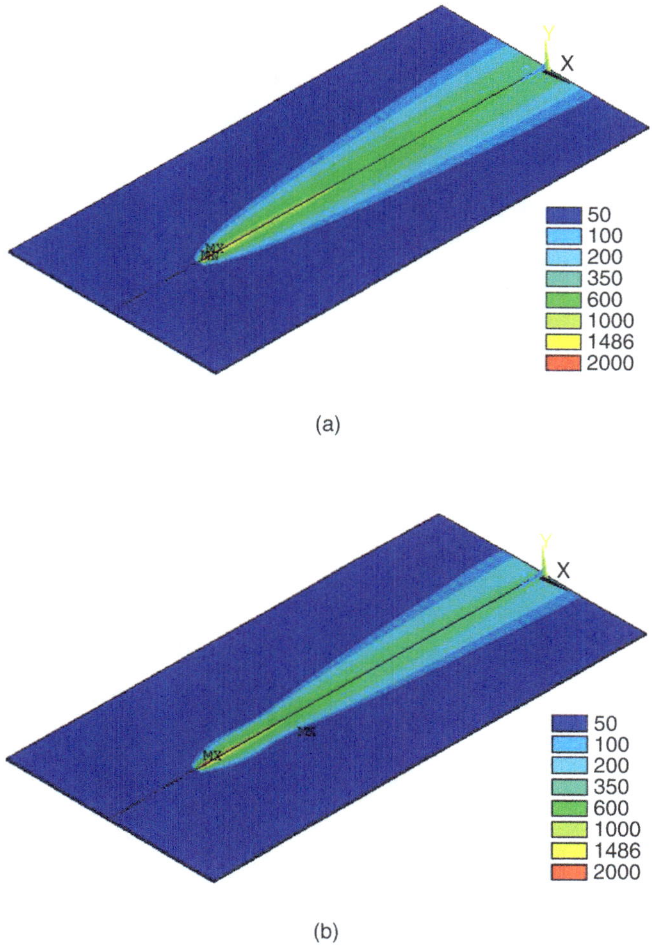

12.36 Thermal contour plots in butt-welded plate (°C): (a) welding without cooling; (b) welding with cryogenic cooling.

12.7.5 Distortion and residual stresses

Figure 12.40 shows the predictions of out-of-plane deformations in butt-welded plates, according to the different structural computation strategies. All results show the classic hogging pattern longitudinally and the normal positive, V-shaped transverse angular contraction. As the shapes are regular and consistent, it is reasonable to characterise the distortion in terms of the deformation range, measured between the lowest point of the V-shape at the starting end and the highest point at the centre of the outside edge.

Firstly, in the cases without cooling, the CE approach predicts a much smaller deformation than the other two strategies. Careful comparison

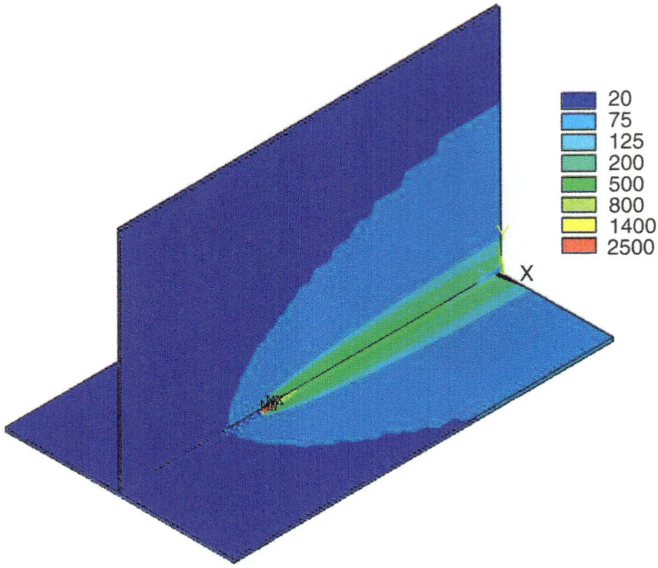

(a)

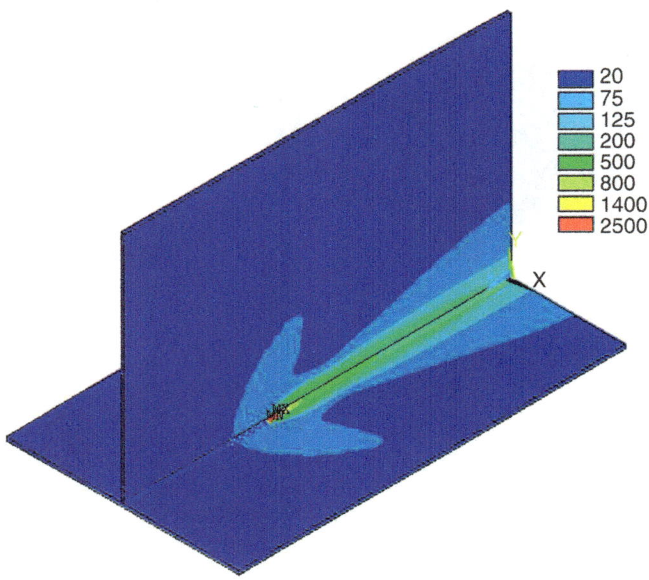

(b)

12.37 Thermal contour plots in fillet-welded plate and stiffener (°C): (a) welding without cooling; (b) welding with cryogenic cooling.

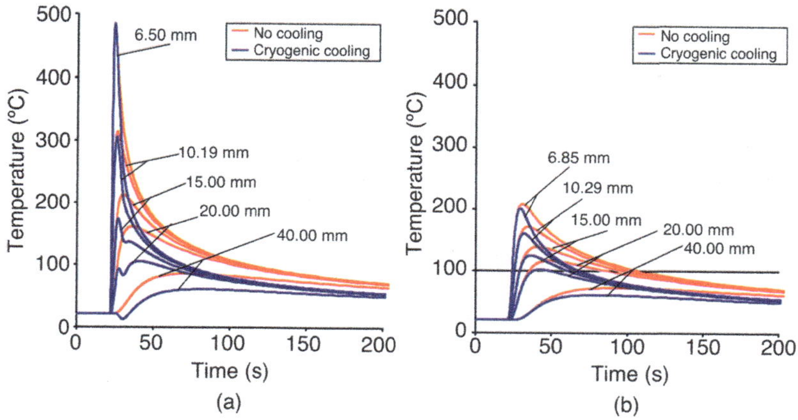

12.38 Thermal transients in fillet-welded case: (a) welded side of stiffener; (b) opposite side of stiffener from weld.

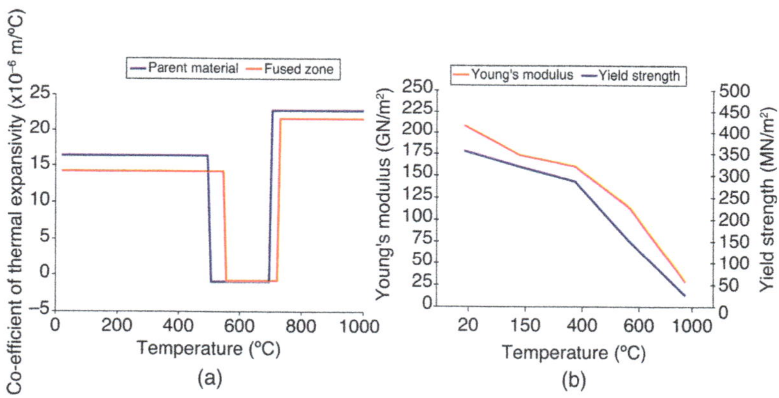

12.39 Material properties used in welding simulation of cryogenic cooling: (a) non-linear contraction coefficient; (b) non-linear strength properties.

shows that the main source of the difference lies in the prediction of angular deformation, which is much less in the CE approach. This result exposes a weakness of the method – it does not model the effect of gravity on transverse deformation when using widely spaced supports. The assumption in the transverse deformation algorithm is that the material outside the fusion zone is elastic and, inside the zone, strain is governed entirely by the transverse expansivity with no contribution of plastic strain. The same is not true of the longitudinal deformation where the structural model reflects the tension due to contraction and also the flattening effect of gravity. However, the longitudinal bowing in the case of the CE model is again somewhat less

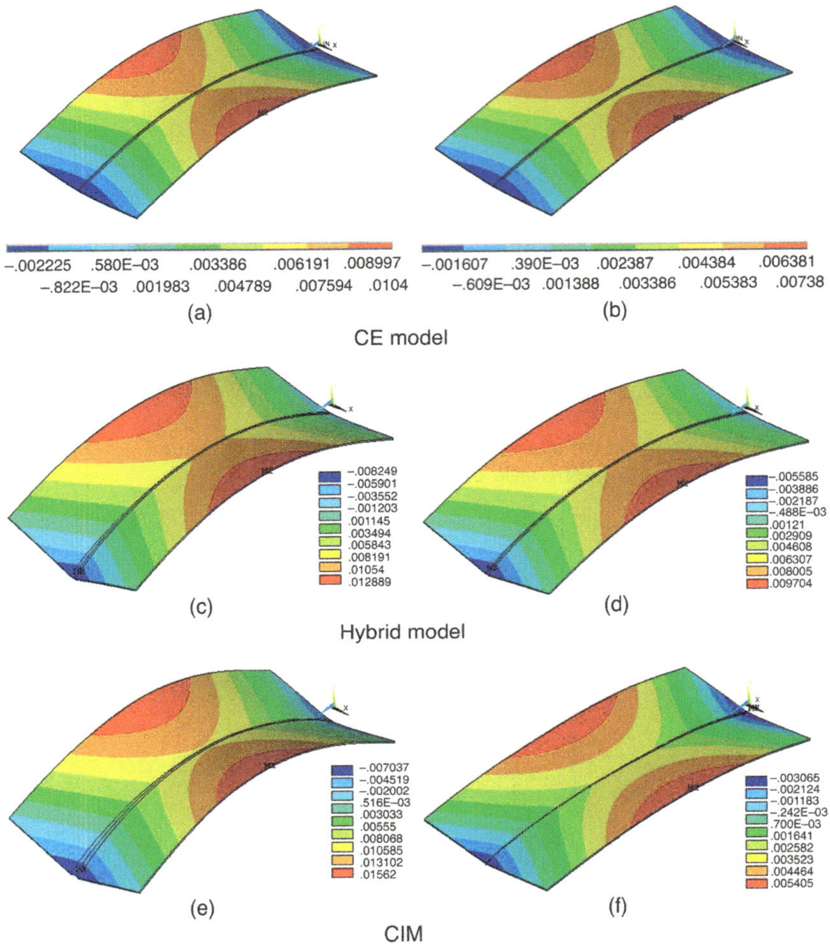

12.40 Predicted out-of-plane deformation with and without cryogenic cooling. Computationally efficient model: (a) without cooling (range 12.6 mm) and (b) with cooling (range 9.0 mm). Hybrid model: (c) without cooling (range 21.1 mm) and (d) with cooling (range 15.3 mm). CIM thermo-elastic-plastic model: (e) without cooling (range 22.7 mm) and (f) with cooling (range 8.5 mm).

than for the other two, as the longitudinal contraction force has less offset from the neutral axis of the V-shape.

These considerations should also apply to the results of the 'with cooling' analyses. The result that shows the CE analysis to have a numerically similar outcome to the CIM thermo-elastic-plastic model is essentially fortuitous, as the shapes are different as regards the balance between angular and longitudinal deformations.

Aside from the gravity effect on transverse deformation, all strategies

Table 12.14 Predicted out-of-plane deformation for fillet-welded fabrications

	Out-of-plane deformation (mm)	
	Without cooling	With cooling
Computationally efficient model	0.985	0.974
Hybrid model	0.634	0.638
Thermo-elastic-plastic model	1.951	1.877

predict a substantial reduction in distortion due to cooling. This improvement is largely due to a reduction in longitudinal bending and reflects a reduction in the width of yielded zones and contraction force. The hybrid method underestimates the improvement relative to the thermo-elastic-plastic model; this is probably due to the simplification of the thermal description, which is based on the maximum temperatures reached at transverse distances and ignores the transient variation of temperature in the longitudinal direction. The hybrid and full analyses both show a small reduction in transverse angular deformation when cooling is applied, but it is not clear if this is a direct result of the cooling effect or if it is due to the large-displacement coupling of longitudinal and transverse actions. If the argument is correct that transverse contraction depends directly on thermal contraction of the fusion zone and if the fusion zone size and configuration are unaffected by cooling, then cooling should not be expected to influence transverse contraction.

In the case of the fillet-welded, stiffened fabrications, the main component of distortion is angular bending of the base plate (see Fig. 12.41). These fabrications are relatively stiff in longitudinal bending and little bowing is evident. As a consequence, the effect of cooling is very small (see Table 12.14). In this case, the angular deformation arises from local bending of the plate. The lower prediction of distortion range in the CE analyses derives from the algorithmic approach, which is of uncertain accuracy relative to the full analysis. The hybrid method predicts even smaller deformations and negligible effect of cooling. The error in this case is also due to poor prediction of base-plate transverse bending local to the fillet weld and, again, the deficiency is likely to be that the true longitudinal transient temperature pattern is replaced in the reduced solutions by maximum temperature profiles. Nevertheless, the contraction force is reduced by cooling in all cases, as confirmed by the residual stress patterns shown in Fig. 12.42.

12.7.6 Conclusion

These studies suggest that cryogenic cooling, using the parameters given, should have a significant effect on out-of-plane deformation. However, the

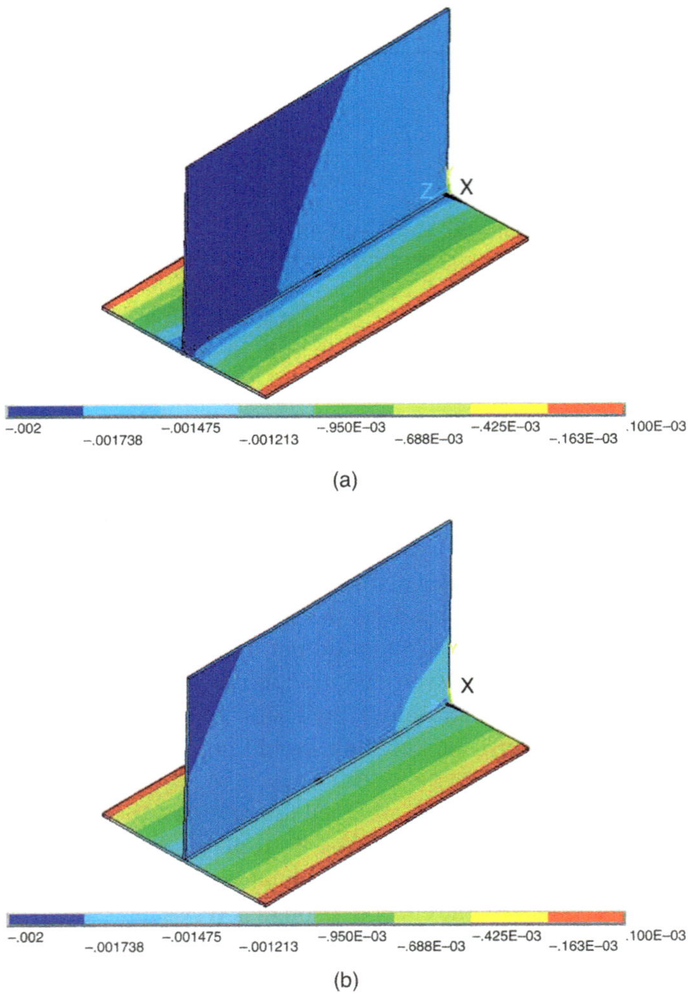

12.41 Out-of-plane distortion predicted by CIM thermo-elastic-plastic model: (a) without cooling; (b) with cooling.

effect is predominantly related to reduction of the longitudinal contraction force and hence reduction of longitudinal bending or bowing. Transverse deformations are less affected by cooling as they relate mainly to transverse contraction of the fused zone in the case of butt welds and bending of the base plate in the case of the stiffened fabrications.

If the thermo-elastic-plastic solutions are accepted as accurate, the reduced solutions underestimate the deformations in both cases, with and without cooling. The problem in the CE case is related in this example to inadequate modelling of the gravity effect on transverse deformation. Both reduced solutions are deficient in terms of modelling the complex pattern of

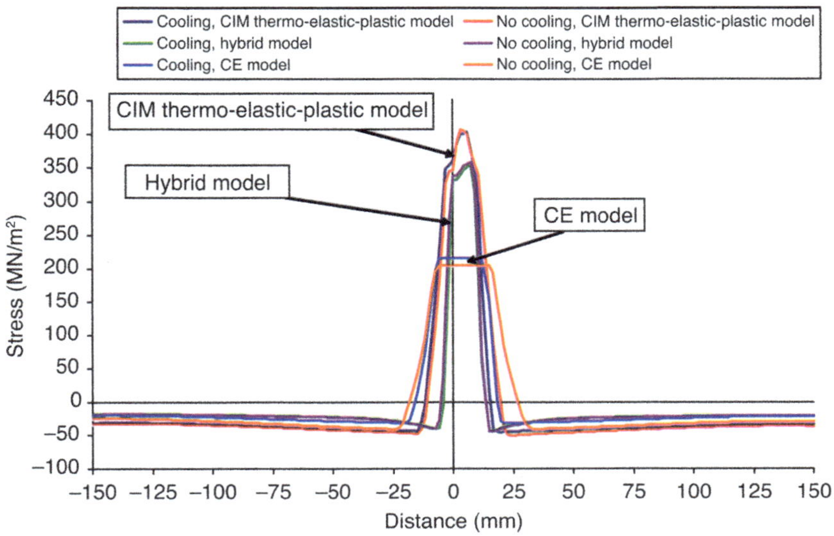

12.42 Longitudinal residual stresses at mid-length in fillet-welded fabrications.

longitudinal thermal distribution, as they are based on a time-independent assumption in the use of a maximum temperature profile. Nevertheless, they both show reductions in deformation when cooling is applied and the simpler CE model reduces the computational budget by a factor of approximately 4000. The reduced solutions therefore have potential in terms of screening a proposal to use cryogenic cooling at a given level, to see whether any benefit is likely. A more accurate thermo-elastic-plastic analysis may then be applied to improve prediction.

12.8 Conclusion

Computational simulations have been used to investigate various design and manufacturing strategies for fabricating stiffened plates. Some of the simulations are supported by experimental studies and the strengths and weaknesses of different computational strategies have been highlighted.

Stiffener configurations were examined in the design-related sections (sections 12.2 and 12.3). The increased distortion effect of multiple attachment welds in close proximity was considered relative to the increased structural efficiency of more frequent stiffening. Table 12.2 showed experimental and computational results for fabrications with alternative two-stiffener and three-stiffener layouts under different support conditions. As for plain butt-welded fabrications, initial flatness and uniform support were shown to be critical factors. It was also noted that the non-linear effects of assembly and welding sequence depend principally on the interaction of residual stress systems generated by the build-up of welds on the fabrication (see Fig. 12.12). The application of twin-arc welding systems for stiffener welding was explored in section 12.4 through simulation and it was shown that if the trailing arc is positioned too closely to the leading arc, thereby increasing the effective heat input, conditions for buckling are easily reached (see Fig. 12.15).

Various techniques aimed at reducing distortion in the fabrication phase were investigated in sections 12.5–12.7. Transient thermal tensioning was examined computationally for a particular case in section 12.5 and it was shown that this technique can reduce the longitudinal bowing of fabrications by inducing additional longitudinal tensile stress zones in the fabrication. However, the spacing of the heated zones is critical to success, as is the quantity of heat supplied. A case was made for the application of computational methods to optimise the thermal tensioning layout. The use of low-transformation-temperature welding wire was examined computationally in section 12.6, using expansivity data determined for actual filler materials. Modest improvements were shown for butt welds in flat plate, using available filler materials (see Fig. 12.29 and Table 12.13). However, it is clear that informative simulation of this technique will require detailed material property data and realistic information on the different welding process conditions applied for proposed filler materials. Finally, the use of cryogenic cooling applied in the trailing zone of the weld was examined through computational studies in section 12.7.

12.9 References

1. Young, W.C., 1989, *Roark's formulas for stress and strain*, 6th edn, McGraw-Hill International Editions, New York, USA.
2. Timoshenko, S., 1936, *Theory of elastic stability*, McGraw-Hill, New York, USA, pp 394–401.
3. Guan, Q., 2005, Control of buckling distortion in plates and shells, in *Processes and mechanisms of welding residual stress and distortion*, Feng, Z., Ed., Woodhead Publishing Ltd, Cambridge, UK.
4. Deo, M.V., 2011, Minimization of bowing distortion in welded stiffeners using differential heating, in *Minimization of welding distortion and buckling*, Michaleris, P., Ed., Woodhead Publishing Ltd, Cambridge, UK.
5. Michaleris, P., 2011, Introduction to welding residual stress and distortion, in *Minimization of welding distortion and buckling*, Michaleris, P., Ed., Woodhead Publishing Ltd, Cambridge, UK.
6. Michaleris, P. and Sun, X., 1997, Finite element analysis of thermal tensioning techniques mitigating weld buckling distortion, *Welding J.*, 76, 22, pp 451s–457s.
7. Dull, R.M., Dydo, J.R., Russell, J.J. and Shanghvi, J., 2005, Method of reducing distortion by transient thermal tensioning, US Patent 6861617.

8. Nagy, T., Williams, S., Colegrove, P., Ikeagu, C. and Fafiolu, I., 2008, Distortion mitigation in welded ship panels, *International Workshop on Thermal Forming and Welding Distortion, Bremen, Germany.*
9. Scicluna, M. and Camilleri, D., 2009, Numerical modelling of welded structures subject to thermal tensioning, University of Malta, Msida, Malta.
10. Fafiolu, I.O., 2007, *Evaluating the heating methods to reduce the distortion of 4 mm thick DH36 ship panels*, MSc thesis, Cranfield University, Cranfield, UK.
11. Henwood, C., Bibby, M.J., Goldak, J.A. and Watt, D.F., 1988, Coupled transient heat transfer microstructure weld computations, *Acta Metall.*, 36, pp 3037–3046.
12. Bhadeisha, H., 1989, Modelling the microstructural development in weld heat affected zones, *Proc. Int. Conf. Trends in Welding Technology, Gatlinburg, TN, USA.*
13. Mikami, Y., Morikage, Y., Mochizuki, M. and Toyoda, M., 2009, Angular distortion of fillet welded T-joint using low transformation temperature welding wire, *Sci. Tech. Weld. Join.*, 14, pp 97–105.
14. Murakawa, H., Beres, M., Vega, A., Rashed, S., Davies, C. M., Dye, D. and Nikbin, K. M., 2008, Effect of phase transformation onset temperature on residual stress in welded thin steel plates, *Trans. JWRI*, 37, pp 75–80.
15. Shirzadi A.A., Bhadeshia H., Karlsson, L. and Withers, P.J., 2009, Stainless steel weld metal designed to mitigate residual stresses, *Sci. Tech. Weld. Join.*, 14, pp 559–565.
16. Li, J., 2011, Minimising buckling distortion in welding by weld cooling, in *Minimization of welding distortion and buckling*, Michaleris, P., Ed., Woodhead Publishing Ltd, Cambridge, UK.
17. Gabzdyl, J., Johnson, A., Williams, S. and Price, D., 2002, Laser weld distortion control by cryogenic cooling, *Proc. SPIE, Osaka, Japan*, 4831, pp 269–274.
18. Camilleri, D., Gray, T.G.F. and Nash, D., 2008, Mitigation of welding distortion and residual stresses via cryogenic CO_2 cooling – a numerical investigation, *17th International Conference and Tutorial Workshop – Computer Technology in Welding and Manufacturing Cranfield, UK.*

Index

CPSIA information can be obtained at www.ICGtesting.com
Printed in the USA
BVOW10*1629130414

350245BV00001BA/6/P

9 780857 090478